FOOD PROTEINS
AND
PEPTIDES
CHEMISTRY, FUNCTIONALITY, INTERACTIONS,
AND COMMERCIALIZATION

FOOD PROTEINS
AND
PEPTIDES
CHEMISTRY, FUNCTIONALITY, INTERACTIONS, AND COMMERCIALIZATION

Edited by
Navam S. Hettiarachchy
Associate Editors
Kenji Sato • Maurice R. Marshall • Arvind Kannan

CRC Press
Taylor & Francis Group
Boca Raton London New York

CRC Press is an imprint of the
Taylor & Francis Group, an **informa** business

CRC Press
Taylor & Francis Group
6000 Broken Sound Parkway NW, Suite 300
Boca Raton, FL 33487-2742

First issued in paperback 2016

© 2012 by Taylor & Francis Group, LLC
CRC Press is an imprint of Taylor & Francis Group, an Informa business

No claim to original U.S. Government works

ISBN 13: 978-1-138-19900-2 (pbk)
ISBN 13: 978-1-4200-9341-4 (hbk)

Visit the Taylor & Francis Web site at
http://www.taylorandfrancis.com

and the CRC Press Web site at
http://www.crcpress.com

Contents

v

Preface

This book features a comprehensive review of food proteins and peptides, covering a wide variety of chapters on the chemistry, structure–function relationships, functionalities, characterization, technological advances, industrial perspectives, and commercialization trends. The book has been designed to dwell deep into the basic chemistry, structures, and functions of food proteins and peptides, to give the reader an understanding of the physicochemical and biochemical factors that govern functionality. This will give the reader an overall perspective of functional foods in relation to their biochemical and physical aspects, making this book a distinct one that is multidisciplinary. It also provides a comprehensive resource for researchers and students working in biochemistry, biotechnology, food science and technology, nutrition, and medicine.

The contributors to this book edition review the literature and also present research from their own work, emphasizing both the philosophy and significance that have led to a deeper understanding of the basic subjects covered. Accordingly, this book should be of benefit to those following the basic principles governing the study, isolation, structure, chemistry, and functionality of food proteins and peptides.

Editors

Dr. Navam Hettiarachchy is a university professor in the Department of Food Science and is also associated with the Institute of Food Science and Engineering at the University of Arkansas, Fayetteville, Arkansas. She earned her MS and PhD degrees in molecular biochemistry at the University of Edinburgh Medical School, Edinburgh, Scotland, and the University of Hull, England, UK, respectively. She also earned her postgraduate diploma in nutrition from the Indian Council of Medical Research, Hyderabad, India. She has been a professor at three major universities in the United States. Prior to this, she was a faculty member in the department of biochemistry in a major medical school where her research was focused on nutrition in health and disease. In 2003, she was elected as an Institute of Food Technologists (IFT) fellow.

Her research program focuses on an integrated approach to nutraceuticals and protein chemistry. She has published more than 100 refereed peer-reviewed journal articles and more than 300 abstracts and presentations, and has been awarded more than $3.5 million as the principal investigator for her research program from competitive federal and other grant funding sources. Dr. Hettiarachchy is internationally recognized for her research program on cereals, particularly rice. Three patents have been filed on her research on rice nutrients and nutraceuticals. She has made several invited presentations at both the national and international levels. Her tireless research on rice bran resulted in the formation of a new company, Nutraceutical Innovations, with the University of Arkansas in 2007. The focus of this company is to generate value-added nutraceuticals from rice bran with a total system approach. In her efforts to find alternate uses for anticancer drugs, she has been working with plant-derived proteins and peptides. Her research program has pioneered anticancer research on rice and soybean peptides and demonstrated potent anticancer activities in various types of cancer cell lines. This major breakthrough is the result of focusing on the fundamental research on rice bran that can have a healthier impact on people's lives.

Professor Maurice Marshall is a faculty member in the Food Science and Human Nutrition Department, University of Florida, where he has taught and researched food chemistry/biochemistry for 30 years. He received his bachelor's degree in biology from Duquesne University, and both his master's degree from Rutgers University and his PhD degree from Ohio State University are in the area of food science, specializing in food chemistry/biochemistry. His research focuses on the analysis of food constituents including biogenic amines, antioxidants, and pesticide residues. He is director of the Southern Region USDA-IR-4 Project, which assists specialty crop growers to obtain registered low-risk crop protection tools. His additional research focuses on food enzymes, specifically polyphenol oxidases in fruits and vegetables; proteases from stomachless marine organisms; and unique browning inhibitors from insects, mussels, and crustaceans. He also works on the hydrolysis of waste seafood proteins for bioactive peptide generation and their action as antioxidants and as antidiabetic, anticancer, and high-blood-pressure regulators.

Professor Kenji Sato has been a faculty member for 22 years at the Division of Applied Life Sciences, Graduate School of Life and Environmental Sciences, Kyoto Prefectural University at Kyoto City, where he is currently a professor. He obtained his undergraduate degree and PhD degree in agriculture from Kyoto University. After spending 1 year at Kyoto University as a postdoctoral fellow, he joined the faculty at Kyoto Prefectural University. His current research interest focuses on bioactive peptides, especially on the metabolic fate of these peptides. His lab performs research on food-derived peptides to demonstrate their absorption, metabolic fate, and efficacy and elucidate their molecular mechanisms of action.

Dr. Arvind Kannan is currently working as a postdoctoral fellow in the food science department at the University of Arkansas. He received his bachelor's degree in biochemistry from Madras University, India, and a master's degree in medical laboratory technology from Birla Institute of Technology & Science (BITS), India. He then worked as a research associate at the Madras Diabetes Research Foundation, India, studying the markers of coronary artery disease and diabetes, before getting overseas experience as a research associate at the National Tsing-Hua University in Taiwan. He worked on bacterial gene and protein expression and biophysical experiments related to protein structure in the Structural Proteomics program of Prof. Chin Yu. He obtained his PhD degree in cell and molecular biology from the University of Arkansas in 2009. His research was on characterizing a bioactive peptide from rice bran for cancer antiproliferative activity using human cancer cell lines. He is currently furthering his studies on bioactive peptides and is involved in projects dealing with bioactive proteins and peptides. He also serves as a scientist at Nutraceutical Innovations, LLC, a startup firm that focuses on the fermentation of rice bran to generate bioactives for value-added functional and nutraceutical uses.

Contributors

Małgorzata Darewicz
Chair of Food Biochemistry
University of Warmia and Mazury in
    Olsztyn
Olsztyn, Poland

Bartłomiej Dziuba
Chair of Industrial and Food
    Microbiology
University of Warmia and Mazury in
    Olsztyn
Olsztyn, Poland

Dina Fernandez
Department of Food Science and
    Human Nutrition
University of Illinois
Urbana, Illinois

Ryuichi Fujisawa
Graduate School of Dental Medicine
Hokkaido University
Sapporo, Japan

Elvira Gonzalez de Mejia
Department of Food Science and
    Human Nutrition
University of Illinois
Urbana, Illinois

Sundaram Gunasekaran
Biological Systems Engineering
    Department
Food and Bioprocess Engineering
    Laboratory
University of Wisconsin-Madison
Madison, Wisconsin

Kaori Hashimoto
Division of Applied Biosciences
Graduate School of Life and
    Environmental Sciences
Kyoto Prefectural University
Kyoto, Japan

Navam Hettiarachchy
Department of Food Science
University of Arkansas
Fayetteville, Arkansas

Geoffrey Hunt
Center for Bioethics and Emerging
    Technologies
St Mary's University College
London, United Kingdom

Arvind Kannan
Department of Food Science
University of Arkansas
Fayetteville, Arkansas

Se-Kwon Kim
Department of Chemistry
and
Marine Bioprocess Research Center
Pukyong National University
Busan, Republic of Korea

Yoshinori Kuboki
Graduate School of Dental Medicine
Hokkaido University
Sapporo, Japan

Hitomi Kumagai
Department of Chemistry and Life
  Science
College of Bioresource Sciences
Nihon University
Fujisawa-shi, Kanagawa, Japan

Motonaka Kuroda
Institute for Health Fundamentals
Ajinomoto Co, Inc
Kawasaki city, Kanagawa, Japan

Maurice Marshall
Department of Food Science and
  Human Nutrition
University of Florida
Gainesville, Florida

Cristina Martinez-Villaluenga
Department of Food Science and
  Human Nutrition
University of Illinois
Urbana, Illinois

Masami Matsuda
Department of Health Nutrition
Tokyo Kasei-gakuin University
Tokyo, Japan

Kentaro Matsumiya
Laboratory of Quality Analysis and
  Assessment
Division of Agronomy and
  Horticultural Science
Graduate School of Agriculture
Kyoto University
Kyoto, Japan

Yasuki Matsumura
Laboratory of Quality Analysis and
  Assessment
Division of Agronomy and
  Horticultural Science
Graduate School of Agriculture
Kyoto University
Kyoto, Japan

Piotr Minkiewicz
Chair of Food Biochemistry
University of Warmia and Mazury in
  Olsztyn
Olsztyn, Poland

Satoshi Nagaoka
Department of Applied Life Science
Faculty of Applied Biological Sciences
Gifu University
Gifu, Japan

Toshio Ogino
Department of Electrical and
  Computer Engineering
Yokohama National University
Yokohama, Japan

Hiroki Saeki
Faculty of Fisheries Sciences
Hokkaido University
Hakodate, Hokkaido, Japan

Rachel L. Sammons
School of Dentistry
University of Birmingham
Birmingham, United Kingdom

Shridhar K. Sathe
Department of Nutrition, Food and
   Exercise Sciences
College of Human Sciences
Florida State University
Tallahassee, Florida

Kenji Sato
Division of Applied Life Sciences
Graduate School of Life and
   Environmental Sciences
Kyoto Prefectural University
Kyoto, Japan

Toshio Shimizu
School of Health and Human Life
Nagoya-bunri University
Inazawa-city, Aichi Prefecture, Japan

Oscar Solar
Biological Systems Engineering
   Department
Food and Bioprocess Engineering
   Laboratory
University of Wisconsin-Madison
Madison, Wisconsin

Daisuke Urado
Division of Applied Life Sciences
Graduate School of Life and
   Environmental Sciences
Kyoto Prefectural University
Kyoto, Japan

Fumio Watari
Department of Biomedical, Dental
   Materials and Engineering
Graduate School of Dental Medicine
Hokkaido University
Sapporo, Japan

Isuru Wijesekara
Department of Chemistry
Pukyong National University
Busan, Republic of Korea

Abbreviations

ANNs	Artificial neural networks
BSA	Bovine serum albumin
CATH	Class architecture topology homology
Dha	Didehydroalanine
Dhb	Didehydroaminobutyric acid
DMA	Dynamic mechanical analysis
DSC	Differential scanning calorimetry
DSSP	*Dictionary of Protein Secondary Structure*
DTA	Differential thermal analysis
EAI	Emulsifying activity index
ESI	Electrospray ionization
FAB	Fast atom bombardment
GSH	Glutathione
HAT	Histone acetyltransferase
HDAC	Histone deacetylase
HEWL	Hen egg-white lysozyme
HMW	High molecular weight
HPLC	High-performance liquid chromatography
IEF	Isoelectric focusing
ITAMs	Immunoreceptor tyrosine-based activation motifs
ITT	Inverse temperature transition
Lan	Lanthionine
lg	Lactoglobulin
LMW	Low molecular weight
MALDI	Matrix-assisted laser desorption/ionization
MeLan	Methyllanthionine
mRNA	Messenger ribonucleic acid
MS	Mass spectrometry
Mw	Molecular weight
NF	Nanofiltration
NMR	Nuclear magnetic resonance
pI	Isoelectric point
$POCl_3$	Phosphorous oxychloride

PrP^C	Cellular prion protein
PSD	Post source decay
SDS-PAGE	Sodium dodecyl sulfate polyacrylamide gel electrophoresis
SEC	Size exclusion chromatography
SELDI	Surface-enhanced laser desorption/ionization
SIB	Swiss Institute of Bioinformatics
SPE	Solid-phase extraction
STMP	Sodium trimetaphosphate
TG	Thermogravimetry
TMA	Thermal mechanical analysis
UF	Ultrafiltration

Chapter 1

Food Proteins–Peptides: Chemistry and Structure

Arvind Kannan, Navam Hettiarachchy,
and Maurice Marshall

Contents

1.1 Introduction

The concept of food being a very important source of high-quality proteins and peptides has caught the attention of many research scientists in this decade. Proteins are important biological entities that not only provide health benefits but also help sustain the overall growth, metabolism, balance, and function of cells in any system. They are ubiquitous and hence readily available in most tissues. Amino acids make up the building blocks of proteins. Although only 20 in number, their ability to bond with each other in an orderly yet permutational manner has helped in the generation of millions of protein molecules (Brocchieri and Karlin 2005). With the evolution of proteomic and genomic tools, new protein molecules are gaining access into the enormously growing protein databases. Understanding the behavior of proteins, amino acids, and peptides and their chemistry, structure, and properties helps design and categorize proteins on the basis of the functional significance of their properties.

The overall function of a protein is governed by a variety of forces. Molecular bonding, forces of attractions, environment, and structural integrity are the major ones among scores of other biophysical entities. Nuclear magnetic resonance (NMR) and x-ray crystallography have proved to be successful tools for determining structures at high resolutions. Structure–function relationships have gained a new dimension with these tools. The study of protein chemistry and structural biology has never been more intense and accurate. As newer tools and state-of-the-art technological practices develop, biologists find it easier to isolate and characterize the proteins of interest from virtually any cell or substance. Basic scientists have mostly delved deep into protein structure, chemistry, and function, while clinical scientists use the information in clinical practices.

Embarking upon a society that has become seemingly conscious about low-calorie booms, health risks, and genetic predispositions, research is not restricted to only chemical drugs, chemotherapy, conventional medicines, homeopathy, and others when it comes to blocking disease manifestations. A whole array of new fields and subjects has appeared as exploration continues. The science of food is one such area that has interested many young and aspiring scientists. Food proteins have attained the status of being a functional ingredient or a health-promoting food-derived substance (Klotzbach-Shimomura 2001), a status attributed by peptides or amino acid chains encrypted within each protein. Hence, food proteins and peptides may be considered the next generation of natural supplements and treatment molecules that may aid in arresting disease propensities. Several of them already promote anticancerous, antimicrobial, antihypertensive, and antimutagenic properties.

Chapter 1 provides basic information on the structure and chemistry of amino acids, peptides, and proteins—important factors that relate structure and chemistry, thereby leading to a structural hierarchy—and their dimensions and complexities. This chapter is intended to help the reader gain a basic knowledge of protein chemistry. This knowledge will help understand other complex phenomena describing the structure and functional relationships of food proteins and peptides in the accompanying chapters.

1.2 Amino Acids: Structures and Chemistry

A molecule having an amino (NH_2) functional group as well as a carboxyl (COOH) functional group is referred to as an amino acid. The general formula for an amino acid is represented in Figure 1.1. Since amino acids bear both amino (basic) and carboxyl (acidic) groups that can react, they are said to form an inner salt called a zwitterion and hence are referred to as being zwitterionic at neutral pH. The zwitterion has no net charge; there is one positive (NH_3^+) and one negative (COO^-) charge. Typical ionization of the charges in the functional groups present at physiological pH within an amino acid is shown in Figure 1.2.

All amino acid types possess the characteristic α-amino acid moiety. The carbon atom found in the central position of amino acids links both the amino and the carboxyl functional groups. It is referred to as the C_α-group, represented in Figure 1.1. Except for glycine, all amino acids possess a chiral carbon atom placed adjacent to the carboxyl group, allowing possible stereoisomeric enantiomers. The most natural configuration seen in protein amino acids is the L-configuration about their chiral carbon atoms (Creighton 1993; Berg, Tymoczk, and Stryer 2002). Amino acids with a D-configuration do not occur in proteins. However, they are said to be associated with certain bacterial structure and metabolism.

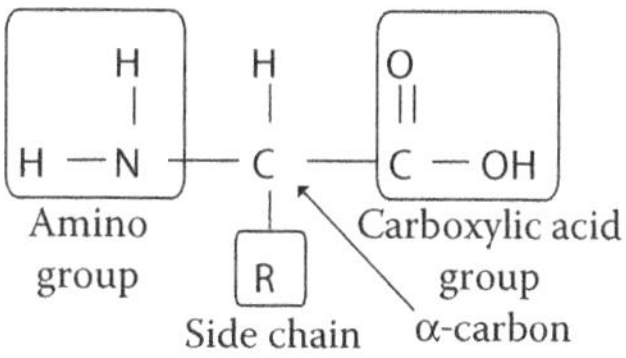

Figure 1.1 General structure of an amino acid where the central carbon atom, denoted as C_α, is shown (pointed arrow).

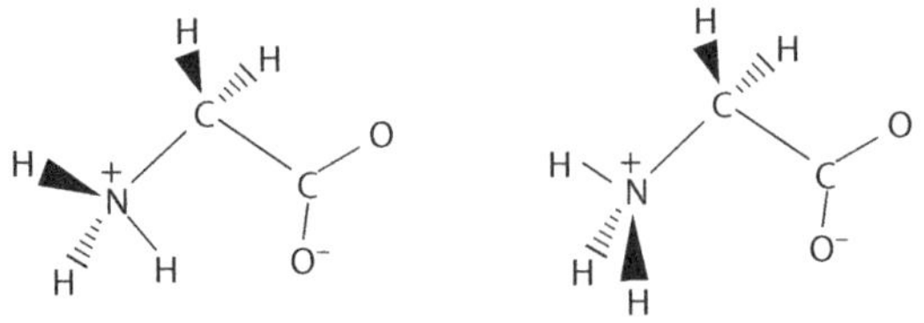

Figure 1.2   Zwitterion state of amino acid glycine (cis and trans).

1.2.1 Side Chains

Basically, the side chains (R) of amino acids vary in the number of atoms. From a single hydrogen (glycine) to a complex heterocyclic group (tryptophan), amino acid side chains give each amino acid its unique structure and property. The side-chain complexity is characteristic of the property and charge of the side chain, ranging from polar to nonpolar and charged to uncharged. The side chain confers uniqueness to each amino acid. The α-carbon also binds to a side chain carbon specific for each type of amino acid. Proline does not conform to this unique identity because its hydrogen atom is replaced by a bond to the side-chain. It is also chiral since the carbon atom is bound to four different groups. Glycine, however, is not chiral since it contains a hydrogen atom as its side chain (Berg, Tymoczk, and Stryer 2002).

1.2.2 Polar, Uncharged Side Chains

There are several amino acids with polar, uncharged side chains. Serine and threonine have hydroxyl groups. Tyrosine has a phenolic side chain. Tryptophan has a heterocyclic aromatic amine side chain, as does histidine but it confers a positive charge (Figure 1.3).

1.2.3 Nonpolar Side Chains

There are eight amino acids with nonpolar side chains. Glycine, alanine, and proline have small, nonpolar side chains and are all weakly hydrophobic. Phenylalanine, valine, leucine, isoleucine, and methionine have larger side chains and are more strongly hydrophobic (Figure 1.3).

1.2.4 Charged Side Chains

There are four amino acids with charged side chains. Aspartic acid and glutamic acid have carboxyl groups on their side chains. Each acid is fully ionized at pH 7.4. Arginine and lysine have side chains with amino groups. Histidine also renders a positive charge. Their side chains are fully protonated at pH 7.4 (Figure 1.3).

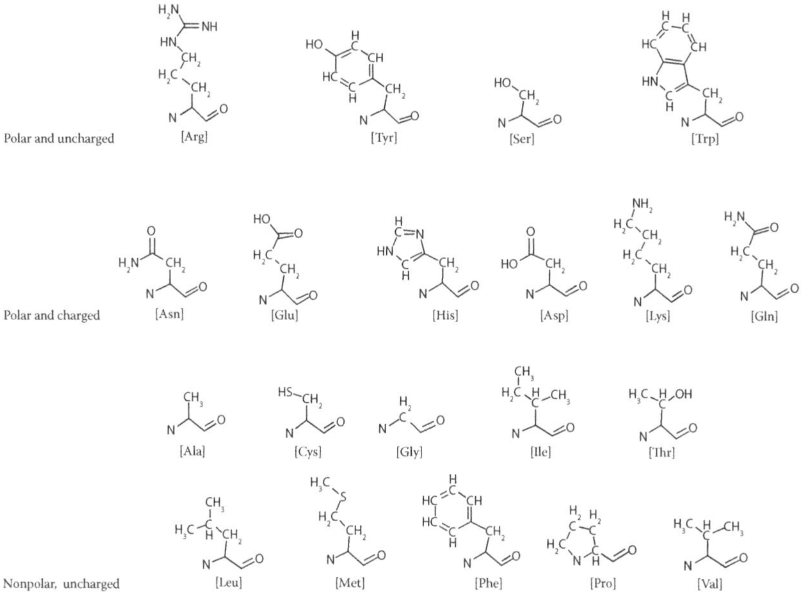

Figure 1.3 **Classification of amino acids based on their side-chain structure and properties. N, neutral; +, positive; –, negative; Np, nonpolar; P, polar. (From Campbell, Department of Biology, Davidson College, Davidson, NC.)**

1.2.5 Hydrophobic and Hydrophilic Amino Acids

Hydrophobicity and hydrophilicity are governed by varying degrees of polarity of the side chains of amino acids. The ability of the amino acids to interact is influenced by the extent of polarity, which imparts importance to a protein's physical properties. The arrangement and distribution of hydrophobic and hydrophilic amino acids in a polypeptide chain also help determine the tertiary structure of a protein (Creighton and Thomas 1993). For example, soluble proteins have polar amino acids such as serine and threonine on their surfaces, while integral membrane proteins tend to have an outer ring of hydrophobic amino acids. These amino acids help to anchor the protein into the membrane lipid bilayer. Likewise, charge specificities dictate as to which residues to bind to which proteins. Proteins that bind to positively charged molecules have surfaces abundant with negatively charged amino such as glutamate and aspartate, while proteins that should bind to negatively charged molecules have abundant positively charged chains such as lysine and arginine (Nelson and Cox 2008). This charge-specific property of proteins has also been applied to purify proteins from a given mixture using ion-exchange chromatography columns. The concept is discussed in detail in the following chapters.

There are several proposed hydrophobicity scales for amino acids. This property has been the key driving force in understanding protein folding behavior and

Table 1.1   Hydrophobicity Scale Comparing Amino Acid Hydrophobicities

Kyte and Doolittle (1982)	Rose et al. (1985)	Wolfenden et al. (1981)	Janin (1979)
Ile	Cys	Gly, eu, Ile	Cys
Val		Val, Ala	Ile
	Phe, Ile		Val
Leu	Val	Phe	Leu, Phe
	Leu, Met, Trp	Cys	Met
Phe		Met	Ala, Gly, Trp
Cys			
Met, Ala	His	Thr, Ser	
	Tyr	Trp, Tyr	His, Ser
Gly	Ala		Thr
Thr, Ser	Gly		Pro
Trp, Tyr	Thr		Tyr
Pro			Asn
		Asp, Lys, Gln	Asp
His	Ser	Glu, His	Gln, Glu
Asn, Gln	Pro, Arg	Asp	
Asp, Glu	Asn		
Lys	Gln, Asp, Glu		Arg
Arg	Lys	Arg	Lys

Comparison of amino acids based on hydrophobicity (top to bottom: higher to lower hydrophobicity) on four different hydrophobicity scales proposed.

states. The four representative scales (Table 1.1) proposed differ to an extent only compounded by the differences in methods used to determine the scaling pattern (Janin 1979; Rose et al. 1985; Wolfenden et al. 1981; Kyte and Doolite 1982). While separation between largely hydrophobic, moderately hydrophobic, and slightly hydrophobic amino acids exists in all four scales, subtle differences within the same group (largely, moderately, and slightly hydrophobic) of amino acids exist, probably owing to the nature of the methods used.

More recently, scientists have proposed a scale of hydrophobicity based on the free energy of hydrophobic association (Urry 2004). The change in Gibbs free

Table 1.2 Classification of Amino Acids Based on Their Side-Chain Structure and Properties

Name	Abbreviation	Charge	Polarity	Linear Structure
Alanine[ne]	ala A	N	Np	CH_3–CH(NH_2)–COOH
Arginine[ne]	arg R	+	P	HN=C(NH_2)–NH–$(CH_2)_3$–CH(NH_2)–COOH
Asparagine[ne]	asn N	–	P	H_2N–CO–CH_2–CH(NH_2)–COOH
Aspartic acid[ne]	asp D	N	P	HOOC–CH_2–CH(NH_2)–COOH
Cysteine[ne]	cys C	N	Np	HS–CH_2–CH(NH_2)–COOH
Glutamic acid[ne]	glu E	–	P	HOOC–$(CH_2)_2$–CH(NH_2)–COOH
Glutamine[ne]	gln Q	N	P	H_2N–CO–$(CH_2)_2$–CH(NH_2)–COOH
Glycine[ne]	gly G	N	Np	NH_2–CH_2–COOH
Histidine[e]	his H	+	P	NH–CH=N–CH=C–CH_2–CH(NH_2)–COOH
Isoleucine[e]	ile I	N	Np	CH_3–CH_2–CH(CH_3)–CH(NH_2)–COOH
Leucine[e]	leu L	N	Np	$(CH_3)_2$–CH–CH_2–CH(NH_2)–COOH
Lysine[e]	lys K	+		H_2N–$(CH_2)_4$–CH(NH_2)–COOH
Methionine[e]	met M	N	Np	CH_3–S–$(CH_2)_2$–CH(NH_2)–COOH
Phenylalanine[e,*]	phe F	N	Np	Ph–CH_2–CH(NH_2)–COOH
Proline[ne]	pro P	N	Np	NH–$(CH_2)_3$–CH–COOH
Serine[ne]	ser S	N	P	HO–CH_2–CH(NH_2)–COOH

(*continued*)

Table 1.2 (Continued) Classification of Amino Acids Based on Their Side-Chain Structure and Properties

Name	Abbreviation	Charge	Polarity	Linear Structure
Threonine[e]	thr T	N	Np	CH_3–CH(OH)–$CH(NH_2)$–COOH
Tryptophan[e*]	trp W	N	P	Ph–NH–CH=C–CH_2–$CH(NH_2)$–COOH
Tyrosine*	tyr Y	N	P	HO–Ph–CH_2–$CH(NH_2)$–COOH
Valine[e]	val V	N	Np	$(CH_3)_2$–CH–$CH(NH_2)$–COOH

N, neutral; +, positive; –, negative; Np, nonpolar; P, polar; ph, phenyl ring; *, aromatic; e, essential amino acid; ne, nonessential amino acid.

energy resulting from changes in the temperature (T_t) and heat (ΔH_t) of an inverse temperature transition (ITT) of hydrophobic association was derived and utilized to develop a hydrophobicity scale for amino acid residues within an elastic model protein $(GVGVP)_n$. When using this scale, on raising the temperature, hydrophobic residues go from water to the hydrophobic association.

Having elucidated the importance of the hydrophobic character of amino acid residues, it is imperative to keep in mind that interactions due to hydrophobicity or hydrophilicity need not always depend on the side-chain conformity. Posttranslational modifications can help associate certain chains to the proteins. Examples include lipoproteins (Magee and Seabra 2005) and glycoproteins (Pilobello and Mahal 2007).

1.2.6 Amino Acids Classification

Amino acids are mainly classified according to their physical groups via side-chain properties. Thus, amino acids are classified under indexes that include charge, size, hydrophilic–hydrophobic, and functional groups. The IUPAC–IUB joint commission has set standard symbols and nomenclature for amino acids and peptides. Apart from this classification, from a nutritional viewpoint, amino acids are classified into essential and nonessential amino acids.

Essential amino acids are those that cannot be synthesized by the body de novo and hence must be supplied in the diet. Essential amino acids include histidine, isoleucine, leucine, lysine, methionine, phenylalanine, threonine, tryptophan, and valine. Nonessential amino acids are those that the body can synthesize through various metabolic pathways and so need not be provided in the diet. Nonessential amino acids include alanine, arginine, aspartic acid, asparagine, cysteine, cystine, glutamic acid, glycine, glutamine, proline, serine, and tyrosine (Reeds 2000). Table 1.2 lists the 20 amino acids and classifies them according to their properties.

1.3 Peptides Formation, Structure, Chemistry, and Classification

1.3.1 Peptide Bond

Condensation of two or more amino acids results in the formation of a peptide bond with the release of water. The resultant molecules are known as peptides, which can range from di-, tri- to even 100 amino acid residue peptides (Nelson and Cox 2008). Figure 1.4 describes the formation of a peptide by the condensation of two amino acids.

The reaction between an amino group of one amino acid and a carboxyl group of an adjacent or other amino acid with the liberation of a water molecule forms a peptide bond. This condensation reaction forms the CO–NH bond called the peptide bond. This reaction is said to be catalyzed by the ribosome, and the process is called translation. In other cases, specific enzymes can synthesize peptides. For example, the tripeptide glutathione (GSH) is synthesized in two steps from free amino acids involving two enzyme systems, γ-glutamylcysteine synthetase and GSH synthetase (Meister 1988). Figure 1.5 shows the structure of the tripeptide GSH.

1.3.2 Chemistry

The resonance forms of the peptide (amide) bond enable the bond to be less reactive (resonance stabilization) and confer a partial double bond character. The partial double bond nature imparts planarity to the amide group, allowing either the cis or the trans isomer. However, in the folded state, only the trans form is favored. Although stabilized by resonance, the less reactive peptide bonds can be broken down through chemical reactions favoring the breaking of the carbonyl double bond and forming an intermediate, as observed in several enzymatic proteolysis reactions (Berg, Tymoczk, and Stryer 2002; Nelson and Cox 2008).

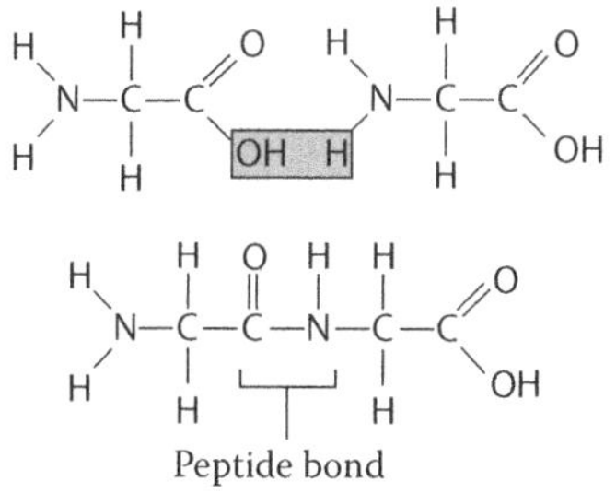

Figure 1.4 Formation of a peptide bond. The COOH moiety from one amino acid condenses with the NH_2 moiety of another amino acid with the liberation of water to form a peptide bond.

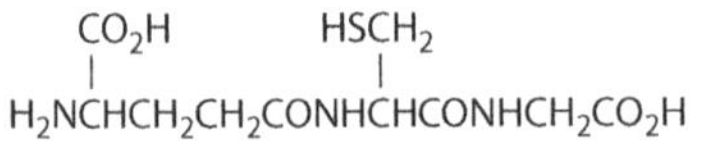

$$\begin{array}{cc} CO_2H & HSCH_2 \\ | & | \\ H_2NCHCH_2CH_2CONHCHCONHCH_2CO_2H \end{array}$$

Figure 1.5   Structure of glutathione.

The peptide bond is planar due to the delocalization of electrons from the double bond. The rigid peptide dihedral angle ω(the bond between C_1 and N) is always close to 180°. The dihedral angles φ(the bond between N and C_α) and ψ(the bond between C_α and C_1) can have a certain range of possible values. Figure 1.6 shows the various angles involved in the peptide bond. These angles are represented as degrees of freedom of a protein, which control the protein's three-dimensional structures. They are restricted by geometry to allowed ranges typical for particular secondary structure elements, and are represented in a Ramachandran plot (Ramachandran, Ramakrishnan, and Sasisekharan 1963).

1.3.3 Classification

Peptides are classified into ribosomal and nonribosomal classes. Ribosomal peptides are synthesized by the translation of messenger RNA (mRNA), whereas nonribosomal peptides are arranged by enzymes rather than by the ribosome. Ribosomal peptides can undergo posttranslational modifications, unlike nonribosomal peptides (Nelson and Cox 2008). Several other peptide classes exist based on function, including vasoactive intestinal peptides, opioid peptides, calcitonin peptides, pancreatic polypeptide-related peptides, and tachikinin peptides. Some very common peptides bearing biological significance are GSH and nisin. GSH is a tripeptide. It contains an unusual peptide linkage between the amine group of cysteine and the carboxyl group of the glutamate side chain (Figure 1.5). Nisin is an antibacterial polycyclic peptide with 34 amino acid residues used as a food preservative. It contains the uncommon amino acids

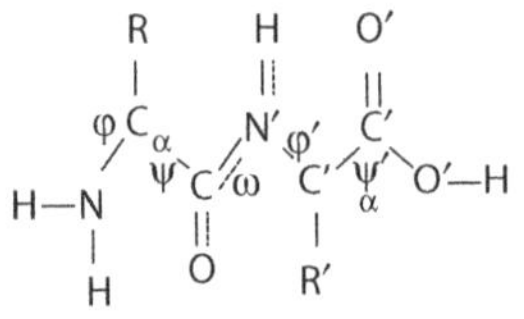

Figure 1.6   Bond angles for ψ and ω represented in valine. The rigid peptide dihedral angle ω(the bond between C_1 and N) is always close to 180°. The dihedral angles φ(the bond between N and C_α) and ψ(the bond between C_α and C_1) can have a certain range of possible values. (From Renfrew, D. P., et al., *Proteins*, 71, 1637–46, 2008.)

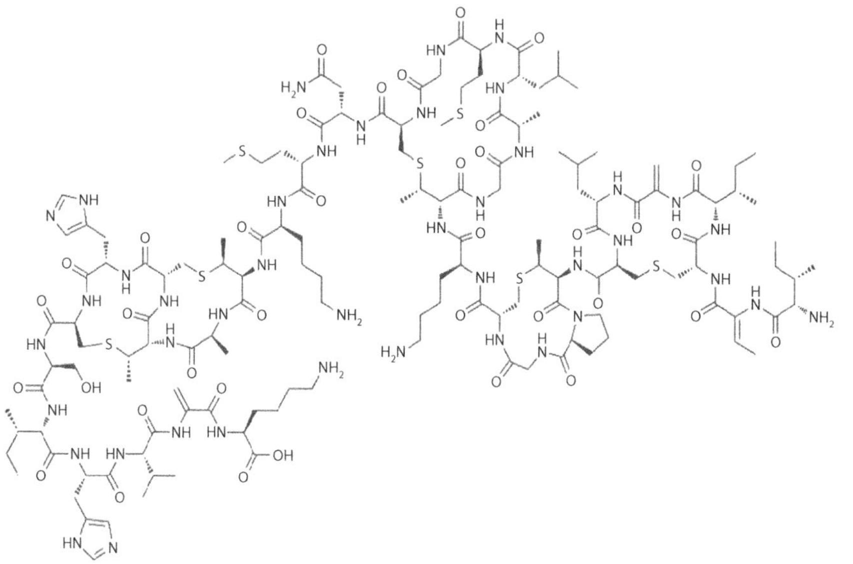

Figure 1.7   Structure of nisin. (From Li, B., et al., *Science*, 311, 1464–67, 2006.)

lanthionine (Lan), methyllanthionine (MeLan), didehydroalanine (Dha), and didehydroaminobutyric acid (Dhb). These unusual amino acids are introduced by the posttranslational modification of the precursor peptide. Nisin is produced by fermentation using the bacterium *Lactococcus lactis*. Commercially, it is obtained from the culturing of *L. lactis* on natural substrates, such as milk or dextrose, and is not chemically synthesized. The structure of nisin is represented in Figure 1.7.

1.4 Protein Structure, Chemistry, and Function

1.4.1 Structural Hierarchy

Proteins are an important class of biological macromolecules present in all biological organisms made up of such elements as carbon, hydrogen, nitrogen, oxygen, and sulfur. All proteins are polymers of amino acids. If classified by their physical size, proteins are nanoparticles (definition: 1–100 nm). The polymers, also known as polypeptides, consist of a sequence of 20 different L-α-amino acids, also referred to as residues. For chains under 40 residues, the term "peptide" is frequently used instead of protein. To be able to perform their biological functions, proteins fold into one, or more, specific spatial conformations, driven by a number of noncovalent interactions, such as hydrogen bonding, ionic interactions, van der Waals forces, and hydrophobic packing. In order to understand the functions of proteins at a molecular level, it is often necessary to determine the three-dimensional structure of proteins.

1.4.2 Primary Structure

The primary structure of a protein constitutes the specific composition of the polypeptide chain. Several amino acids bind together through peptide bonds to form the backbone of a polypeptide chain. Each protein has a unique polypeptide chain comprising a highly specific arrangement of an amino acid sequence (Pauling, Corey, and Branson 1951). A linear arrangement of amino acids forming a peptide chain is represented in Figure 1.8 and is referred to as the primary structure of proteins.

In the early nineteenth century, several hypotheses about the polypeptide chain arrangement began to accumulate. The colloidal protein hypothesis, cyclol hypothesis, came into existence when Frederick Sanger successfully sequenced the insulin molecule, disproving many of the models and hypotheses proposed (Sanger 1952). It was later known that the primary structure of the protein plays an important role in determining the three-dimensional shape of the protein, known as the tertiary structure. Understanding the relationship between the primary structure and other structures needed a wider understanding of the protein folding and complex phenomenon, and it was difficult to predict the secondary or the tertiary structures. However, knowledge of the structure of a homologous protein sequence

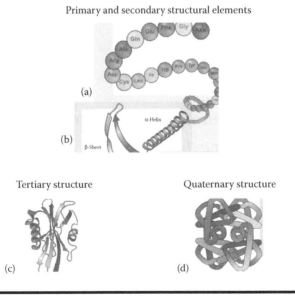

Figure 1.8 (a–d) Levels of protein structure. (From http://commons.wikimedia. org/wiki/File:Main_protein_structure_levels_en.svg.)

obtained from the same family can help identify the tertiary structure. Sequence families are often determined by structural genomics and sequence clustering.

1.4.3 Modifications to Primary Structure

Although the linear sequence of amino acids along their backbone makes the polypeptide unbranched, a cross-linkage using disulfide bonds is not uncommon. Racemization of L-amino acids to D-amino acids at the C_α-atom can occur in proteins where L-amino acids exist, preventing proteolytic cleavage. Posttranslational modifications like acylation, amidation, myristoylation, phosphorylation, hydroxylation, methylation, glycosylation, ADP-ribosylation, ubiquitination, carboxylation, and sulfation can occur to the polypeptide chains (Berg, Tymoczk, and Stryer 2002; Nelson and Cox 2008). Posttranslational modifications are covalent changing events, processing the polypeptide to undergo signal cascades, localization, and interactions with other proteins. Mann and Jensen (2003) have reviewed posttranslational modifications to the proteome in the new era whereby more than phosphorylation, other modifications are becoming common and more understandable.

Phosphorylation happens to be the most important chemical modification of proteins and plays a chief role in signal transduction (Schreiber et al. 2008). A phosphate group can be attached to the side chain hydroxyl group of serine, threonine, and tyrosine residues, adding a negative charge at that site. Such reactions are catalyzed by kinases that add phosphate groups, and the reverse reaction is catalyzed by phosphatases that remove phosphate groups. One of the earliest events is phosphorylation of protein tyrosines in order for efficient and complete signal transduction to occur (Pribluda, Pribluda, and Metzger 1994). While phosphorylated tyrosines can help proteins bind to one another, the phosphorylation of Ser-Thr often induces conformational changes, presumably because of the introduced negative charge (Berg, Tymoczk, and Stryer 2002).

To understand these effects and to study in depth about phosphorylating residues, scientists simulate phosphorylation in the laboratory by mutating the Ser-Thr residues. For example, even though several mechanisms for signal transduction exist, immunoreceptor tyrosine-based activation motifs (ITAMs) that may be present in receptor complexes aid in signaling beyond the immune system (Fodor, Jakus, and Mócsai 2006). Upon the ligand binding to the receptors, tyrosines within the ITAM sequence are phosphorylated by Src family kinases, leading to an SH2 domain-mediated recruitment and activation of the Syk or the related ZAP-70 tyrosine kinase. These kinases then initiate further downstream signaling events. The authors also observed that ITAMs play significant roles in signaling other types of cells, including osteoclasts and blood platelets (Phillips et al. 2001).

In addition to phosphorylation, the most important modification of the primary structure is peptide cleavage. Proteins are often synthesized in an inactive form; typically, an N-terminal or C-terminal segment blocks the active site of the protein, inhibiting its function. The protein then needs to be activated from its precursor form by cleaving off the inhibitory peptide (Berg, Tymoczk, and Stryer 2002).

Proteins tend to cleave within their sites. A typical N–O acyl shift can be cited as an example in which the hydroxyl group of a serine attacks the carbonyl carbon of a preceding peptide bond, thereby forming a tetrahedral intermediate. This intermediate tends to revert to the amide form, expelling the attacking group, since the amide form is usually favored by free energy. However, additional molecular interactions may render the amide form less stable; the amino group is expelled instead, resulting in an ester (Ser-Thr) or thioester (Cys) bond in place of the peptide bond (Nelson and Cox 2008).

Other posttranslational modifications include acylation–acetylation, the addition of an acetyl group, either at the N-terminus of the protein (Polevoda and Sherman 2003) or at lysine residues (Yang and Seto 2008). α- and β-Tubulins form heterodimers and serve as building blocks for microtubules. α-Tubulin was among the first acetylated nonhistone proteins to be discovered (Westermann and Weber 2003). The acetylation site was mapped to Lys40, which is absent in β-tubulin and is located on the lumen of microtubules.

Histone acetylation (Bártová et al. 2008; Glozak 2005) is also an important modification involving enzyme catalysis by histone acetyltransferase (HAT) or histone deacetylase (HDAC) activity, although it should be noted that HATs and HDACs can modify the acetylation status of nonhistone proteins as well. Yang and Seto (2008) have also linked protein acetylation to other modifications, including phosphorylation, methylation, and ubiquitination.

1.4.4 Primary Structure Determination

Since the primary structure comprises chiefly the amino acids making up the polypeptide chain, the determination of amino acid composition and sequence will form the primary means of determination of the primary structure. Prior to the determination of amino acids within a protein, all possible hydrolysis (acid, alkali, etc.) should be undertaken to establish the full amino acid composition of the protein. Spectrophotometric and fluorometric analyses with the conventional ninhydrin method can help identify a majority of amino acids, although suitable digestion procedures need to be followed. Techniques like reducing and nonreducing electrophoresis can reveal the presence of possible disulfide bonds.

It is important that while sequencing, the protein or peptide needs to be free from other contaminant proteins or substances used in the characterization steps. Most often, the Edman degradation method is used, but recently, the use of proteomic tools like mass spectrometers coupled in tandem with the help of software can predict the amino acid sequence on the basis of the ionization of individual residues. Chapter 2 of this book reviews the methods used for determining protein structure, amino acid analysis, and sequencing.

1.4.5 Secondary Structure

The secondary structures of proteins, the α-helices and β-sheets, help form the three-dimensional protein structure via inter–amino acid residue interactions of the polypeptide backbone.

α-Helix: A common motif in the secondary structure of proteins, the α-helix is a right- or left-handed coiled conformation in which every backbone N–H group donates a hydrogen bond to the backbone C=O group of the amino acid, four residues earlier as an iteration (i + 4 – i hydrogen bonding). The amino acids in an α-helix pattern are arranged in a right-handed helical structure where each amino acid corresponds to a 100° turn in the helix (i.e., the helix has 3.6 residues per turn) and a translation of 1.5 Å (= 0.15 nm) along the helical axis. The pitch of the helix (the vertical distance between two points on the helix) is 5.4 Å (= 0.54 nm), which is the product of 1.5 and 3.6. Figure 1.9 shows hemoglobin, a full helically structured protein.

β-Sheet: The β-sheet (also β-pleated sheet) is the second form of a regular secondary structure in proteins, consisting of β-strands connected laterally by five or more hydrogen bonds, forming a generally twisted, pleated sheet. A β-strand is a stretch of amino acids, typically 5–10 amino acids long, whose peptide backbones are almost fully extended. The association of β-sheets has been implicated in the formation of protein aggregates and fibrils observed in many human diseases, notably the amyloidoses. Figure 1.10 shows a β-pleated amyloid fibril three-dimensional structure.

β-Sheets are present in all -β,α + β, and α/β domains and in many peptides or small proteins with poorly defined overall architecture. All-β domains may form β-barrels, β-sandwiches, β-prisms, β-propellers, and β-helices (Voet and Voet 2004). In rare cases, posttranslational mechanisms cause a typical α-helical protein to convert into an entirely β-sheeted protein, as in the case of prion proteins. The normal, cellular prion protein (PrP^C) is converted into PrP^{Sc} through a posttranslational process during which it acquires high β-sheet content (Prusiner 1997).

Figure 1.9   Three-dimensional structure of hemoglobin. Quaternary structure of hemoglobin. (From http://commons.wikimedia.org/wiki/File:Main_protein_structure_levels_en.svg.)

Figure 1.10 Three-dimensional structure of β-pleated amyloid Aβ (1–42) fibril. (From Lührs, T., et al., *Proceedings of the National Academy of Sciences,* **1, 17342–47, 2005.)**

The side chains of amino acids play crucial roles in designing the three-dimensional conformation. Several intermolecular forces with the available torsion angles that rotate the bond about them are required to stabilize the secondary structures into a three-dimensional fold. In proteins, the secondary structure is denoted largely by patterns of hydrogen bonds between backbone amide and carboxyl groups, where the *Dictionary of Protein Secondary Structure* (DSSP) definition of a hydrogen bond is used (Kabsch and Sander 1983).

1.4.6 *Super Secondary Structure*

The super secondary structures are compact three-dimensional structures of several adjacent elements of a secondary structure that is smaller than a protein's domain or a subunit. Super secondary structures can act as nucleations in the process of protein folding. Examples include β-hairpins, α-helix hairpins, and β–α–β motifs. Protein structural motifs often include loops of variable length and unspecified structure. Examples of super secondary structures include the following (Chiang et al. 2007):

β-Ribbon: This occurs when two antiparallel β-strands connect by a tight turn of a few amino acids between them.

Greek key: Four β-strands folded over into a sandwich shape.

Omega loop: A loop in which residues that make up the beginning and end of the loop are placed close together.

Helix-loop-helix: Consists of α-helices bound by a looping stretch of amino acids.

Zinc finger: Two β-strands with an α-helix end folded over to bind a zinc ion. This motif is seen in transcription factors.

1.4.7 *Secondary Structure Determination and Prediction*

The rough secondary structure content of a protein containing approximately 40% α-helix and 20% β-sheet can often be estimated using spectroscopy–spectrometry methods. For proteins, a common method is far-ultraviolet (far-UV, 170–250 nm) circular dichroism. A pronounced double intensity at 208 and 222 nm indicates an α-helical structure, whereas a single intensity minimum at 204 or 217 nm denotes a random coil or a β-sheet structure, respectively. Infrared spectroscopy, which detects differences in the bond oscillations of amide groups due to hydrogen bonding, is less common as a tool to study secondary structures (Branden and Tooze 1999). Using NMR spectroscopy, secondary structural elements can be accurately estimated based on chemical shifts, thereby assigning residue-specific chemical shifts to the NMR spectrum.

The eight-state DSSP code is actually a simplification of the 20 amino acid residues present in a protein, but the majority of secondary structure prediction methods still use the simple three dominant states: helix, sheet, and coil. Early methods of secondary structure prediction were based on the helix- or sheet-forming propensities of individual amino acids and were considered inaccurate in predicting the actual state that the protein adopts (helix/sheet/coil).

By using multiple sequence alignment, wherein the alignment of multiple sequences is usually three or more assuming an evolutionary relationship among the sequences, structural propensities proximate to a given residue position can be determined (Duret and Abdeddaim 2000). For example, glycine at a particular position might indicate the possibility of a random coil. However, in homologous proteins, the same position may reveal a helix formation due to the presence of helix-favoring amino acids by multiple sequence alignment experiments. Furthermore, this alignment may also suggest factors like hydrophobicity at this position, influencing solvent accessibility. These suggest that even though the presence of glycine at that position is established, other factors considered would make that position adopt a helical conformation, rather than a random coil. Several types of methods, including modern prediction methods, are used to combine all the available data to provide a confidence score for their predictions at every position.

1.4.8 Tertiary Structure

The three-dimensional structure of protein represents the tertiary structure. Determinants of a tertiary structure include folding, hydrophobicity, and disulfide bonds (for secreted proteins). Largely, proper folding of the protein in its native form constitutes the tertiary structure. Several physical and chemical factors govern native protein folding. Interactions stabilizing the tertiary structure include disulfide bonds, hydrophobic interactions, hydrogen bonds, and ionic interactions (Nelson and Cox 2008). Within the cell, several protein-folding assisting molecules, called chaperones, enable proper folding. Of course, there are several disordered or randomly coiled proteins that do not have a well-defined tertiary structure.

In globular proteins, interactions at the tertiary structure are stabilized by the presence of hydrophobic amino acid residues in the protein core, from which water is excluded, and by the resultant accumulation of charged or hydrophilic residues on the protein's water-exposed surface. In secreted proteins, disulfide bonds between cysteine residues help maintain the protein's tertiary structure (Berg, Tymoczk, and Stryer 2002). A variety of common and stable tertiary structures appear in a large number of proteins that are unrelated in both function and evolution, for example, many proteins are shaped like a TIM barrel, named after the enzyme triosephosphateisomerase. Another common structure is a highly stable dimer that takes up a coiled-coil structure composed of two to seven α-helices. Proteins are classified by the folds they represent in databases like SCOP (structural classification of proteins) and CATH (class architecture topology homology).

1.4.9 Determination of Tertiary Structure

The majority of protein structures known to date have been solved with the experimental technique of x-ray crystallography, which typically provides data of high resolution but lacks information about dynamic conformational flexibility. The second best method to solve protein structures is NMR spectroscopy, which provides moderate resolution data but is limited to relatively small proteins. Most importantly, it can provide real-time information about the motion of a protein in solution. For determining the overall conformation and conformational changes in surface-captured proteins, dual polarization interferometry may be used. In general, more is known about the tertiary structural features of soluble globular proteins than about membrane proteins because membrane proteins are extremely difficult to study using these methods.

1.4.10 Quaternary Structure

Multiple subunits coalesce into a protein's native environment to form a quaternary structure. For example, enzymes composed of subunits (holoenzymes) need to interact to activate the regulatory or catalytic subunit within. Hemoglobin and DNA polymerase are typical examples of proteins exhibiting quaternary structures (Berg, Tymoczk, and Stryer 2002). Protein–protein interactions can also be grouped under this category. Depending on the number of subunits that are arranged into an oligomeric protein complex, proteins can be named as monomers, dimers, trimers, tetramers, and so on. Extremely complex oligomeric structures may comprise more than 20 subunits. For example, certain viral capsids contain multiples of 60 proteins.

Examples of proteins with a quaternary structure include hemoglobin, DNA polymerase, and ion channels. Other assemblies, referred to as "multiprotein complexes," also possess a quaternary structure. Examples include nucleosomes and microtubules. Changes in a quaternary structure can occur through conformational changes within individual subunits or through relative reorientation of the subunits. It is through such changes, which underlie cooperativity and allostery in "multimeric" enzymes, that many proteins undergo regulation and perform their physiological function.

Many proteins have a quaternary structure, which consists of several polypeptide chains that associate to form an oligomer. Each polypeptide chain is referred to as a subunit. Hemoglobin, for example, consists of two α-subunits and two β-subunits. Each of the four chains has an all-α globin fold with a specific heme pocket. Domain swapping is a mechanism for forming oligomeric assemblies (Bennett, Schlunegger, and Eisenberg 1995). In domain swapping, a secondary or tertiary element of a monomeric protein is replaced by the same element of another protein. Domain swapping can range from secondary structure elements to whole structural domains. It also represents a model of evolution for functional adaptation by oligomerization, for example, oligomeric enzymes that have their active site at subunit interfaces (Heringa and Taylor 1997).

1.4.11 Determination of Quaternary Structure

A protein quaternary structure can be determined using a variety of experimental techniques that require a protein sample in a variety of experimental conditions. The experiments often provide an estimate of the mass of the native protein and, together with the knowledge of the masses and/or stoichiometry of the subunits, allow the quaternary structure to be predicted with a given accuracy. For a variety of reasons, it is not always possible to obtain a precise determination of the subunit composition.

The number of subunits in a protein complex can often be determined by measuring the hydrodynamic molecular volume or mass of the intact complex, which requires native proteins in solution. The mass can then be inferred from its volume using the partial specific volume of 0.73 mL/g. However, for unfolding proteins, volume measurements are less accurate than mass measurements, and additional experiments are required to determine whether a protein is unfolded or has formed an oligomer.

1.5 Dimensions and Complexity of Protein Structure

Owing to the innumerous adaptations that proteins can organize, studying and understanding these complex dimensions and folding can be very challenging. As protein structures are solved step-by-step, early prediction of key structural aspects becomes possible. With this information, structures can be solved using high-resolution spectroscopy or x-ray crystallography; of course, the protein needs to be engineered in such a way that its structure can be solved using either of the methods. Once solved, examination of the key structural elements with other regions (prosthetic group) can be done to get a clear picture of the protein's overall fold and quaternary structure. Hence, let us focus on determining the protein structure.

If a protein was to attain its correctly folded configuration by sequentially sampling all possible conformations, it would require an enormous amount of time to attain its correct native conformation. This is true, even if conformations are sampled at rapid (nanosecond or picosecond) rates. The "paradox," however, is that most small proteins fold spontaneously on a millisecond or even microsecond timescale (Levinthal 1968). This paradox is usually noted in the context of computational approaches to protein structure prediction. Levinthal, however, suggested that the stable structure could have a higher energy and that protein folding will be accelerated and guided by the rapid formation of local interactions. These then determine the further folding of the peptide, suggesting local amino acid sequences that form stable interactions and serve as nucleation spots in the folding process.

1.5.1 Protein Structure Prediction and Determination

Around 90% of protein structures available in the Protein Data Bank (PDB) have been determined by x-ray crystallography (Zhang 2008). This method allows one to measure the three-dimensional density distribution of electrons in protein (in the crystallized state) and thereby infer the three-dimensional coordinates of all atoms to be determined to a certain resolution. On the contrary, only 9% of known protein structures have been obtained by NMR techniques, which can also be used to determine the secondary structure. Aspects of the secondary structure as a whole can be determined via other biochemical techniques, such as circular dichroism.

The secondary structure can also be predicted with a high degree of accuracy. Cryo-electron microscopy has recently come into vogue for determining protein structures to a high resolution (<5 Å or 0.5 nm) and is anticipated to provide high-resolution data in the next decade (Brocchieri and Karlin 2005). This technique is still a valuable resource for researchers working with very large protein complexes, such as viral coat proteins and amyloid fibers.

Higher organization and study of structures have become possible with the discovery of higher evolutionary functional groups associated with protein structures. Modules, families, folds, and higher organizational domains have enabled the determination of the relationship between primary and tertiary structures (Wheelan, Marchler-Bauer, and Bryant 2000). Domains could be autonomously folding units present within protein structures. They are used to generate new sequences and those that are repeatedly found in diverse proteins are often referred to as modules; examples can be found among extracellular proteins associated with clotting, fibrinolysis, complement, the extracellular matrix, cell surface adhesion molecules, and cytokine receptors.

There are currently about 45,000 experimentally determined three-dimensional protein structures present within the PDB. However, this set contains a lot of identical or very similar structures. All proteins should be classified according to structural families to understand their evolutionary relationships. Structural comparisons are best achieved at the domain level. For this reason, many algorithms have been developed to automatically assign domains in proteins with a known three-dimensional structure. For example, a CATH domain database classifies domains into approximately 800 fold families (Orengo et al. 1997), 10 of these folds are referred to as "super-folds" based on their densities. Super-folds are defined as folds for which there are at least three structures without significant sequence similarity. The most populated is the α/β-barrel super-fold.

The simplest multidomain organization seen in proteins is that of a single domain repeated in tandem. The domains may interact with each other or remain isolated, like beads on a string. Sometimes, as seen in serine proteases, a gene duplication event can lead to the formation of two β-barrel domain enzymes.

Modules often show connectivity relationships, as observed by kinesins and ABC transporters. The kinesin motor domain can be at either end of a polypeptide chain that includes a coiled-coil region and a cargo domain. ABC transporters are built with up to four domains consisting of two unrelated modules, an ATP-binding cassette, and an integral membrane module, arranged in various combinations.

Not only do domains recombine, but they also insert into one another. Sequence or structural similarities to other domains demonstrate that homologs of inserted and parent domains can exist independently. Several examples exist to illustrate this effect. Zinc fingers and fingers inserted into the Pol I family can be cited (Krishna, Majumdar, and Grishin 2003).

A number of factors exist that make protein structure prediction a very difficult task. The two main problems are that the number of possible protein structures is extremely large and that the physical basis of protein structural stability is not fully understood. As a result, any protein structure prediction method needs a way to explore the space of possible structures efficiently and a way to identify the most plausible structure. The tertiary structure of a native protein may not be readily formed or determined without the aid of additional agents. For example, proteins known as chaperones are required for some proteins to properly fold. Other proteins cannot fold properly without modifications, such as glycosylation. A particular protein may be influenced by its chemical environment to attain multiple conformations. The biologically active conformation of the protein may not be the most thermodynamically favorable one.

Due to the increase in computer software, databases, and especially new algorithms, much progress is being made to overcome these problems. However, routine de novo prediction of protein structures, even for small proteins, is still not achieved.

1.6 Conclusion

There are currently about 45,000 experimentally determined three-dimensional protein structures present within the PDB. With the discovery of higher organizational forms and domains, structure prediction and determination have become more challenging. Yet, newer algorithms and programs are being used to decipher the structures to precision in order to determine the function and functionality of a protein. Thus, knowing the basic chemistry behind amino acids, peptide bonds, and primary and secondary structural elements can help understand how a protein's structure can be determined and understood. Deciphering the structure will eventually help determine the protein's function. Understanding amino acids' properties, their chemistry, and the forces involved will help elucidate the structures of peptides and proteins.

References

Bártová, E., Krejcí, J., Harnicarová, A., Galiová, G., and Kozubek, S. 2008. Histone modifications and nuclear architecture: A review. *Journal of Histochemistry and Cytochemistry* 56: 711–21.

Bennett, M. J., Schlunegger, M. P., and Eisenberg, D. 1995. 3D domain swapping: A mechanism for oligomer assembly. *Protein Science* 4: 2455–68.

Berg, J. M., Tymoczk, J. L., and Stryer, L. 2002. Protein structure and function. In *Biochemistry*, ed. L. Stryer, 5th edn, 693–98. San Francisco: W. H. Freeman & Company.

Branden, C. and Tooze, J. 1999. *Introduction to Protein Structure*, 2nd edn. New York, NY: Garland Publishing.

Brocchieri, L. and Karlin, S. 2005. Protein length in eukaryotic and prokaryotic proteomes. *Nucleic Acids Research* 33: 3390–3400.

Campbell, M. 2007. Amino acids. www.bio.davidson.edu.

Chiang, Y. S., Gelfand, T. I., Kister, A. E., and Gelfand, I. M. 2007. New classification of supersecondary structures of sandwich-like proteins uncovers strict patterns of strand assemblage. *Proteins* 68(4): 915–21.

Creighton, T. H. 1993. *Proteins: Structures and Molecular Properties*, Chapter 1. San Francisco: W. H. Freeman.

Duret, L. and Abdeddaim, S. 2000. Multiple alignment for structural functional or phylogenetic analyses of homologous sequences. In *Bioinformatics Sequence Structure and Databanks*, ed. D. Higgins and W. Taylor, 1–14. Oxford University Press.

Fodor, S., Jakus, Z., and Mócsai, A. 2006. ITAM-based signaling beyond the adaptive immune response. *Immunology Letters* 104: 29–37.

Glozak, M. A., Sengupta, N., Zhang, X., and Seto, E. 2005. Acetylation and deacetylation of non-histone proteins. *Gene* 363: 15–23.

Heringa, J. and Taylor, W. R. 1997. Three-dimensional domain duplication, swapping and stealing. *Current Opinion in Structural Biology* 7: 416–21.

Janin, J. 1979. Surface and inside volumes in globular proteins. *Nature* 277: 491–92.

Kabsch, W. and Sander, C. 1983. Dictionary of protein secondary structure: Pattern recognition of hydrogen-bonded and geometrical features. *Biopolymers* 22: 2577–637.

Klotzbach-Shimomura, K. 2001. Functional foods: The role of physiologically active compounds in relation to disease. *Topics in Clinical Nutrition* 16: 68–78.

Krishna, S. E., Majumdar, I., and Grishin, N. V. 2003. Survey and summary: Structural classification of zinc fingers. *Nucleic Acids Research* 31: 532–50.

Kyte, J. and Doolite, R. 1982. A simple method for displaying the hydropathic character of a protein. *Journal of Molecular Biology* 157: 105–32.

Levinthal, C. 1968. Are there pathways for protein folding? *Journal de Chimie Physique et de Physico-Chimie Biologique* 65: 44–45.

Li, B., Yu, J. P., Brunzelle, J. S., Moll, G. N., van der Donk, W. A., and Nair, S. K. 2006. Structure and mechanism of the lantibiotic cyclase involved in nisin biosynthesis. *Science* 311: 1464–67.

Lührs, T., Ritter, C., Adrian, M., Riek-Loher, D., Bohrmann, B., Döbeli, H., Schubert, D., and Riek, R. 2005. 3D structure of Alzheimer's amyloid-β (1–42) fibrils. *Proceedings of the National Academy of Sciences* 1: 17342–47.

Magee, T. and Seabra, M. C. 2005. Fatty acylation and prenylation of proteins: What's hot in fat. *Current Opinion in Cell Biology* 17: 190–96.

Mann, M. and Jensen, N. 2003. Proteomic analysis of post-translational modifications. *Nature Biotechnology* 21: 255–61.

Meister, A. 1988. Glutathione metabolism and its selective modification. *Journal of Biological Chemistry* 263: 17205–208.

Nelson, D. and Cox, M. 2008. *Lehninger Principles of Biochemistry*, 4th edn. New Jersey: W. H. Freeman and Co.

Orengo, C. A., Michie, A. D., Jones, S., Jones, D. T., Swindells, M. B., and Thornton, J. M. 1997. CATH – a hierarchic classification of protein domain structures. *Structure* 5: 1093–1108.

Pauling, L., Corey, R. B., and Branson, H. R. 1951. The structure of proteins: Two hydrogen-bonded helical configurations of the polypeptide chain. *Proceedings of the National Academy of Sciences USA* 37: 205–11.

Phillips, D. R., Prasad, K. S., Manganello, J., Bao, M., and Nannizzi-Alaimo, L. 2001. Integrin tyrosine phosphorylation in platelet signaling. *Current Opinion in Cell Biology* 13: 546–54.

Pilobello, K. T. and Mahal, L. K. 2007. Deciphering the glycocode: The complexity and analytical challenge of glycomics. *Current Opinion in Chemical Biology* 11: 300–305.

Polevoda, B. and Sherman, F. 2003. N-terminal acetyltransferases and sequence requirements for N-terminal acetylation of eukaryotic proteins. *Journal of Molecular Biology* 325: 595–622.

Pribluda, V. S., Pribluda, C., and Metzger, H. 1994. Transphosphorylation as the mechanism by which the high-affinity receptor for IgE is phosphorylated upon aggregation. *Proceedings of the National Academy of Sciences USA* 91: 11246–50.

Prusiner, S. B. 1997. Prion diseases and the BSE crisis. *Science* 278: 245–51.

Ramachandran, G. N., Ramakrishnan, C., and Sasisekharan, V. 1963. Stereochemistry of polypeptide chain configurations. *Journal of Molecular Biology* 7: 95–99.

Reeds, P. J. 2000. Dispensable and indispensable amino acids for humans. *Journal of Nutrition* 130(7): 1835S–40S.

Renfrew, D. P., Butterfoss, G., and Kuhlman, B. 2008. Using quantum mechanics to improve estimates of amino acid side chain rotamer energies. *Proteins* 71: 1637–46.

Rose, G., Geselowitz, A., Lesser, G., Lee, R., and Zehfus, M. 1985. Hydrophobicity of amino acid residues in globular proteins. *Science* 229: 834–38.

Sanger, F. 1952. The arrangement of amino acids in proteins. *Advances in Protein Chemistry* 7: 1–67.

Schreiber, T. B., Mäusbacher, N., Breitkopf, S. B., Grundner-Culemann, K., and Daub, H. 2008. Quantitative phosphoproteomics – an emerging key technology in signal-transduction research. *Proteomics* 8: 4416–32.

Urry, D. W. 2004. The change in Gibbs free energy for hydrophobic association – Derivation and evaluation by means of inverse temperature transitions. *Chemical Physics Letters* 399: 177–83.

Voet, D. and Voet, J. G. 2004. *Biochemistry*, 3rd edn, 227–31. Hoboken, NJ: Wiley.

Westermann, S. and Weber, K. 2003. Post-translational modifications regulate microtubule function. *Nature Reviews Molecular Cell Biology* 4: 938–47.

Wheelan, S. J., Marchler-Bauer, A., and Bryant, S. H. 2000. Domain size distributions can predict domain boundaries. *Bioinformatics* 16: 613–18.

Wolfenden, R., Andersson, L., Cullis, P., and Southgate, C. 1981. Affinities of amino acid side chains for solvent water. *Biochemistry* 20: 849–55.

Yang, X. J. and Seto, E. 2008. Lysine acetylation: Codified crosstalk with other posttranslational modifications. *Molecular Cell* 31: 449–61.

Zhang, Y. 2008. Progress and challenges in protein structure prediction. *Current Opinion in Structural Biology* 18: 342–48.

Chapter 2

Food Proteins–Peptides: Determination and Characterization

Arvind Kannan, Navam Hettiarachchy, and Maurice Marshall

Contents

2.1 Introduction

With the advent of the advancement in characterization tools and modeling programs, several proteins have been characterized successfully. Using advanced proteomic tools, it has become possible to study the key structural elements of proteins and their behavior and the way their biological activity can be modulated in relation to structure. Structural proteomics has advanced to the extent that predicting a protein's tertiary structure from its primary structure has become possible. By precisely determining the structural characteristics using in vitro experiments and supplementing information into proteomic tools, scientists can fully characterize a protein or a peptide.

This chapter attempts to explore the modes of characterizing a protein–peptide from a mixture, eventually identifying the structural components using spectrometry coupled with mass spectrometry (MS). Several models of characterization steps involving membrane separation processes, purification, sequencing, and mass determination in fully characterizing food proteins and peptides have been discussed. Emphasis is given to the order of the chapter's contents, so that a systemic approach can be adopted for characterization. At the end, the reader will be able to assimilate the various methods and choose one or more that are needed to separate, determine, and purify the protein–peptide of interest and essentially follow a stepwise paradigm to fully characterize the protein–peptide of interest. A review of the modern tools used in combination with conventional in vitro equipment for high-resolution characterization has been done, highlighting the software and structure prediction programs. Overall, this chapter will not only cover the basic principles of protein–peptide structural determination and characterization but also delve into sophisticated in vitro tools to explore the structures in depth.

2.1.1 Determination of Proteins and Peptides in Foods

Specific peptide sequences with specific amino acid sequences are thought to be modulators of biological function. Several bioactive peptides have been fully characterized from food proteins using a systematic approach involving several steps and methods. Characterization generally involves the separation, purification, and detection of the structural components of the peptide or protein under investigation. The separation of bioactive peptides from enzymatic hydrolysates provides the scope to create functional foods that can slow disease progression. Proper analytical separation, detection, and purification techniques can identify specific amino acid sequences and hence peptides as important biological modulators. A number of separation techniques are available. As the subsequent stage in characterization, purity becomes an important criterion to concentrate all of the bioactivity within a pure compound. The purification of peptides ideally requires a preparative mode of chromatography that relies on ion exchange, hydrophobic interaction, size exclusion, or affinity. These are generally employed as initial purification steps, where the peptides of interest are isolated out from a given pool that had exhibited bioactivity as a whole fraction. Following this, reversed-phase high-performance liquid chromatography (HPLC) has been widely employed with or without coupling to ion-exchange chromatography for efficient purification of the peptide of interest (Leonil et al. 2000).

To assess the purity and further characterize the structural components within the bioactive peptide, various proteomic tools can be used. Several food peptides have been characterized using MS, a versatile tool not only for determining accurate molecular sizes but also for providing the structural information of the pure compound. MS is by far the most informative technique for determining the purity of protein–peptide samples. The mass spectrum provides the number of components present and an indication about the length of the proteins–peptides. It has been widely used as a modern tool to characterize peptides and proteins present in foods that can impart bioactive properties. MS in conjunction with database searching is playing an increasingly important role in the characterization of peptides. Of the ionization modes, matrix-assisted laser desorption ionization (MALDI) has made it possible to not only determine the accurate mass of peptides obtained in a hydrolysate of food proteins but also identify the purity of a pure protein or peptide, determine its fragmenting patterns, and also determine the de novo sequence of the amino acids.

The shorter the peptides, the more accurate will be the prediction by MS-MALDI of the sequence; nevertheless, the score for predicting the best hit for sequences present in longer peptides can enable the identification of C-terminal amino acids to a reasonable accuracy. Although MS was applied more popularly to biologically important peptides, its use is growing in food peptide characterization as well. Precise characterization of peptides therefore becomes a valuable tool in ascribing each peptide to its biofunctional role. Because peptides fragment in a sequence-specific way, the original amino acid sequence can be derived from the fragment ion signals. Post source decay (PSD) has been used successfully to obtain amino acid sequence information for many peptides (Spengler 1997).

2.2 Characterization Principles and Methods

Characterization involves determining the physical properties as well as the chemical composition. The physical characterization of the isolated protein or peptide mostly employs the procedures that determine the apparent molecular weight and charge, size, and shape of the molecule (Nandi 1997). The chemical characterization involves amino acid composition and sequence analysis. The physical properties can determine, to a great extent, the methodology needed to purify the protein of interest. For example, a protein with a net negative charge can be purified from a mixture, using affinity or ion-exchange chromatography on positively charged resin. Similarly a known molecular weight protein can be separated by gel filtration chromatography using selective pore sizes within the gel matrix to allow or retain the compound of interest.

Determining the size or molecular weight of the molecule is based on (a) the composition of the molecule; (b) the colligative properties, such as vapor and osmotic pressure, that influence the protein molecule; (c) the transport of molecules within a solution under the effects of mechanical, electrical, or centrifugal force; and (d) the ability of the protein molecule to scatter the incident light.

Composition: The estimation of the molecular mass of the protein or peptide can be made from measuring its dry weight; however, for accuracy and reproducibility, protein concentration is normally determined by refractometry or spectrophotometry. Amino acid analysis also gives accurate information about the molecular mass of the protein or peptide in question based on the composition. Often, the composition analysis is coupled with other techniques like electrophoresis to determine the accurate mass of the protein, particularly when low amounts of samples are being used.

Since proteins are made up of amino acids, the properties of certain amino acids and their ability to absorb a definite wavelength of light have been sought to determine the protein concentration. Protein determination methods include the bicinchoninic acid assay (also known as the BCA assay or Smith assay), which determines the total level of protein in a solution, similar to the Lowry protein assay, Bradford protein assay, or biuret reagent. The total protein concentration is exhibited by a color change of the sample solution in proportion to the protein concentration, which can then be measured using colorimetric techniques.

Colligative properties and transport properties: The colligative properties are a result of the modification of the chemical potential of a solvent by the presence of the solute. Boiling point elevation, freezing point depression, and osmotic pressure all depend on the amount of solute present and can therefore aid in measuring the concentration of the molecule.

Transport properties involve sedimentation methods that help determine both the mass and the shape of the molecule. Sedimentation equilibrium can be used to determine the shape of the protein molecule, and sedimentation velocity is used to determine the hydrodynamic properties of the molecule (Nandi 1997).

In sedimentation equilibrium, the sample is subjected to high-speed ultracentrifugation in an analytical ultracentrifuge to force the protein toward the outside of the rotor, but not high enough to cause the sample to form a residual pellet. As the centrifugal force produces a gradient in the protein concentration across the centrifuge cell, diffusion acts to oppose this concentration gradient. Eventually, an exact balance is reached between sedimentation and diffusion, and the concentration distribution reaches equilibrium. This equilibrium concentration distribution across the cell is then measured while the sample is spinning, using either absorbance or refractive index detection.

The sedimentation velocity, on the other hand, uses the sedimentation rate and provides information about both the molecular mass and the shape of the molecules. In the sedimentation velocity method, a sample is spun at very high speed in an analytical ultracentrifuge. The high centrifugal force rapidly depletes all the protein from the region nearest the center of the rotor (the meniscus region), forming a boundary, which moves toward the outside of the rotor with time until, finally, all the protein forms a residual pellet at the outer layer of the cell. The concentration distribution across the cell at various times during the experiment is measured while the sample is spinning, using either absorbance or refractive index detection.

Amino acid analysis: Amino acid composition usually provides a method for quantitatively characterizing the component amino acids in the protein or peptide. Generally, hydrolysis using acid (6N HCl) is carried out, but certain amino acids undergo modifications under such hydrolyzing conditions (Ozols 1990). For example, glutamine and asparagine are hydrolyzed to glutamic acid and aspartic acid, respectively. Tryptophan is generally lost during the acid hydrolysis process, but alkaline hydrolysis can be performed to recover tryptophan (Edelhoch 1967). The addition of 2-mercaptoethanol to the hydrolyzing mixture helps capture methionine by preventing its oxidation to methionine sulfoxide. Determination of cysteine can be achieved by oxidation using performic acid. The extent of hydrolysis also plays a role in capturing certain peptide bonds like ala-ala, Ile-Ile, val-Ile, and ala-val.

Care must be taken to ensure that there is no loss of any amino acid, particularly when there is no clue about the composition of the probable amino acids in the peptide. Spectrophotometric and fluorometric analyses with the conventional ninhydrin method can identify, to some extent, a few amino acids, and information from these experiments can be used to ascertain the type of hydrolysis required for conducting amino acid analysis.

Sequencing: To characterize the structural and functional domains of the protein or peptide, N-terminal amino acid sequencing can be performed. It is important that the protein or peptide that needs to be sequenced is free from other contaminant proteins or substances used in the characterization steps. In cases where a sufficient signal is not generated, there could be a block in the N-terminal end. Unblocking the terminus by cleaving the protein can help determine the N-terminal sequence (Matsudaira 1990). Most often, Edman degradation method is used, but recently, the use of proteomic tools like mass spectrometers coupled in tandem with the help of software can predict the amino acid sequence based on the ionization of individual residues. The section on characterization using proteomic tools covers this aspect in more detail.

2.2.1 Fractionation and Separation Techniques

Proteins and hence their constituent amino acids are present abundantly in many food sources. The isolation and identification of proteins or peptides of interest for bioactive functional roles are very important steps. However, characterization has proven to be a challenging task because many proteins–peptides can possess closely related physicochemical properties.

The separation of bioactive peptides from enzymatic hydrolysates provides the scope to create functional foods that can slow disease progression. Proper analytical separation and purification techniques alone can identify specific amino acid sequences and hence peptides as important biological modulators. A number of separation techniques are available. Fractionation and purification techniques are normally performed based on where and for what purpose the final product or molecule is rendered applicable. Achieving purity generally requires the use of expensive methods, such as liquid and gas chromatography, particularly for industrial applications. As the first step in the characterization, even before bioactivity can be determined, the generated peptide hydrolysates, for example, from a food source after extraction, are normally subject to at least one of several membrane filtration procedures to fractionate the peptides into molecular sizes for ease of identification of bioactive fractions and also for further characterization.

2.2.1.1 Membrane Separation Processes

A range of filtration material is used to fractionate molecules based on the membrane pore sizes. The choice of the appropriate filtration system depends on the type of product needed to be fractionated. For example, an ultrafiltration (UF) membrane typically consisting of pore sizes between 1 and 500 nm is used to fractionate proteins and peptides from a protein hydrolysate (Figure 2.1).

Nanofiltration (NF) materials are used to separate smaller molecules, such as amino acids and salts, based on size and charge. They usually operate with membranes of pore sizes between 0.1 and 1 nm. These filtration methods and some others must operate under a consistent pressure, ranging from 0.1 up to 50 bars. Table 2.1 lists the comparative features of membrane filtration methods.

The membranes are normally made up of polymers and bear configurations of hollow fibers, flat sheets, tubules, or spiral windings. Owing to the subtle differences in the physicochemical properties among mixtures of peptides, a fractionation process, such as UF or NF, is usually preferred over other conventional separation methods.

To cite examples of using membrane filtration methods, Turgeon and Gauthier used hollow fiber polysulfone UF membranes having molecular weight cutoffs (MWCO) of 30 and 1 kDa. They were able to obtain a pool of polypeptides that improved the overall composition compared with the ratio of commercially available whey protein concentrate to higher protein content. They also obtained peptides lower than 5000 Da, which could be used as emulsifiers in the food industry (Turgeon and Gauthier 1990). Nau et al. (1995) used the NF process to fractionate β-casein peptides. With 0.1–1 nm pore size ranges, Nau et al. showed separation casein peptides based on both electric charges and size exclusion. Kimura and Tamano (1986) used charged UF membranes to separate amino acids between

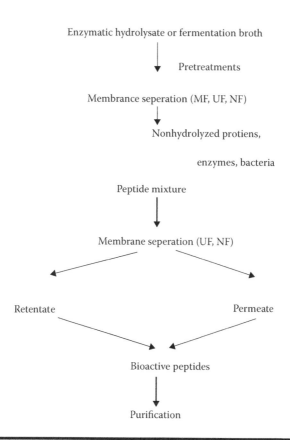

Enzymatic hydrolysate or fermentation broth

↓ Pretreatments

Membrance seperation (MF, UF, NF)

↓

Nonhydrolyzed protiens,

enzymes, bacteria

Peptide mixture

↓

Membrane seperation (UF, NF)

Retentate   Permeate

Bioactive peptides

↓

Purification

Figure 2.1   Production of peptides using membrane filtration methods.

75 and 200 kDa. The separation of a complex mixture of 15 amino acids using two membranes with molecular cutoffs of 500 and 1000 Da, respectively, was studied by Garem et al. (1998). It was observed that separation with the high MWCO membrane was pH dependent and that maximum selectivity was obtained at pH 10 with a low ionic strength. Garem et al. also used NF membranes to fractionate β-casein peptides into acidic and basic peptides at pH 8.0.

Peptides with different amino acid sequences were obtained from soybean protein, after hydrolysis with enzymes and UF. They possessed various biological activities based on the initial protein source, enzyme used, and processing conditions. Different enzymes were used for hydrolyzing the soy protein that yielded peptides with antioxidant peptides (Pena-Ramos and Xiong 2002), peptides with anticancer properties (Kim et al. 2000), or peptides with hypotensive activity (Wu and Ding 2001). Proteases from bacterial cultures were used to hydrolyze milk protein to obtain peptides, which were fractionated by UF and were shown to have significant antioxidant properties (Hogan et al. 2009).

Table 2.1   Membrane Filtration Methods

	Microfiltration (MF)	Ultrafiltration (UF)	Nanofiltration (NF)	Reverse Osmosis (RO)
Pore size	>0.1 μm	1–500 nm	0.1–1 nm	<0.1 nm
Sieving mechanism	Size	Size, charge	Size, charge	Size
Separation products	Particles, globules (>0.1 μm)	Proteins/ peptides (1–300 kDa)	Amino acids, salts, solutes	Monovalent ions (K, Cl)
Operating pressure	0.1–2 bars	1–10 bars	15–30 bars	30–50 bars

Soy protein hydrolysates were fractionated using UF membranes where the effect of the membrane pore sizes on the molecular size distribution as well as the functional properties of the protein hydrolysate was studied (Deeslie and Cherian 2006). UF has also been used to obtain bioactive peptides from soy for angiotensin-converting enzyme (ACE)-I inhibitory activity (Cha and Park 2005). β-Lactoglobulin from whey was isolated using a UF membrane enzymic reactor, where the authors were able to retain β-lactoglobulin in the reactor while the peptides generated from the hydrolysis of α-lactalbumin and serum albumin permeated through the membrane (Sannier, Bordenave, and Piot 2000).

2.2.2 Purification Techniques

Liquid chromatography is the most widely used method for the analytic as well as preparative separation of food-derived proteins and peptides. Coupling the chromatographic system to conventional UV and fluorescence detectors enables the detection of the proteins and peptides of interest. Furthermore, the structural characterization of purified proteins and peptides can be achieved by methods such as amino acid analysis, MS, and sodium dodecyl sulfate (SDS)-electrophoresis. The recent advent of ionization techniques, starting in 1981 with the introduction of fast atom bombardment (FAB), followed by electrospray ionization (ESI) and MALDI, has made such analyses by MS possible for proteins and peptides, with (0.01%) high accuracy. The latter two methods have become important tools for the detection and characterization of large biomolecules because of their sensitivity, high mass range, and capacity to isolate from complex mixtures. In addition, MS analysis can be extended to include interfaces for a MALDI source to enable precise structural information and characterization and even locate events such as post-translational modifications.

2.2.2.1 Liquid Chromatography Techniques

Several preparative chromatographic methods have been used to purify proteins and peptides from a complex mixture. Ion-exchange chromatography, hydrophobic interaction chromatography, affinity chromatography, and size exclusion chromatography (SEC) are the most commonly used chromatographic techniques. These are generally employed as initial separation steps, where the peptides of interest are isolated out from a given pool that had exhibited bioactivity as a whole fraction. Following this, reversed-phase HPLC has been widely employed with or without coupling to ion-exchange chromatography for efficient purification of the peptides of interest (Leonil et al. 2000).

Several models have been proposed in favor of the quantitative characterization of the structure of a peptide. Researchers also studied the prediction of a peptide gradient in RP-HPLC retention under optimal conditions. The amino acid composition of the peptides influences their chromatographic behavior. Specific retention coefficients were derived by artificial neural networks (ANNs) using a set of peptide retentions (Palmblad 2007). Scientists have presented various approaches for peptide elution time prediction in RP-HPLC, and the contribution of sequence-dependent parameters and peptide physicochemical descriptors, which have been shown to affect the peptide retention time in liquid chromatography, was analyzed. Hence, efficient analysis of the chromatographic data by modeling techniques could have direct use for proteomics (Baczek and Kaliszan 2009). Saz and Marina's (2007) review article presented a comprehensive focus on the analytical methodologies developed using HPLC and capillary electrophoresis in soybean protein characterization and analysis. They also presented the relationship between the nutritional, functional, and biomedical properties and the use of analytical techniques in the characterization of soybean protein fractions and their hydrolysates. The review also studied the application of proteomic methodologies in soybean food technology (Saz and Marina 2007).

Ion-exchange chromatography: Proteins have numerous functional groups that can have both positive and negative charges. Ion-exchange chromatography separates proteins according to their net charge, which is dependent on the composition of the mobile phase. By adjusting the pH or the ionic concentration of the mobile phase, various protein molecules can be separated. For example, if a protein has a net positive charge at pH 7, then it will bind to a column of negatively charged beads, whereas a negatively charged protein would not. By altering the pH, the net charge on the protein can be made negative, and the protein can be eluted (Gjerde and Fritz 2000).

Elution by changing the ionic strength of the mobile phase works when ions from the mobile phase interact with the immobilized ions in preference over those from the stationary phase. This masks the stationary phase from the protein, allowing the protein to elute (Gjerde and Fritz 2000).

Cationic and anionic exchangers (resins) are used to separate molecules based on charge. The retention of the molecule in the matrix (column) depends on the intrinsic charge and also the induced electrostatic interactions. For proteins and peptides, the pI (isoelectric point) is the important property for retention. Several proteins and peptides have been subjected to ion-exchange chromatography for effective purification. Lactoferricin, a peptide from the protein lactoferrin, was separated with high purity using cation-exchange chromatography (Dyonisius and Milne 1997). Another peptide, luffacylin, was isolated from sponge guard seeds and characterized with the help of DEAE-cellulose ion-exchange chromatography (Parkash, Ng, and Tso 2002).

An ACE inhibitor was identified from buckwheat to be a tripeptide, Gly-Pro-Pro, utilizing ion-exchange chromatography Sephadex C-25 ion-exchange column (Ma et al. 2006). Similarly, the ACE-I inhibitory peptide from lactic acid–fermented peanut flour was purified and characterized using ion-exchange SP-Sephadex C-25 column (Liu et al. 2000). In both cases, novel peptides were purified using ion-exchange resins followed by HPLC systems.

ACE inhibitory peptides have been identified in various plant food sources including soybean, sunflower, rice, corn, wheat, buckwheat, mushroom, garlic, spinach, and wine (Guang and Phillips 2009). The supernatants of solvent-extracted protein hydrolysates obtained from each plant source are generally applied first to a column containing a cation-exchange resin such as Dowex 50 (polystyrene-divinylbenzene beads with sulfonic acid groups), which is washed with deionized water to remove impurities, and the desired peptides are eluted with ammonia solution (Suetsuna 1998). Resin can be adopted to fractionate protein hydrolysates and peptide extracts before or after membrane separation.

Jeong et al. (2007) purified lunasin, a 43-amino acid peptide, from soybean to characterize and later test its anticancer properties on laboratory rats. They used Ni-NTA chelating agarose CL-60 ion-exchange chromatography for purification of the peptide, which was tested for inhibition of core histone acetylation. HPLC was used for the analytical comparison of lunasin from various soybean varieties (Jeong et al. 2007).

Size exclusion chromatography: SEC can be used to separate relatively large peptides, as the gel matrix is not suited for small peptides of <1 kDa. The matrix consists of a gel of a specific pore size through which molecules may pass or be retained on the column owing to their molecular sizes. In general, smaller molecules diffuse faster and better into pores than larger molecules, resulting in differential retention times. Parkash, Ng, and Tso (2002) prepared protein hydrolysates from rice bran, employing SEC coupled with HPLC. The molecular fragments ranged between 1 and 150 kDa after hydrolysis with alcalase and flavourzyme. Further, anion-exchange chromatography of the extracted rice bran hydrolysates was performed. A comparison between both the alcalase and the flavourzyme treatments for deamidation showed on the chromatogram that the protein hydrolysates treated with flavourzyme had a longer retention time, suggesting the occurrence of more negative charges due to deamidation.

For certain proteins and bacteriocins, a combination of chromatographic techniques was used for efficient recovery. Stoffels, Sahl, and Gudmundsdtottir (1993) developed a two-step process of chromatographic separation: first, based on adsorption onto hydrophobic interaction chromatography resin and second, based on a cation-exchange resin. Stoffels, Sahl, and Gudmundsdtottir used this two-way process for 100% recovery of nisin. In another similar study by Uteng et al. (2002), bacteriocins were initially bound to sepharose cation-exchanger resins and eluted with NaCl followed by an RP-HPLC process with the application of low pressure. Thus, they were able to obtain 90% pure fraction of bacteriocins.

The characterization of proteins and peptides can thus be performed using several chromatographic methods individually or in combination. Efficient purification generally requires manifold optimization trials and hence proves to be a challenging task. However, sophisticated instrumentation and automation aid in faster and less cumbersome protein recoveries.

Studies show that SEC provides a significant enrichment of N-linked glycopeptides relative to nonglycosylated peptides. This is because N-linked glycans expressed on tryptic glycopeptides tend to contribute substantially to their mass. LC-MS/MS was later used to identify the glycosylated peptides. In vitro analyses of human serum showed an enhanced number of glycopeptides identified due to the SEC glycopeptide isolation procedure. This demonstrated that a simple, nonselective, rapid method is an effective tool to facilitate the identification of peptides with N-linked glycosylation sites (Alvarez-Manilla et al. 2006).

RP-HPLC for the purification of bioactive peptides from food: HPLC/RP-HPLC has replaced classical methods as the technique for the separation and purification of peptides. Efficient recovery of biological activity is possible with most low molecular weight peptides. Nowadays, the most popular column material is the reversed-phase column, in which separation is achieved through partition and adsorption by unprotected silanol groups. In reversed-phase chromatography, the stationary phase is nonpolar (or less polar than the mobile phase) and the analytes are retained until they are eluted with a sufficiently polar solvent or solvent mixture (in the case of a mobile-phase gradient). Large peptides (>4000 Da) are generally analyzed with reversed-phase columns having particles with greater pore diameters (300 Å) so that the molecules will have greater access to the alkyl chains. On the other hand, smaller molecular pore sizes (60–100 Å) are preferred for separating smaller peptides. The physicochemical characteristics of peptides greatly influence the choice of stationary phase, for example, C4 chains are suited for hydrophobic peptides while C8 chains are preferred for hydrophilic or relatively less hydrophobic peptides. The most widely used organic modifier for the separation of peptides has been acetonitrile, because of its high transparency in UV detection at around 200–220 nm. Ion-pairing reagents have been used to increase efficiency in the separation of peptides. Thus, ion-pairing organic solvents operating in reversed-phase conditions are proven to yield highly pure peptides (Vanhoute et al. 2009).

The formation of opioid peptides by in vitro proteolysis of bovine hemoglobin using pepsin was investigated by RP-HPLC and the subsequent offline identification of collected fractions by FAB-MS (Piot et al. 1992). The same technique was used to isolate opioid peptides from milk fermented with a strain of *Lactobacillus helveticus* and to identify some bioactive peptides released by pepsin and trypsin digestion of UHT milk fermented by *L. casei*. Using RP-HPLC off line with ESI-MS allowed Dyonisius and Milne (1997) to identify peptides with antimicrobial activity in the tryptic hydrolysate of bovine lactoferrin. Working with the same protein, Shimazaki et al. (1998) identified a heparin-binding peptide after pepsin hydrolysis, RP-HPLC separation, and MALDI-TOF-MS measurements. RP-HPLC–ESI-MS was also reported to be useful to monitor the kinetics of a bioactive production during continuous hydrolysis of β-casein with chymosin in a membrane reactor (Leonil et al. 2000). Table 2.2 illustrates the types of HPLC used for the separation of food peptides, their separation mechanism, and characteristics of separation.

2.2.2.2 Electrophoresis

Electrophoresis is one of the most widely used analytical techniques for the separation, identification, and characterization of proteins. Both one- and two-dimensional electrophoresis can obtain information about the protein's physical properties like size and charge. One-dimensional separation of complex protein mixtures can be based on differential electrophoretic motilities, which in turn are based on the pH (isoelectric focusing, IEF), size (SDS polyacrylamide gel electrophoresis, SDS-PAGE), charge-to-size ratio (native PAGE), or ligand affinity (affinity electrophoresis). The one-dimensional electrophoretic runs can be coupled to the second dimension, most often seen with IEF (as a first-dimension run) and SDS-PAGE (as a second-dimension run).

Several food proteins have been characterized using electrophoretic procedures. Guo and Yao (2006) successfully characterized proteins from buckwheat flour. Protein fractions (albumin, globulin, prolamin, and glutelin) subjected to electrophoresis under reductive and nonreductive conditions revealed the presence of disulfide bonds within the four fractions. Nonreduced albumin showed major bands at 64, 57, 41, and 38 kDa and globulin at 57, 28, 23, 19, and 15 kDa. The reduced albumin and globulin shared two common bands at 41 and 38 kDa (Figure 2.2).

Quantitative studies of food-derived proteins using proteomics were first done on dairy proteins. It was observed that several bacterial proteins were released into Swiss cheese through the lysis of lactic acid starter cultures. These proteins were characterized using two-dimensional gel electrophoresis (Gagnaire et al. 2009). The proteins included peptidases and glycolytic enzymes.

Costa et al. (2007) used capillary electrophoresis to separate an ACE mixture with inhibition and with the addition of whey protein hydrolysate and were able to characterize the whey protein hydrolysate with the highest ACE inhibitory activity. Likewise, several food proteins have been and are being characterized using electrophoretic techniques alone or in combination with proteomic tools. Peptide

Table 2.2   HPLC Methods Involved with Food Peptides

HPLC Type	Separation Mechanism	Characteristics
RP-HPLC	Hydrophobic interaction of the peptides with the stationary phase	High resolution
		High speed
		Standard separation procedures
		Prediction of retention times
		Good detection and reproducibility
		Quantitative recovery using volatile buffers
		Moderate capacity in analytical columns
IE-HPLC	Ionic interaction of peptides with anion exchangers or cation exchangers as the stationary phase	Moderate resolution
		High speed
		High capacity
		Prefractionation of peptides by charge
		Wide range of pH
SE-HPLC	Size exclusion of peptides on hydrophilic polymers as the stationary phase	Moderate or low resolution
		High speed
		Very high capacity
		Desalting
		Good prefractionation by size
		Molecular weight determination
		Presence of nonspecific interactions

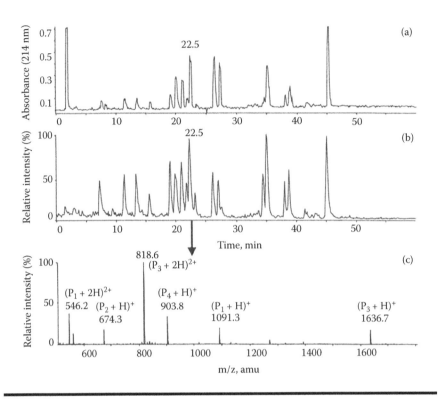

Figure 2.2 (a–c) On-line HPLC and MS combined separation of casein proteins. (From Leonil et al. 2000.)

mapping is one such example that involves a combination of techniques including electrophoresis.

Peptide mapping: Peptide mapping involves a series of processes that help determine information regarding the sequence of proteins and peptides. This is usually a combination of techniques such as electrophoresis, labeling, and chromatography. This technique will help identify and separate from a mixture of peptides using specific endoproteases and antibodies need be (Nandi 1997). First, purification of the proteins or peptides will be done following labeling using radioiodination. Specific endopeptidases will be allowed to act on the mixture to yield cleaved fragments that can be analyzed by electrophoresis or chromatography.

An increase in MALDI-TOF-MS sensitivity toward lysine-terminated peptides was reported due to derivatization reaction and guanidine treatment. The sequence information conveyed in this application can be used as a parameter in peptide mass mapping database searches. Karty et al. (2002) reported a systematic study of the impact of guanidination on proteomic analysis, and a novel computer algorithm was developed to analyze the data. This study showed the advantages of guanidination using peptide mapping data (Karty et al. 2002).

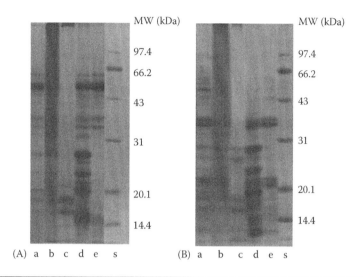

Figure 2.3   SDS-PAGE distribution of buckwheat flour proteins. SDS-PAGE under nonreductive conditions (A) and under reductive conditions (B): (a) defatted tartary flour; (b) glutelin; (c) prolamin; (d) globulin; (e) albumin; and (s) molecular weight marker. (From Guo and Yao 2006.)

Peptide mapping using RP-HPLC of a casein hydrolysate was done with two different proteolytic enzymes, a bacterial proteinase and a fungal peptidase (Kanekanian, Gallagher, and Evans 2000) (Figure 2.3).

The neutral bacterial protease extracted from *Bacillus subtilis* was used to produce large peptides, which imparted a bitter taste to the hydrolysate. The bitterness is usually linked to the presence of peptides with a high proportion of hydrophobic, side-chain amino acids. The fungal protease extracted from *Aspergillus oryzae* has high amino and carboxy peptidase activities and was used to debitter the hydrolysate. A comparison of the peptide map following treatment reveals the changes that had taken place after each hydrolysis.

2.2.2.3 Proteomic Tools

With the advent of modern tools that aid in precise and quick identification and detection of structural components within a complex as well as a pure mixture, characterization protocols have been made more simple, quick, and reliable. Proteomic tools have been developed keeping in mind several factors involved in characterizing a compound for purity and to retain functionality. Chromatographic separations are being coupled to spectrometers in tandem to enhance the separation as well as mass determining capabilities at the same time. Such features not only decrease time but also increase sensitivities. Of the several tools available, the most popular and increasingly diverse, especially with proteins, is the mass spectrometer and its accessory coupling instruments available for characterization studies (Ball et al. 2002; Barnidge et al. 2003).

Identification and characterization with peptide mass fingerprinting data, identification and characterization with MS–MS data, and identification with the isoelectric point, molecular weight, and/or amino acid composition are common proteome tools listed by ExPASy as popular tools dedicated to protein analysis. These include programs like (a) MASCOT—peptide mass fingerprint; (b) PepMapper—peptide mass fingerprinting tool; (c) PepFrag—searches known protein sequences with peptide fragment mass information; and (d) TagIdent—identifies proteins with isoelectric point (pI), molecular weight (Mw), and sequence tag or generates a list of proteins close to a given pI and Mw (Beardsley and Reilly 2003).

Other tools include the following: (a) ProtParam—computes the physicochemical parameters of a protein sequence (amino acid and atomic compositions, isoelectric point, and extinction coefficient); (b) PeptideCutter—predicts the potential protease and cleavage sites and the sites cleaved by chemicals in a given protein sequence; (c) PeptideMass—calculates the masses of peptides and their posttranslational modifications for a UniProtKB/Swiss-Prot or UniProtKB/TrEMBL entry or for a user sequence; (d) MALDIPepQuant—quantifies MALDI peptides (SILAC); and (e) a range of similar searches like BLAST, PropSearch, SEQUEROME, and BLAST2FASTA. Conrads et al. have reviewed a perspective on the development of proteomic patterns for the diagnosis of early-stage cancers (Conrads et al. 2001). A whole range of available tools and software programs for use in proteomic tools are given and some are even offered by the ExPASy (Expert Protein Analysis System) proteomics server of the Swiss Institute of Bioinformatics (SIB) dedicated to the analysis of protein sequences and structures as well as two-dimensional PAGE.

2.3 Scale-Up Separation and Purification of Food Proteins and Peptides

There is an increasing an commercial interest in the production of bioactive peptides from various sources. Industrial-scale production of such peptides is limited due to the lack of scalable technologies. There is a need to develop technologies that retain or even enhance the activity of bioactive peptides in food systems. Over the last 20 years, considerable progress has been made in technologies aimed at separation, fractionation, and isolation in a purified form of many interesting proteins occurring in bovine colostrum and milk (Korhonen and Pihlanto 2007). Industrial-scale methods have been developed for native whey proteins, such as immunoglobulins, lactoferrin, lactoperoxidase, α-lactalbumin, and β-lactoglobulin. Commercially viable products obtained from bioactive peptides, particularly those derived from milk proteins, have hit the market.

Advances in technologies will enable the recovery of bioactive peptides with minimal destruction, thereby enabling the utilization of these active peptides in functional foods or specific nutraceutical applications.

Approaches toward the large-scale production of proteins from cereal grains have also been practiced. For example, rice grain can be processed to obtain oil, starch, and proteins for various food and industrial applications, which may require further processing or conversion to the desired product. Rice protein content is low (6–10 wt%) compared to other grains, but rice flour has a high nutritional value and has been used to formulate various food products, such as pudding, instant milk, and baby foods. A biocatalysis process to produce starch-derived and high-protein products from rice has been successfully developed (Akoh et al. 2008) (Figure 2.4).

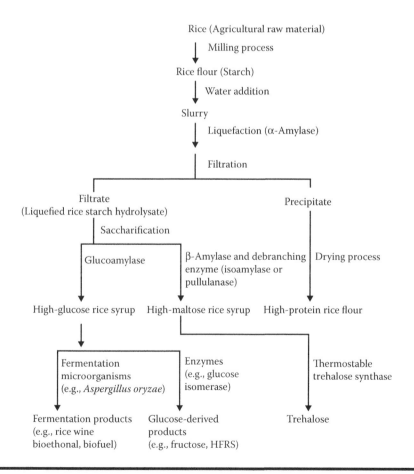

Figure 2.4 A flow diagram for the process of converting raw rice starch into high-protein rice flour and other industrial and functional food products.

Enzymatic hydrolysis has the potential to be an alternative successful approach for the production of protein-enriched flour from cereal grains with improved nutrition (Casimir et al. 2008). The application of enzymes to convert low-value, underutilized agricultural raw produce into high, value-added commercial products, including functional foods and bioactive compounds, needs to be oriented toward technologically amenable protocols. Gene cloning and expression, recombinant enzyme biotechnology, and transgenic plant biotechnology are expected to further develop and sustain the improved application potential of enzymes for producing high-value products.

2.4 Recent Advances in Characterization Techniques

2.4.1 Mass Spectrometry

MS is an analytical technique for determining the elemental composition of a sample or molecule. It is also used for elucidating the chemical structures of molecules, such as peptides and other chemical compounds. MS is based on the principle of ionizing the chemical compounds to generate charged molecules or molecule fragments and measuring their mass-to-charge ratios (Sparkman 2000). In a typical MS procedure, a sample is loaded onto the MS instrument, where its compounds are ionized by different methods (e.g., by impacting them with an electron beam), resulting in the formation of charged particles (ions). The mass-to-charge ratio of the particles is then calculated from the motion of the ions as they transit through the electromagnetic fields.

The two primary methods for the ionization of whole proteins are ESI and MALDI. MALDI prefers thermolabile, nonvolatile organic compounds, especially those of high molecular mass, and is used successfully in the field of biochemistry for the analysis of proteins, peptides, glycoproteins, oligosaccharides, and oligonucleotides (Hillenkamp et al. 1991). It is relatively straightforward and reasonably acceptable to use buffers and other additives. The mass accuracy depends on the type and performance of the analyzer of the mass spectrometer, but most modern instruments are designed to achieve mass measurement accuracies to within 0.01% of the molecular mass of the sample, at least up to 40,000 Da. Two approaches are used for characterizing proteins. In the first approach, intact proteins are ionized by ESI or MALDI and then introduced into a mass analyzer.

In the second approach, proteins are enzymatically digested into smaller peptides using specific proteases, such as trypsin or pepsin, either in a solution or in a gel after electrophoretic separation. A collection of peptide pools is then introduced in to the mass analyzer. Either by peptide mass fingerprinting (peptides used for protein identification) or by de novo sequencing (sequence data from tandem MS–MS analysis), full characterization of the protein can be achieved (Schuchardt and Sickmann 2007).

MALDI is based on the bombardment of sample molecules with a laser light to bring about sample ionization. The sample is premixed with a highly absorbing matrix compound for the most consistent and reliable results, and a low concentration ratio of sample to matrix works best. The matrix transforms the laser energy into excitation energy for the sample, which leads to sputtering of the analyte and matrix ions from the surface of the mixture. In this way, energy transfer is efficient and also the analyte molecules are prevented from decomposition due to excessive direct energy.

MS–MS is used to generate structural information about a compound by fragmenting specific sample ions inside the mass spectrometer and identifying the resulting fragment ions. This information can then be put together to generate structural information regarding the intact molecule. MS–MS also enables specific compounds to be detected in complex mixtures on account of their specific and characteristic fragmentation patterns.

Peptides fragment in a reasonably well-documented manner. The protonated molecules fragment along the peptide backbone and also show certain side-chain fragmentation. Three different types of bonds can fragment along the amino acid backbone: the NH–CH, CH–CO, and CO–NH bonds. Each bond breakage gives rise to two species, one neutral and one charged, and only the charged species is monitored by the mass spectrometer. The charge can stay on either of the two fragments, depending on the chemistry and relative proton affinity of the two species. The most common cleavage sites are at the CO–NH bonds, which give rise to the "b" and/or the "y" ions. The mass difference between two adjacent b ions, or y ions, indicates the presence of a particular amino acid residue.

Bioactive peptides have not only been quantified in food products but there are also a few attempts dealing with the quantification of peptides in biological fluids using HPLC/MS. Characterization of a fraction from β-lactoglobulin (β-Lg) tryptic digest resulted in the identification of an ACE inhibitory peptide, β-Lg f (142–148) by Mullally, Meisel, and FitzGerald (1997). Curtis et al. (2002) developed a straightforward quantitative method for measuring the antihypertensive peptide LKPNM in bonito muscle hydrolysate using one step of the solid-phase extraction (SPE) followed by quantification using HPLC/MS/MS with a hybrid QTOF spectrometer. Kuwata et al. (1998) evaluated the in vivo generation of lactoferricin after the ingestion of bovine lactoferrin by surface-enhanced laser desorption or ionization (SELDI).

MALDI techniques have also contributed to the bioavailability studies of antihypertensive peptides. Using MALDI-TOF-MS, the presence of the antihypertensive peptide, ALPMHIR, derived from a tryptic digestion of β-lactoglobulin, was demonstrated in both the mucosal and the serosal sides of a Caco-2 cell monolayer (Vermeirssen et al. 2002). The heptapeptide was transported intact through the monolayer but in concentrations too low to exert an ACE inhibitory activity. A recent promising technique for food-derived peptides, particularly the phosphopeptides, involves the direct detection and sequence of IMAC-enriched peptides through MALDI-MS/MS on an orthogonal injection QTOF mass spectrometer (Raska et al. 2002).

2.4.2 Peptide Identification Using Tandem Mass Spectra

Peptide identification algorithms fall into two broad classes: database search and de novo search. The former search takes place against a database containing all amino acid sequences assumed to be present in the analyzed sample, whereas the latter infers peptide sequences without the knowledge of the genomic data.

2.4.3 Database Search Programs

SEQUEST, Mascot, Phenyx, and X!Tandem are all database software programs that collect information from tandem mass spectra and run database searches to identify the protein or peptide of interest, while SPIDER and PEAKS are de novo search programs.

SEQUEST identifies each tandem mass spectrum individually. The software evaluates protein sequences from a database to compute the list of peptides that could result from each (Eng et al. 1994). The peptide's intact mass is known from the mass spectrum, and SEQUEST uses this information to determine the set of candidate peptide sequences that could be compared to the spectrum near the peptide ion. For each candidate peptide, SEQUEST projects theoretical tandem mass spectra and compares these spectra to the observed tandem mass spectrum. The candidate sequence with the best matching theoretical tandem mass spectrum is reported as the best identification for this spectrum.

Mascot is a proprietary identification program available from Matrix Science. It performs MS data analysis through a statistical evaluation of matches between the observed and the projected peptide fragments (Perkins et al. 1999).

Phenyx incorporates OLAV, a family of statistical scoring models, to generate and optimize scoring schemes that can be programmed for all kinds of instruments (Colinge et al. 2003). It computes a score to evaluate the quality of a match between a theoretical and an experimental peak list. The basic peptide score is ultimately transformed into a normalized z-score and a p-value. A basic peptide score is the sum of raw scores for up to 12 physicochemical properties.

To identify proteins using MS–MS, the de novo sequencing software computes one or several possible amino acid sequences (sequence tags) for each MS–MS spectrum (Han, Ma, and Zhang 2005). These tags are then used to match, accounting for amino acid mutations, the sequences in a protein database. If the de novo sequencing gives correct tags, the homologs of the proteins can be identified by this approach, and software such as MS-BLAST is available for the matching. The most common error is that a segment of amino acids is replaced by another segment with approximately the same mass. The SPIDER algorithm matches the sequence tags with the errors to database sequences for the purpose of protein and peptide identification, and is designed to solve the problems arising from BLAST.

PEAKS identifies unknown proteins and provides a protein identification search engine (Ma et al. 2003). This software is one of the earliest and successful tools for de novo sequencing (both automated and manual) and sequence tag–based searching (SPIDER). In short, de novo sequencing is peptide sequencing performed without prior knowledge of the amino acid sequence, approximately 1 spectra per second or a run of 1000 spectra in about 20 min.

Currently, a database search is more popular and is considered to produce higher quality results for most uses. With increasing instrument precision, however, a de novo search may become increasingly usable and attractive.

2.5 Conclusion

It is apparent that proper analytical separation, detection, and purification techniques can identify the precise structures of proteins and peptides and thus enable the identification of function. As more proteins and peptides are being identified with bioactive properties, the need to explore and test the technologies for precise characterization and commercialization has arisen. Furthermore, the use of advanced technologies, including gene cloning and expression, recombinant enzyme biotechnology, and transgenic plant biotechnology, is expected to further develop as preparation or production tools for the effective isolation of bioactive proteins and peptides.

References

Akoh, C. C., Chang, S. W., Lee, G. C., and Shaw, J. F. 2008. Biocatalysis for the production of industrial products and functional foods from rice and other agricultural produce. *Journal of Agricultural and Food Chemistry* 56(22): 10445–51.

Alvarez-Manilla, G., Atwood, J. III, Guo, Y., Warren, N. L., Orlando, R., and Pierce, M. 2006. Tools for glycoproteomic analysis: Size exclusion chromatography facilitates identification of tryptic glycopeptides with N-linked glycosylation sites. *Journal of Proteome Research* 5(3): 701–8.

Baczek, T. and Kaliszan, R. 2009. Predictions of peptides' retention times in reversed-phase liquid chromatography as a new supportive tool to improve protein identification in proteomics. *Proteomics* 9: 835–47.

Ball, G., Mian, S., Holding, F., Allibone, R. O., Lowe, J., Ali, S., Li, G., McCardle, S., Ellis, O., Creaser, C., and Rees, R. C. 2002. An integrated approach utilizing artificial neural networks and SELDI mass spectrometry for the classification of human tumours and rapid identification of potential biomarkers. *Bioinformatics* 18: 395–404.

Barnidge, D. R., Dratz, E. A., Martin, T., Bonilla, L. E., Moran, L. B., and Lindall, A. 2003. Absolute quantification of the G proteincoupled receptor rhodpsin by LC/MS/MS using proteolysis product peptides and synthetic peptide standards. *Analytical Chemistry* 75: 445–51.

Beardsley, R. L. and Reilly, J. P. 2003. Quantitation using enhanced signal tags: A technique for comparative proteomics. *Journal of Proteome Research* 2: 15–21.

Cha, M. and Park, J. R. 2005. Production and characterization of a soy protein-derived angiotensin I-converting enzyme inhibitory hydrolysate. *Journal of Medicinal Food* 8: 305–10.

Colinge, J., Masselot, A., Giron, M., Dessingy, T., and Magnin, J. 2003. OLAV: Towards high-throughput tandem mass spectrometry data identification. *Proteomics* 3: 1454–63.

Conrads, T. P., Alving, K., Veenstra, T. D., Belov, M. E., Anderson, G. A., Anderson, D. J., Lipton, M. S., Pasa-Tolic, L., Udseth, H. R., Chrisler, W. B., Thrall, B. D., and Smith, R. D. 2001. Quantitative analysis of bacterial and mammalian proteomes using a combination of cysteine affinity tags and 15N-metabolic labeling. *Analytical Chemistry* 73: 2132–9.

Costa, El., Gontijo, J. A. R., and Netto, F. M. 2007. Effect of heat and enzymatic treatment on the antihypertensive activity of whey protein hydrolysates. *International Dairy Journal* 17: 632–40.

Curtis, J. M., Dennis, D., Waddell, D. S., MacGillivray, T., and Ewart, H. S. 2002. Determination of angiotensin-converting enzyme inhibitory peptide Leu-Lys-Pro-Asn-Met (LKPNM) in bonito muscle hydrolysates by LC-MS/MS. *Journal of Agricultural and Food Chemistry* 50: 3919–25.

Deeslie, W. D. and Cherian, M. 2006. Fractionation of soy protein hydrolysates using ultrafiltration membranes. *Journal of Food Science* 57: 411–13.

Dyonisius, A. and Milne, J. M. 1997. Antibacterial peptides of bovine lactoferrin: Purification and characterization. *Journal of Dairy Science* 80: 667–74.

Edelhoch, H. 1967. Spectroscopic determination of tryptophan and tyrosine in proteins. *Biochemistry* 6: 1948.

Eng, J. K., et al. 1994. An approach to correlate tandem mass spectral data of peptides with amino acid sequences in a protein database. *Journal of the American Society for Mass Spectrometry* 53: 186.

Gagnaire, V., Jardin, J., Jan, G., and Lortal, S. 2009. Invited review: Proteomics of milk and bacteria used in fermented dairy products: From qualitative to quantitative advances. *Journal of Dairy Science* 92: 811–25.

Garem, A., Daufin, G., Maubois, J. L., Chaufer, B., and Leonil, J. 1998. Ionic interactions in nanofiltration of β-casein peptides. *Biotechnology and Bioengineering* 57: 109–17.

Gjerde, D. T. and Fritz, J. S. 2000. *Ion Chromatography*. Weinheim: Wiley-VCH.

Guang, C. and Phillips, R. D. 2009. Plant food-derived angiotensin I converting enzyme inhibitory peptides. *Journal of Agricultural and Food Chemistry* 57: 5113–20.

Guo, X. and Yao, H. 2006. Fractionation and characterization of tartary buckwheat flour proteins. *Food Chemistry* 98(1): 90–4.

Han, Y., Ma, B., and Zhang, K. 2005. SPIDER: Software for protein identification from sequence tags containing de novo sequencing error. *Journal of Bioinformatics and Computational Biology* 3: 697–716.

Hillenkamp, F., Karas, M., Beavis, R. C., and Chait, B. T. 1991. Matrix-assisted laser desorption/ionization mass spectrometry of biopolymers. *Analytical Chemistry* 63: 1193A–1203A.

Hogan, S., Zhang, L., Li, J. R., Wang, H. J., and Zhou, K. Q. 2009. Development of antioxidant rich peptides from milk protein by microbial proteases and analysis of their effects on lipid peroxidation in cooked beef. *Food Chemistry* 117: 438–43.

Jeong, J. B., Jeong, H. J., Park, J. H., Lee, S. H., Lee, J. R., Lee, H. K., Chung, G. Y., Choi, J. D., and de Lumen, B. O. 2007. Cancer-preventive peptide lunasin from *Solanum nigrum* L. inhibits acetylation of core histones H3 and H4 and phosphorylation of retinoblastoma protein (Rb). *Journal of Agricultural and Food Chemistry* 55(26): 10707–13.

Kanekanian, A., Gallagher, J., and Evans, P. E. 2000. Casein hydrolysis and peptide mapping. *International Journal of Dairy Technology* 53: 1–5.

Karty, J. A., Ireland, M. M. E., Brun, Y. V., and Reilly, J. P. 2002. http://www.sciencedirect.com/science/article/pii/S1570023202005500 - COR1Artifacts and unassigned masses encountered in peptide mass mapping. *Journal of Chromatography B* 782(1–2): 363–83.

Kim, Se., Kim, H. H., and Kim, J. Y. 2000. Anticancer activity of hydrophobic peptides from soy proteins. *Biofactors* 12(1–4): 151–5.

Kimura, S. and Tamano, A. 1986. Separation of amino acids by charged ultrafiltration membranes. In *Membranes and Membrane Processes*, eds. E. Droli and N. Nakagaki, 191. New York: Plenum Press.

Korhonen, H. and Pihlanto, A. 2007. Technological options for the production of heatth-promoting proteins and peptides derived from milk and colostrum. *Current Pharmaceutical Design* 13: 829–43.

Kuwata, H., Yip, T. T., Yip, C. L., Tomita, M., and Hutchens, W. 1998. Bactericidal domain of lactoferrin: Detection, quantitation, and characterization of lactoferricin in serum by SELDI affinity mass spectrometry. http://en.wikipedia.org/wiki/International_Standard_Book_Number. *Biochemical and Biophysical Research Communications* 245: 764–73.

Léonil, J., Gagnaire, V., Mollé, D., Pezennec, S., and Bouhallab, S. 2000. Application of chromatography and mass spectrometry to the characterization of food proteins and derived peptides. *Journal of Chromatography A* 881(1–2): 1–21.

Liu, L.-N., Zhang, S. J., and He, D.-P. 2000. Detection of an angiotensin converting enzyme inhibitory peptide from peanut protein isolate and peanut polypeptides by western blot and dot blot hybridization. *European Food Research and Technology* 230(1): 89–94.

Ma, B., Zhang, K., Hendrie, C., Liang, C., Li, M., Doherty-Kirby, A., and Lajoie, G. 2003. PEAKS: Powerful software for peptide de novo sequencing by MS/MS. *Rapid Communications in Mass Spectrometry* 17: 2337–42.

Ma, M., Bae, I. Y., Lee, H. G., and Yang, C. 2006. Purification and identification of angiotensin I-converting enzyme inhibitory peptide from buckwheat (*Fagopyrum esculentum* Moench). *Food Chemistry* 96: 36–42.

Matsudaira, P. 1990. Limited N-terminal sequence analysis. In *Methods in Enzymology: Guide to Protein Purification*, ed. M. P. Deutscher, p. 602. New York: Academic Press.

Mullally, M. M., Meisel, H., and FitzGerald, R. J. 1997. *FEBS Letters* 402: 99–101.

Nai-Fu, W., Chun-Yang, L., and Zheng, Y. 2008. Purification and characterization of angiotensin I-converting enzyme inhibitory peptide from lactic acid fermented peanut flour. *Journal of Biotechnology* 136: S723.

Nandi, P. K. 1997. Chemical and physical methods for the characterization of proteins. In *Food Proteins and Their Applications*, eds. S. Damodaran and A. Paraf, 597–602. New York: Marcel Dekker.

Nau, F., Kerherve, F., Leonil, J., and Daufin, G. 1995. Selective separation of tryptic b-casein peptides through ultrafiltration membranes: Influence of ionic interactions. *Biotechnology and Bioengineering* 46: 246–53.

Ozols, J. 1990. Amino acid analysis. In *Methods in Enzymology: Guide to Protein Purification*, ed. M. P. Deutscher, p. 587. New York: Academic Press.

Palmblad, M., 2007. In *Mass Spectrometry Data Analysis in Proteomics*, ed. Matthiesen, R., 195–207. Totowa, NJ: Humana Press.

Parkash, A., Ng, T. B., and Tso, W. W. 2002. Purification and characterization of charantin, anapin-like ribosome-inactivating peptide from bitter gourd (*Momordica charantia*) seeds. *Journal of Peptide Research* 59: 197–202.

Pena-Ramos, E. A. and Xiong, Y. L. 2002. Antioxidant activity of soy protein hydrolyzates in a liposomial system. *Journal of Food Science* 67(8): 2952–6.

Perkins, D. N., Darryl, J., Pappin, C., Creasy, D. M., and Cottrell, J. S. 1999. Probability-based protein identification by searching sequence databases using mass spectrometry data. *Electrophoresis* 20: 3551–67.

Piot, J.-M., Zhao, Q., Guillochon, D., Ricart, G., and Thomas, D. 1992. Isolation and characterization of a bradykinin-potentiating peptide from a bovine peptic hemoglobin hydrolysate. *FEBS Letters* 299: 75–79.

Raska, C. S., Parker, C. E., Glish, G. L., Dominski, Z., Marzluff, W. F., Pope, R. M., and Borchers, C. H. 2002. Direct MALDI-MS/MS of Phosphopeptides Affinity-Bound to IMAC Beads. *Analytical Chemistry* 74: 3429–33.

Sannier, F., Bordenave, S., and Piot, J. 2000. Purification of goat *b*-lactoglobulin from whey by an ultrafiltration membrane enzymic reactor. *Journal of Dairy Research* 67: 43–51.

Saz, J. M. and Marina, M. L. 2007. High performance liquid chromatography and capillary electrophoresis in the analysis of soybean proteins and peptides in foodstuffs. *Journal of Separation Science* 30(4): 431–51.

Schuchardt, S. and Sickmann, A. 2007. Protein identification using mass spectrometry: A method overview. *Plant Systems Biology* 97: 141–70.

Shimazaki, K., Tazume, T., Uji, K., Tanaka, M., Kumura, H., Mikawa, K., and Shimo-Oka, T. 1998. Properties of a heparin-binding peptide derived from bovine lactoferrin. *Journal of Dairy Science* 81: 2841–49.

Sparkman, O. D. 2000. *Mass Spectrometry Desk Reference*. Pittsburgh, PA: Global View Pub.

Spengler, B. 1997. Post-source decay analysis in matrix-assisted laser desorption/ionization mass spectrometry of biomolecules. *Journal of Mass Spectrometry* 32: 1019–36.

Stoffels, G., Sahl, H. G., and Gudmundsdtottir, A. 1993. Carnocin U149, a biopreservative produced by *Carnobacterium pisicola*: Large scale purification and activity against various Gram-positive bacteria including *Listeria* sp. *International Journal of Food Microbiology* 20: 199–210.

Suetsuna, K. 1998. Isolation and characterization of angiotensin I-converting enzyme inhibitor dipeptides derived from *Allium sativum* L (garlic). *Journal of Nutritional Biochemistry* 9: 415–19.

Turgeon, S. L. and Gauthier, S. F. 1990. Whey peptide fractions obtained with a two-step ultrafiltration process: Production and characterization. *Journal of Food Science* 55: 106–10.

Uteng, M., Hauge, H. H., Brondz, I., Nissen-Meyer, J., and Fimland, G. 2002. Rapid two-step procedure for large-scale purification of pediocin-like bacteriocins and other cationic antimicrobial peptides from complex culture medium. *Applied and Environmental Microbiology* 68: 952–56.

Vanhoute, M., Froidevaux, R., Vanvlassenbroeck, A., Lecouturier, D., Dhulster, P., and Guillochon, D. 2009. Ion-pairing separation of bioactive peptides using an aqueous/octan-1-ol micro-extraction system from bovine haemoglobin complex hydrolysates. *Journal of Chromatography B* 877: 1683–88.

Vermeirssen, V., Deplancke, B., Tappenden, K. A., Van Camp, J., Gaskins, H. R., and Verstraete, W. 2002. Intestinal transport of the lactokinin Ala-Leu-Pro-Met-His-Ile-Arg through a Caco-2 Bbe monolayer. *Journal of Peptide Science* 8(3): 95–100.

Wu, J. and Ding, X. 2001. Hypotensive and physiological effect of angiotensin converting enzyme inhibitory peptides derived from soy proteins on spontaneously hypertensive rats. *Journal of Agricultural and Food Chemistry* 49: 501–6.

Chapter 3

Food Proteins and Peptides: Structure– Function Relationship

Arvind Kannan, Navam Hettiarachchy, Kenji Sato, and Maurice Marshall

Contents

3.1 Introduction: Definition of Functionality

Food proteins constitute a diverse and complex collection of biological macromolecules. The expression of these functional properties during the preparation, processing, and different states of foods is largely influenced by the structural properties of the proteins involved. Kinsella (1976) defined the functional properties as "those physical and chemical properties which affect the behavior of proteins in food systems during processing, storage, preparation and consumption." The properties that influence the function of a protein in food systems include size, shape, net charge, polarity, structure, composition, sequence, interaction with other food components, and physical parameters such as pH and changing chemical environments. Predicting the functional properties using several factors that influence protein function has been unsuccessful possibly due to changes in the protein conformation or arrangement. Hence, denaturation has become a prerequisite for protein functionality (Kinsella 1976).

This chapter attempts to delve into the structure and functional relationships of food proteins–peptides from a variety of sources. It is by means of this review that the structural changes and properties associated with food proteins as they are processed into foods or as a whole functional ingredient can be understood. Recent advances in the field of protein chemistry have significantly enhanced our understanding of the possible intermediates that may cause denaturation, or protein folding and unfolding. In particular, studies on α-lactalbumin have led to the theory that the molten globule state may be one possible intermediate in the folding of many proteins. The molten globule state is characterized by a compact structure, a high degree of hydration and side chain flexibility, a significant amount of native secondary structure but little tertiary structure, and the ability to react with chaperones (Farrel et al. 2002). Other partially folded conformations (e.g., the premolten globule) have also been found. Many proteins, known as natively unfolded, intrinsically unstructured, or intrinsically disordered, were shown to be highly flexible under physiological conditions. By taking advantage of this dimension of protein folding, and applying these concepts to engineered macromolecules and food proteins, it may be possible to generate new and useful forms of proteins for efficient or improved functionality in food systems as food ingredients and in the nutraceutical industries (Damodaran and Paraf 1997).

Protein structures depend on composition, whereas functional properties depend on the behavior of the protein molecules within a solvent or when exposed to several modifications that are needed to achieve a particular food matrix. Heat treatment of a protein hydrolysate, for example, can cause structural changes that can modify or affect the functionality of the protein hydrolysate. Extrapolating this to a commercial scale can cause a different behavior. Hence, considering the physicochemical properties and the pattern of behavior of known proteins in systems can help predict or model systems whereby functionality loss would be very minimal and can be reproduced on a large scale.

Numerous functional foods have evolved in the recent decade whereby alternative forms of treatment for certain chronic conditions have been made possible. These proteins could be bioactive if they are proven to be physiologically stable. For example, the Bowman–Birk inhibitor of soy has been studied for its biochemical and functional properties with respect to its structure and its quaternary structural–molecular interactions (Losso 2008). The author has also reviewed the chemical modifications that Bowman–Birk inhibitors can undergo to improve their biofunctional properties.

Various processes, interactions, and structural modifications need to be undertaken to design a functional food that is not only beneficial without side effects but can also be used as an ingredient to improve the functional properties of the whole food.

3.2 Classification of Functional Food Proteins and Peptides

Functional food proteins and peptides can be classified on the basis of their functions. With structural factors influencing the functional role of food proteins and peptides, it will be understandable to classify the proteins and peptides according to their functionalities, such as their emulsifying properties, foaming, gelation, and solubility. Although some proteins, especially those from milk, may share good functional properties with other proteins, it will be relatively easier to select from a functional class, such as the proteins that are needed for emulsification or foaming. In terms of solubility and viscosity, whey proteins and gelatin are said to be superior and are classified under this functional category. The water-binding and gelation functional properties consist mostly of muscle and egg proteins, with a contribution from milk proteins. The cohesion, adhesion, and elasticity functional parameters can be applied to the proteins belonging to muscle, egg, and whey and to some of the cereal proteins. The role of emulsification may be categorized for egg and milk proteins, while the fat- and flavor-binding properties can be ascribed to the proteins derived from milk and egg and to cereal proteins (Damodaran and Paraf 1997). This classification not only helps us understand the structure–function relationship of different sources and classes of proteins, but also enables us to better understand the mechanism of action underlying the structure–function relationship.

Following protein preparation, characterization of the protein is needed for a better understanding of its physicochemical properties and functionalities. Information on the basic physicochemical characteristics of proteins is essential for their application as functional ingredients in food systems. These characteristics include surface hydrophobicity, molecular size, and thermal properties.

Surface hydrophobicity: Surface hydrophobicity is used for protein characterization in relation to the excess free energy of the exposed amino acids that can be transferred from water to a hydrophobic solvent (Bigelow 1967). This hydrophobicity measurement is based on a comparison of the excessive free energy in these two solvents. Information on the hydrophobicity of the protein explains the propensities for protein solubility as well as the emulsifying properties. Several methods have been reported in the literature for hydrophobicity measurements, namely, partition, high performance liquid chromatography (HPLC), binding, contact angle measurement, and the fluorescence probe methods (Nakai and Modler 1996). Among these methods, the fluorescence probe methods is probably the most common method used for this determination due to its simplicity.

The surface hydrophobicity of proteins is an important structural factor that governs the functional properties of proteins, such as the emulsifying and foaming properties (Kinsella and Whitehead 1988). In its native form, the hydrophobic segments of the protein are mostly buried inside the core, causing a better solubility in water, but they are low in functionality, particularly emulsification.

Electrophoresis: The molecular sizes of proteins provide information on food protein identification and characterization. Electrophoresis is a widely used analytical technique to evaluate the molecular sizes of proteins. This technique is based on the movement of charged protein molecules in an applied electric field, and the mobility of the protein molecules is a function of size, charge, and charge-to-size ratio (Van Camp and Dierckx 2003). Electrophoretograms of the proteins provide not only the molecular sizes of the proteins but also the protein cross-linkings, probably caused by intermolecular disulfide bonds.

SDS-polyacrylamide gel electrophoresis (SDS-PAGE) is one of the common techniques used to separate proteins as a function of their molecular weights. SDS-PAGE can determine the proteins in a range of molecular weights from 1 to 1000 kDa. The proteins are usually dissolved in either a nonreducing or reducing buffer. Using a nonreducing buffer would give information on the native protein fractions, while the reducing buffer usually contains mercaptoethanol as the reducing agent, and would provide evidence of multiprotein components being connected by a disulfide bond.

Thermal properties: The thermal properties of proteins give information about their behavioral changes during physical processing and are useful for food processing protocols. The physicochemical changes of proteins in foods as a result of a heat treatment can be measured by a method known as thermal analysis. Some techniques have been developed for thermal analysis; however, only a few are applicable in food protein studies, including differential scanning calorimetry (DSC) or differential thermal analysis (DTA), thermogravimetry (TG), thermal mechanical analysis (TMA), and dynamic mechanical analysis (DMA) (Harwalker and Ma 1989). Thermal property measurements are commonly used to study protein structure–function relationships due to heat. A DSC is used to evaluate the thermal properties of proteins by monitoring the difference in energy input needed for protein denaturation, in comparison with a reference material, usually aluminum oxide, as a function of heat flow or temperature (Harwalker and Ma 1989). These energy changes are recorded as a thermogram that shows the transition temperature/denaturation temperature and enthalpy of denaturation/transition, and are reported as the temperature at maximum heat flow, T_{max}, and enthalpy of transition (ΔH) expressed as calories or joules per gram. These parameters give information on the thermal stability of proteins and are useful for evaluating the reversibility, the thermodynamic parameters, and the kinetics of the reactions (Harwalker and Ma 1996).

Protein functionalities: The functional properties of proteins from various sources have been investigated. These include the proteins from melons (Olaofe et al. 1994), cowpeas (Horax et al. 2004a), rice (Wang, Pan, and Peng 1999; Shih and Daigle 2000; Paraman et al. 2006, 2007), and wheat (Ahmedna et al. 1999; Hettiarachchy et al. 1996). The functional properties of protein isolates in food applications include the extent of foaming (Tang et al. 2003; Horax et al. 2004b; Bamforth and Mi 2004; Deak and Johnson 2007; Lee et al. 2007) and emulsification (Halling 1981; D'Agostina et al. 2006; Jung et al. 2006; Makri and Doxastakis 2006; Zhang et al. 2009).

Protein solubility: Protein solubility is considered a thermodynamic manifestation that provides information on the equilibrium between protein–protein and protein–solvent interactions. This protein–water interaction is related to many of the functionalities of food proteins. The solubility of proteins in water is affected by their peptide bonds and their amino acid side chains (Cheftel, Cuq, and Lorient 1985). The proteins interact with water through their peptide bonds and amino acid side chains by dipole–dipole or hydrogen bonds and interactions with ionized or polar groups, respectively. Higher protein–water interactions result in an increase in the solubility of the proteins, while hydrophobic interactions increase the protein–protein interactions and result in decreases in solubility.

The solubility of proteins is affected by many factors, including the protein source, processing, extraction, the presence of other components needed during the processing, and the pH of the solvent. The solubility of the protein can be defined as the proportion of soluble nitrogen of the protein that is determined after a specifically defined procedure. The pH–solubility profile is often the first functional property determined for a new protein. This profile is determined as the proportion of protein quantified as nitrogen content that is soluble at pH values ranging from 2 to 12. The basic steps involve dispersing the protein in water, adjusting the pH value with acid or alkali, and determining the nitrogen content of the centrifuged supernatant. Solubility is considered as a prerequisite for the major functional properties of proteins, such as the emulsifying, foaming, and gelation properties (Culbertson 2006).

Emulsifying properties: Emulsion can be defined as a two-phase liquid system consisting of two immiscible liquids; one liquid phase is dispersed as droplets in the other liquid (Das and Kinsella 1990). Emulsions can be classified as oil-in-water and water-in-oil emulsions, depending on the dispersed phase. Emulsions are thermodynamically unstable due to a high interfacial tension between water and oil; this can easily cause phase separation with time (Damodaran 1996).

Several methods have been reported in the literature to determine the emulsifying properties of proteins, including the emulsifying activity index (EAI) and the emulsion stability index (ESI). The EAI is used to determine the interfacial area of emulsion generated per gram of protein (Pierce and Kinsella 1978). This property can be determined by a turbidimetric method measured at 500 nm. The emulsion stability index is obtained by determining the time needed for the emulsion to reach one-half of its turbidity value.

Proteins are considered one of the most important surface-active agents or emulsifiers in a food system (Narsimhan 1992). The ability of proteins as an emulsifying agent is due to their amphipathic molecules; hydrophobic groups interact with oil/lipid and hydrophilic groups interact with water in emulsions. Proteins can be used as emulsifiers in food products, such as salad dressings, mayonnaises, and processed meats. The emulsifying properties of some plant proteins, including that of wheat, egg, soybean, rice, cowpea, and rapeseed, have been investigated for their application in food products (Tang et al. 2003; Horax et al. 2004a; D'Agostina et al. 2006; Jung et al. 2006; Makri and Doxastakis 2006; Paraman et al. 2006, 2007; Zhang et al. 2007).

Foaming properties: Foam is usually defined as a dispersion of gas bubbles in a continuous liquid phase that contains a soluble surfactant. The foaming properties of proteins are important in food products, such as sponge cakes, whipped topping, ice cream, mousses, and icings. In foams, gas bubbles are encapsulated by a hydrated thin film in a continuous phase. Similar to emulsions, foams are thermodynamically unstable due to a large air–water interfacial surface (Kinsella and Whitehead 1988). Surface-active agents are needed to maintain the interface against coalescence of the gas bubbles by lowering the interfacial tension, stabilizing the interfacial film against internal and external forces, and forming more elastic films of gas bubbles (Cheftel, Cuq, and Lorient 1985). The foaming properties of proteins are related to the film-forming ability of their secondary and tertiary structures at the air–water interface (Mita et al. 1977).

The foaming properties of proteins are determined based on the increase in volume when a gas is introduced into a protein solution or dispersion. Two methods are widely used in the literature to determine the foaming properties of proteins: shaking the solution in an enclosed container and flowing gas into a solution (Kinsella 1976). These properties are usually expressed as the foaming capacity and the foaming stability. The foaming capacity is defined as the volume of foam formed after a particular time of shaking or gas introduction.

The foaming stability is determined by measuring the volume decrease or density increase with time.

Gelation: Gelation implies the formation of a well-hydrated mass, which is usually insoluble and may often require protein–protein or other interactions. Gelation is governed by water absorption, swelling, water-holding capacity, and wettability. It is also based on conditions such as high ionic strength, relatively high protein concentration, high heating temperature, and neutral pH. Egg proteins and egg substitutes are regarded as very good gelation agents. In general, thermal treatment may aid in the gelation phenomenon, and subsequent cooling in combination with acid treatment may also be helpful. Some protein, however, may gel without heat, preferably aided by mild enzymatic treatments (egg proteins and casein micelles) (Cheftel, Cuq, and Lorient 1985). Although many gels are formed from proteins in solution, certain whey proteins may form gels in sparingly soluble media. Hence, protein solubility may not be an important factor for gel formation, although hydration or the presence of water molecules in the vicinity may help.

Several food proteins and protein–protein interactions have been studied for gel formation in specific foods. With each type of protein, the gelation phenomenon is reviewed in the following sections. To understand the nature of the gelling processes, and the influence of the physical and chemical factors affecting the gel formation, it is useful to review gel characteristics and structure, although the mechanism of the formation of the three-dimensional protein networks that are characteristic of gels is not fully understood.

Protein modification for improved functional properties: Most natural proteins do not exhibit good emulsifying and foaming properties unless they are modified. Several chemical modifications have been done to increase the functionality of the protein ingredients. Modification methods include glycosylation, acylation, alkylation, esterification, phosphorylation, and cross-linking with transglutaminase (Schwenhe 1997; Haard 2001). Glycosylation involves interactions of proteins with reducing sugars; acylation involves acetic and succinic anhydrides to form more hydrophobic and more hydrophilic proteins, respectively; phosphorylation is done using phosphorus oxychloride ($POCl_3$) or sodium trimetaphosphate (STMP); sodium sulfite is used for sulfitolysis; and N-hydroxy succinimidyl esters of fatty acids can be used for alkylation (Damodaran 1996, 2005; Schwenhe 1997; Kato 2002). However, the use of chemical protein modification is limited in the food industries due to the nutritional safety caused by unpleasant reaction conditions. Among these chemical methods, glycosylation of a food protein can be implemented without an adverse effect on the nutrition and safety of the proteins (Damodaran 2005).

Glycosylation: The glycosylation of proteins is chemically conducted to produce more hydrophilic proteins called glycoproteins (Schwenhe 1997). These chemically modified proteins are accomplished by the covalent attachment of monosaccharides or oligosaccharides to the protein structures through the Maillard reaction. Glycosylation can be carried out faster by using aldoses, such as glucose and fructose, because of their higher reducing property (Courthaudon et al. 1989). A controlled and limited glycosylation of a protein could be carried out to achieve the need for a particular function of that protein (Achouri et al. 2005). Many studies have been reported to modify proteins using a controlled and limited glycosylation to improve the emulsifying and foaming properties, water-holding capacity, thermal stability, and solubility of many food proteins (Chevalier et al. 2001; Yeboah et al. 1999, 2000; Aminlari et al. 2005; Achouri et al. 2005; Jiménez-Castaño et al. 2005, 2007).

$$\text{Glu–CHO} + \text{H}_2\text{N–protein} \longrightarrow \text{Glyco–CHO=N–protein}$$
$$\downarrow$$
$$\text{Glyco–CH}_2\text{–NH–protein}$$

Acylation: The acylation of proteins involves the covalent attachment of acyl groups, such as fatty acids and amino acids, to the amino groups of the polypeptide chain. The amino groups of the protein can be acylated by reacting with acid anhydrides, such as acetic anhydride and succinic anhydride. Acetic anhydride will make a protein become more nonpolar, while succinic anhydride makes it more polar (more anionic) (Damodaran 1996).

Protein   Acetic anhydride

Damodaran et al. (1996) have reviewed the acylation of food proteins. Acylation protects the proteins from other chemical modifications induced during food processing. Acylated proteins are generally found to have increased the solubility, compared with native proteins. Charge replacement plays a major role in the acylated proteins attaining an unfolded state owing to its increased electronegativity. This property is used in food proteins to increase their solubility and is hence easily incorporated into certain food matrices without any concern for solubility.

Alkylation: The reactions of the SH and amino groups with iodoacetamide or iodoacetate cause alkylation. The reactions cause lysyl groups to void their positive charges and acquire negative charges at both the lysyl and cysteine residues. Due to the increase in the electronegativity of the alkylated protein, changes to the pH–solubility profile may be encountered, thereby influencing the folding nature of the protein. The hydrophobicity of the protein can be increased if an aldehyde or ketone can be used for reaction purposes. On the other hand, a reducing sugar in the reaction may cause the protein to become hydrophilic (Fennema 1996).

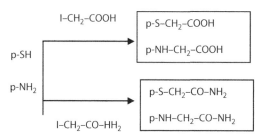

Esterification: The carboxyl groups of proteins can be treated with methanol or ethanol in the presence of an acid catalyst to produce esters. Esterification blocks the negative charges on the carboxyl group, which results in the protein acquiring an increased isoelectric point (pI). Esterification causes conformational changes to the proteins, thereby altering their functional properties (Fennema 1996). β-Lactoglobulin esterification leads to a decrease in the proteins' surface and inter-facial tension. Strong electrostatic interactions of esterified β-lactoglobulin and casein micelles in bovine milk have also been observed. The bioavailability of esteri-fied proteins needs more investigation since studies have shown that the rate of peptide hydrolysis on esterified proteins increases compared with native proteins; however, with β-lactoglobulin it was found to decrease.

$$\text{Protein} \_\_\_\_ COO^- + ROH \xrightarrow{\text{0.02-0.1 M HCl}} \text{Protein} \_\_\_\_ COOR$$

Phosphorylation: Proteins can be phosphorylated by reacting them with $POCl_3$, mainly at the hydroxyl group of serine and threonine residues and at the amino group of lysine residues, to give additional negative charges. Phosphorylation greatly increases the electronegativity of proteins. The addition of two negative charges for each positive charge occurs when phosphorylating the amino groups.

Several foods undergo phosphorylation of their proteins, such as caseins, which may be considered desirable for certain food functionalities. Both enzy-matic and chemical phosphorylation can be achieved. Chardot et al. (2004)

studied enzyme-phosphorylated caseins to show better solubility profiles. The chemically induced phosphorylation of soybean proteins led to an improvement in their solubility, emulsifying activity, and foaming properties (Damodaran 1996).

Cross-linking: Food protein cross-links often result when there are excessive changes to temperature or pH, thus affecting the functional and nutritional properties. The disulfide bonds are the most common types of covalent cross-links found in food protein systems. They are formed by the oxidative coupling of two adjacent cysteine residues within a food protein matrix. The ability of proteins to form intermolecular disulfide bonds during heat treatment is considered to be crucial for the gelation of some food proteins, such as milk proteins, soybeans, and eggs (Zayas 1997).

The reaction of transglutaminase in food systems produces protein cross-links that also have the potential to improve the functionalities, such as elasticity, viscosity, thermal stability, and water-holding capacity of processed foods (Kuraishi, Yamazaki, and Susa 2001). Transglutaminase catalyzes the transfer of the γ-carboxamide group of peptide-bound glutamine residues to the primary amines serving as the acyl acceptor. In addition, the transglutaminase-catalyzed reaction may lead to the formation of both intramolecular and intermolecular isopeptide bonds between the glutamine and lysine residues in proteins. This enzyme has therefore been used for the cross-linking of food proteins with a view to improving their nutritional and/or functional properties (Meade et al. 2005).

$$\text{Protein–CH}_2\text{–CH}_2\text{–CO–NH}_2 \qquad \text{Glutaminyl residue}$$

$$\xrightarrow[\text{H}_2\text{O}]{\text{E}} \quad \text{Protein–CH}_2\text{–CH}_2\text{–CO–OH} \qquad \text{Hydrolase}$$

$$\xrightarrow[\text{H}_2\text{N–R}]{\text{E}} \quad \text{Protein–CH}_2\text{–CH}_2\text{–CO–NHR} \qquad \text{Transferase}$$

$$\xrightarrow[\text{H}_2\text{N–(CH}_2)_4\text{–P}]{\text{E}} \quad \text{Protein–CH}_2\text{–CH}_2\text{–CO–NH–(CH}_2)_4\text{–Protein} \qquad \text{Cross-linking}$$

3.3 Milk Proteins and Peptides

3.3.1 Whey and Casein

Whey: Whey is increasingly being used for human and animal nutrition. Whey protein is said to comprise a mixture of proteins with several diverse functional properties. The main proteins are β-lactoglobulin and α-lactalbumin (Figure 3.1). These constituents influence the behavior of the whey protein–incorporated food systems by hydration, gelling, and other functional properties, such as foaming and emulsification.

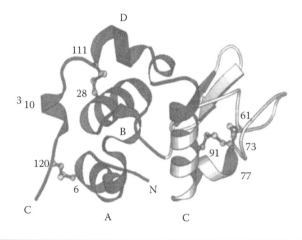

Figure 3.1　Whey proteins. Three-dimensional structure of α-lactoglobulin. (From Horng, J. C., Demarest, S. J., and Raleigh, D. P. 2003. *Proteins* 52(2): 193–202.)

β-*Lactoglobulin*:　β-Lactoglobulin has a molecular mass of 18.3 kDa. It is composed of 162 amino acids with two disulfide bonds. The secondary and tertiary structures of the protein show a greater proportion of β-sheets (43%–50%). The monomer form of β-lactoglobulin has been viewed as close to the retinol-binding protein from blood plasma because of its hydrophobic pocket, which is capable of binding small molecules such as vitamins and fatty acids. There is also a nine β-sheet stack that allows a compact structure that is resistant to complete proteolysis by digestive proteases.

α-*Lactalbumin*:　α-Lactalbumin is a smaller protein of 14.2 kDa with 123 amino acid residues and four disulfide bridges. Unlike β-lactoglobulin, α-lactalbumin has a low content of ordered structures consisting of only 30% helix and 9% β-sheets. Its compact structure is constituted by a bound calcium ion and the four disulfide bridges (Cayot and Lorient 1997). Moatsou et al. (2008) have studied the denaturation phenomena of whey proteins and the effects of high pressure and temperature on total plasmin and plasminogen-derived activity, including that of plasminogen activators. They found that high pressure significantly affected the activity of the proteolytic enzymes and whey protein denaturation in bovine milk. In a similar study, Huppertz, Fox, and Kelly (2004) studied the mechanism of the high-pressure–induced denaturation phenomenon of whey and milk proteins and associated the absence of casein micelles and calcium phosphate with more pronounced denaturation in milk proteins than in whey proteins.

　Two other major whey proteins are albumins and globulins, wherein bovine serum albumins (BSA) and immunoglobulins (mostly glycoproteins) fall under their respective classes.

The structural properties of whey proteins depend on factors such as pH, salts, and other proteins in the surrounding environment. Let us focus first on the physical processing that affects the whey protein structure and thereby its functional properties.

Denaturation: Cooling or heating β-lactoglobulin causes an increase in the reactivity of the thiol group. In the state of cooling, dimers, trimers, or polymers may be induced. With heating, β-lactoglobulin has been observed to dissociate because of a possible destabilizing effect of the hydrogen bonds. With an increase in temperature, the formation of disulfide bonds has been noted to cause β-lactoglobulin polymerization (Azuaga et al. 1992). Certain other globulin proteins, such as soy globulins, also undergo this phenomenon.

Scientists (Goddik 2003) have tried to increase β-lactoglobulin's functionality while minimizing protein aggregation by denaturation. The protein was treated with high pressure for 10 min at 510 and 600 MPa and had the lowest surface tension. In these conditions, protein aggregation was minimized. The surface tension measurements on stored samples were consistent with this trend. CD data indicated secondary structural changes, with the primary changes occurring in the α-helix. Freeze drying the samples minimized the structural changes during their storage.

Experiments conducted at varying pH values to understand the heating phenomenon of β-lactoglobulin monomers have provided considerable information that the SH groups are in fact responsible for polymerization in S–S containing proteins (Cayot and Lorient 1997).

The aggregation of β-lactoglobulin, as evident with increased β-lactoglobulin polymerization due to the pH effect and heating, can also be the result of ions. It has been observed that the addition of calcium salts before or after heating increases the flocculation of β-lactoglobulin, possibly by bridging between the two carboxyl groups. Although it is a general phenomenon that salts induce aggregation, citrate and phosphate have been shown to decrease the aggregation of β-lactoglobulin during heat treatment, possibly due to a chelating effect (Cayot and Lorient 1997).

Lactalbumin, on the other hand, having no SH groups but with four disulfide bridges and the presence of ions, resists heat to a greater extent than β-lactoglobulin. It has the ability to reverse denaturation. However, when α-lactalbumin is heated in the presence of β-lactoglobulin at a neutral pH, α-lactalbumin is readily and irreversibly denatured (Elfgam and Wheelock 1978). The results from experiments studying the denaturing effect of these two proteins do not seem to corroborate well, mainly because of the techniques used.

Emulsifying properties: Heating an emulsion of whey protein induces gelation of the adsorbed film, causing an imminent increase in the mechanical stability. However, when heated for a long time, the emulsifying properties decrease. Similarly, it has been observed that the thermal denaturation of whey protein isolates (WPIs) reduces the stability of the emulsions. Examining the emulsifying properties of WPIs as they are heat-denatured and -undenatured reveals the possibility of a mixture of monomeric and polymeric proteins that form a cohesive interfacial film with an increase in the amount of denatured WPI. It was also observed that the effect of reduction in the disulfide bonds in both proteins causes an increased emulsifying activity (Cayot and Lorient 1976).

Whey and milk proteins have also been the subject of study to understand dehydrated emulsions, generally prepared using encapsulation techniques, including spray-drying methods (Landstrom, Alsins, and Bergenståhl 2000; Millqvist-Fureby, Elofsson, and Bergenståhl 2001; Sliwinski et al. 2003; Vega and Roos 2006). The main disadvantage relative to the use of whey proteins as encapsulants is their susceptibility to heat denaturation and the effects on the emulsion particle size before spray drying and after reconstitution (Figure 3.2) (Sliwinski et al. 2003). An increase in the concentration of the whey protein accelerates the rate and the degree of aggregation, suggesting that the main mechanism is the denaturation and aggregation of the unabsorbed protein (Euston, Finnigan, and Hirst 2000).

Foaming properties: The foaming capacity of α-lactalbumin has been shown to increase with heat, especially at a pH of 7 or 9 with increased stability. Conversely, heating at a pH of 2 or 5 causes a reduction in the foaming properties. β-Lactoglobulin, however, tends to present a negative effect with dispersed pH but a positive effect on foam stability. Moderate heating (50°C–65°C) tends to improve the foaming properties of whey proteins. It also seems that the foaming properties of emulsified whey

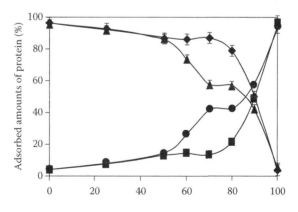

Figure 3.2 Effects of adsorbed sodium caseinates on whey proteins. Percentages of adsorbed sodium caseinate after emulsion preparation (◆) and after spray drying and reconstitution (▲) and that of adsorbed whey protein after emulsion preparation (■) and after spray drying and reconstitution (●) for milk protein–stabilized oil-in-water emulsions with constant protein content and varying ratios of sodium caseinate to whey protein. (From Sliwinski, E. L., et al., 2003. *Colloid and Surface* B 31: 219–29.)

protein concentrate are positively correlated to the number of reactive SH groups (Petonen-Shallaby and Mangino 1986).

It had been established that high-pressure–induced β-lactoglobulin displays lower emulsifying and foaming capacity than native β-lactoglobulin, owing to increased surface hydrophobicity (Pittia et al. 1996). This causes an apparent increase in protein–protein interactions, and hence the protein was observed to form highly viscous films that increase in viscosity on pressurization compared with BSA, which forms less viscous films and weakens after high-pressure treatment (600 MPa). This is the result of decreased surface hydrophobicity, unfolding, and aggregation of the pressurized BSA (Galazka, Dickinson, and Ledward 2000).

There is also evidence that pressure-induced β-lactoglobulin improves the foaming properties, even though it gives emulsions with larger droplets and lower stability. This may be explained based on the different degrees of protein aggregation (Galazka, Dickinson, and Ledward 2000). The pressurization of WPI at a high concentration (2% w/v) or high buffer molarity diminished its foaming properties (Banoglu and Karata 2001). The functional improvement of milk–whey proteins, where both the emulsifying and foaming properties of whey proteins are reviewed, is a useful resource (Lopez-Fandino 2006).

Caseins: Caseins come in a variety of forms, all of which are important in the preparation of foods. They possess different functional properties that may be at least partly attributed to their respective structures.

Bovine milk contains four different types of caseins, denoted as αs1, αs2, β, and κ. When secreted, they are in the form of casein micelles. The two αs-caseins along with the β-casein are said to be calcium sensitive. These caseins are said to share very similar signal peptide sequences and contain one or more phosphate centers. Their binding to phosphate forms a very important biological function (Dalgleish 1997).

Caseins' open structure (Figure 3.3) allows an enzymatic breakdown in the stomach, and the casein micellar fraction makes calcium readily available for absorption. Casein is rich in certain nonessential amino acids, such as glutamine, glutamic acid, serine, and proline. The main biological function of casein is attributed to its structure, and is hypothesized so, because of its open and flexible conformation that can tolerate any mutation, its suboptimal amino acid composition, its inherent ability to be phosphorylated, and its ability to organize calcium and phosphate ions into a milk colloid (Holt and Roginski 2001).

Caseins do not denature at a specific transition temperature because they do not undergo an endothermic enthalpy of denaturation and, therefore, do not attain a molten globule state (Fox and McSweeney 1998). The presence of amino acids such as proline and glutamine enables helix breaking and β-sheet formation. The presence of Cys enables the stabilization of the globular protein, forming cysteine bridges in the casein. Caseins tend to adopt a compact, globular conformation,

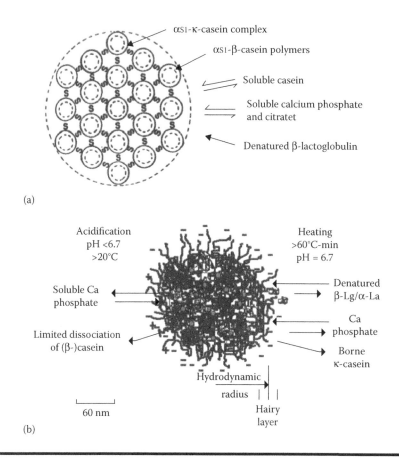

Figure 3.3 Structures of casein micelles. (a) The casein submicelle model. (From Phadungath, 2005, Adapted from Wong, 1988.) (b) The Holt model. (Adapted from Holt and Horne, 1996, from http://www.foodscience.uoguelph.ca/deicon/casein.html)

but other chain sequences, including the C-terminal β-domain, can be made to associate by a hydrophobic effect. Horne (2006) reviewed a detailed account of the nanocluster model of a casein micelle and other models.

Casein consists of a fairly high number of proline peptides, which do not interact. There are also no disulfide bridges. As a result, it has relatively little tertiary structure; therefore, it cannot denature. It is relatively hydrophobic, making it poorly soluble in water. It is found in milk as a suspension of particles called casein micelles, which show some resemblance with surfactant-type micelles in a sense that the hydrophilic parts reside at the surface. The caseins in the micelles are held together by calcium ions and hydrophobic interactions. There are several models that account for the special conformation of the caseins in the micelles (Dalgleish 1997). One of these models proposes that the micellar nucleus is formed by several submicelles, with the periphery consisting of microvellocities of κ-casein (Walstra 1999). Another model suggests that the nucleus is formed by casein-interlinked fibrils (Holt 1992). Finally, the most recent model (Horne 2006) proposes a double link among the caseins to allow gelling to take place. All three models consider micelles as colloidal particles formed by casein aggregates wrapped up in soluble κ-casein molecules.

The pI of casein is 4.6. Since the pH of milk is 6.6, casein has a negative charge in milk. The purified protein is water-insoluble. While it is also insoluble in neutral salt solutions, it is readily dispersible in dilute alkalis and in salt solutions such as sodium oxalate and sodium acetate.

Acid gelation: As milk is treated with acid, the colloidal calcium phosphate dissolves. The calcium ions are displaced from the phosphate centers of the caseins by hydrogen ions. There is a minimal change in the net charge on the protein, and the appearance of the casein micelles does not change initially but changes over time. With time, complex ionic equilibria take place as the pH approaches the pK, and the micelles begin to collapse and aggregate to form a gel (Dalgleish 1997).

Renneting: Several complex processes are involved in renneting. First, the effect of acidification and lactic acid bacteria culture destabilizes the micelle. This is followed by the cleavage of the macropeptide κ-casein. A kinetic reaction monitoring experiment that helps identify the aggregation of the destabilized particles occurs. Eventually, the micelle dynamically reorganizes and causes a disturbance to the whole micellar system (Holt and Roginski 2001).

Coagulation: It was shown that with ethanol treatment, the stability of milk as a function of the pH depends on an effective pK that is sensitive to Ca salts. When different ethanol concentrations were tried, a dielectric constant on the effective pK signified stability differences due to charge variation. Heat-induced coagulation, however, remains very complex to understand. This is due, in part, to the introduction of whey proteins that readily denature and cause a limited coprecipitation of the casein (Holt and Roginski 2001). In the absence of the whey protein, the coagulation time increases sharply around the minimum variation of milk's pH of 6.7. Another observation noted is that κ-casein becomes solubilized on heating to above 110°C.

Emulsifying properties: The emulsifying properties of caseinates arise from the structure of their proteins; mainly because of their hydrophobic regions, they can bind to hydrophobic materials such as oil–water interfaces. Experiments conducted on caseins have shown that they can cover an interface to the extent of 2–3 mg/m^2, and that the proteins present in the interface tend to aggregate followed by dissociation of the aggregates as the protein adsorbs to the oil–water interface (Dalgleish 1997). Casein is capable of extending widely over an oil–water interface, forming a stable emulsion at low protein concentrations, but when the protein concentration increases, the surface coverage increases as the proteins undergo conformational changes. This nature is very unique to caseins.

Emulsions based on caseinates or individual caseins are very stable because adsorbed caseins possess the two important attributes for successful stabilization of colloidal material: being charged and conformationally flexible. Caseins also have the capacity to stabilize emulsions, especially in the presence of substantial concentrations of ethanol, partly due to the collapse of the casein layer with the addition of ethanol in milk. Emulsions formed using casein micelles are not good emulsifiers. This is because the casein aggregates become compact by hydrophobic interactions and the casein micelles are grouped together by calcium phosphate, which cannot resist dissociation of the micelles, causing low emulsifying behavior, as opposed to individual caseins or caseinates (Holt and Roginski 2001).

Foaming properties: Caseins are not widely recommended as foaming agents. As the interactions between caseins are largely hydrophobic, there will be adsorption and prevention of the proteins interacting together, and the hydrophilic regions in the solution may not bind to one another, thereby losing rigidity within the lamella of the foam (Dalgleish 1997). Moreover, intermolecular bonds (disulfides) and interfacial layers are not formed, which are needed to hold the foam together. The foaming capacity is compromised, as opposed to a better foaming capacity seen with β-lactoglobulin, with relatively stronger interfacial layers and an interaction between the molecules through the disulfide bonds. In summary, emulsification and gelation are the major functional characteristics of caseins; other properties, such as foaming and solubility, are less significant.

3.4 Cereal and Legume Proteins and Peptides

3.4.1 Rice and Wheat

Rice: Rice protein is commonly fractioned into four classes based on their solubility in different solvents. Glutelin makes up about 80% of the total protein and is soluble in alkali (Tecson et al. 1971; Padhye and Salunkhe 1979; Juliano 1985); albumin is about 1%–5% and is water-soluble; globulin is soluble in salt and makes up about 4%–15%; and prolamin is 2%–8% and is soluble in alcohol (Cagampang et al. 1966; Houston et al. 1968). However, the relative quantities of each class of proteins are dependent on the genotype, the growing conditions, and the analytical methods employed (Huebner et al. 1990; Krishnan and White 1995).

Glutelin: The high molecular weight (HMW) storage protein of rice glutelin (oryzenin) is the most important protein in the rice kernel. This protein fraction can be extracted with dilute alkali. Gel electrophoresis studies revealed that native oryzenin consists of several polypeptides bound by disulfide bonds. A reduction of oryzenin with 2-mercaptoethanol and alkylation with acrylonitrile drastically reduced the molecular weight. SDS-PAGE studies showed that oryzenin consists of three major subunits with molecular weights of 38, 25, and 16 kDa (Juliano and Boulter 1976).

Albumins and globulins: Albumins and globulins are water- and salt-soluble rice proteins, respectively. The albumins in rice were separated into three to four subfractions using gel filtration chromatography on Sephadex G-100 columns. The molecular weight of the subfractions ranged from 10 to 200 kDa (Cagampang et al. 1976; Iwasaki et al. 1982). Iwasaki et al. (1982) separated about 20 components of the albumins in rice endosperm. The globulins in rice endosperm were also separated by gel filtration chromatography into four subfractions with molecular weights ranging from 16 to 130 kDa (Iwasaki et al. 1975, 1976; Cagampang et al. 1976). The subunits were separated and identified by gel electrophoresis. It was stated that the globulins in rice are highly heterogeneous and have a complicated structure.

Prolamin: The low molecular weight (LMW) storage proteins (prolamins) in rice can be extracted with 70% (v/v) ethanol. Propanol (50%, v/v) can also be used. Electrophoresis studies of rice prolamin by Sgawa et al. (1986) and Kim and Okita (1988) revealed that three different prolamin classes occur in the protein bodies of the lamellar structure. They have molecular weights of 10, 13–15, and 16 kDa.

Rice proteins are extremely insoluble due to their rigid globular structure with excessive intramolecular and intermolecular disulfide bonds and hydrophobic interactions (Tecson et al. 1971). The most abundant amino acids in rice glutelin are glutamine, asparagine, arginine, glycine, and alanine. The amide groups in the glutamine and asparagine side chains promote the aggregation of glutelin (Wen and Luthe 1985). These native structural properties of rice reduce protein molecular flexibility and lead to low solubility and poor surface-active properties. Compared with other plant proteins, rice protein has relatively poor functional properties. Further, the high temperature and pH conditions involved for protein isolation decrease their solubility; thus, they have limited application as a functional ingredient (Shih and Daigle 1997). Very less literature is available on the use of rice endosperm protein as a functional ingredient. A few attempts have been made to improve the functional properties of rice protein. Numerous studies, thus far, have focused on the nutritional and hypoallergenic properties of rice proteins.

Although several methods exist for extracting rice proteins, the use of alkali resulted in purer protein fractions and has been recommended for better functionality (Paraman et al. 2006). The authors showed that alkali-extracted rice proteins have higher emulsifying and foaming properties than those of enzyme-extracted proteins. A relatively more favorable protein composition, a higher solubility, and a lower degree of thermal denaturation of alkali-extracted proteins contributed to higher emulsifying and foaming properties. Paraman et al. (2007) modified rice protein isolates and examined them for better functional properties. They suggest that surface hydrophobicity does not indicate protein solubility prediction and emulsifying properties, but rather solubility and molecular flexibility are the essential factors in achieving good emulsifying properties of rice endosperm protein isolates.

Wheat: Wheat proteins are classified into monomeric and polymeric proteins. Monomeric proteins include gliadins and albumins/globulins. Gliadins are storage proteins, whereas albumins/globulins are metabolic.

Gliadin is a glycoprotein and consists of the following types:

- α-/β-*Gliadins*: soluble in low percentage alcohols
- γ-*Gliadins*: ancestral form of cysteine-rich gliadin with only intrachain disulfide bridges
- ω-*Gliadins*: soluble in higher percentages of 30%–50% acidic acetonitrile

Gliadins are known for their role, along with glutenin, in the formation of gluten. They are slightly soluble in ethanol and contain only intramolecule disulfide links. These proteins are essential for giving bread the ability to rise properly and fix its shape on baking.

Albumins/globulins: Albumins are water-soluble while globulins are salt-soluble, and they are a mixture of LMW compounds. They are quite distinct from the gluten proteins because of their varied amino acid contents. Lysine is present in large amounts of these proteins, while the glutamic acid content is high in gluten proteins (Macritchie and Lafiandra 1997). Polymeric proteins include glutenin, HMW albumin, and tricin.

Glutenin is a protein best known for its role, along with gliadin, in the creation of gluten with its disulfide intermolecular and intramolecular links. It consists of 20% HMW subunits, which are relatively low in sulfur. The other 80% are LMW subunits, which are high in sulfur. It is soluble in dilute acids and bases.

HMW albumins are mainly β-amylases whose subunits do not participate in the interaction with glutenins. Tricins, on the other hand, are globular-type proteins whose ability to interact with glutenin subunits (GS) has not been completely ruled out.

Gluten comprises some 75% protein on a dry weight basis. The vast majority of the proteins are called prolamins, which were initially defined based on their solubility in alcohol–water mixtures (Osborne 1924), typically 60%–70% (v/v) with ethanol. In wheat, these groups of monomeric and polymeric prolamins are known as gliadins and glutenins, respectively, and together form gluten (Shewry et al. 1986). Galili (1997) has proposed that glutens are preferentially retained in the endoplasmic reticulum. Hence, during dough mixing, a gluten network is formed. The precise changes that occur in dough during mixing are still not completely understood. However, an increase in dough stiffness occurs that is generally considered to result from the "optimization" of protein–protein interactions within the gluten network (Anjum et al. 2007).

The HMW-GS accounts for about 5%–10% of the total protein and plays an important role in the process of bread making. The HMW-GS has a higher content of proline and glycine and a low content of lysine with an unusually high content of glutamic acid. Two features of the HMW subunit structure may be relevant to its role in glutenin elastomers: the number and distribution of disulfide bonds and the properties and interactions of the repetitive domains. Although it is now widely accepted that disulfide-linked glutenin chains provide an "elastic backbone" to gluten, evidence from spectroscopic studies (using NMR and FTIR spectroscopy) of HMW subunits and model peptides suggests that noncovalent hydrogen bonding between the GS and polymers may also be important (Gilbert et al. 2000).

Until recently, the scientific literature has shown that disulfide bonds were the only critical covalent cross-links necessary for wheat flour dough formation. However, the participation of nondisulfide bonds, especially those of the tyrosine cross-links, has been discovered to form a molecular basis for gluten structure and function (Tilley et al. 2001). These cross-links occur in wheat flour dough during the mixing and baking processes. Tyrosine bonds can form chemically under baking conditions when synthetic glutenin peptides are used; however, enzymatic mechanisms may also be involved during gluten development in the dough environment. The tyrosine bond structure and its formation during the bread-making process have been documented by HPLC, NMR, and mass spectroscopic analyses. Thus, the HMW-GS are considered key components to understanding the quality and functionality of wheat dough.

3.4.2 Soy

Structure of soy proteins: Albumins and globulins are the main proteins found in soy seeds. Two types of globulins account for 50%–90% of the seed proteins: 7S globulins and 11S globulins (Figure 3.4; Itoh et al. 2006). The ratios between them are between 0.5 and 1.7 in soybeans. The structure of the proteins can be modified to improve specific functional properties. Heat treatment is the most common method, but several other factors, such as the pH, time, protein concentration, ionic strength, and temperature, can be used. Studies have shown that functionality is dependent on the aggregation, denaturation, and degree of dissociation of the 7S and 11S subunits for the soy protein (Arrese et al. 1991). Soy protein isolates (SPIs) have varying levels of soluble and insoluble proteins, which also influence their functional properties.

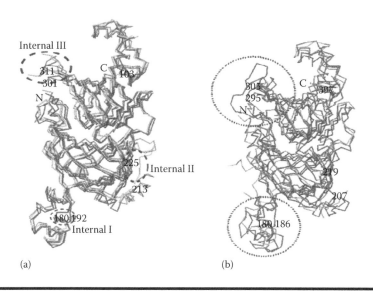

(a)   (b)

Figure 3.4 Superimposed structures of 7s and 11s soy globulins. Structural comparison of overall structures. (a) 7S globulins (soybean β-conglycinin; jack bean canavalin; kidney bean phaseolin) superimposed onto an 8Sα-globulin. (b) Another structurally similar protein and enzyme (soybean 11S proglycinin [A1aB1b]; oxalate decarboxylase Yvrk from *Bacillus subtilis* sp. 168) also superimposed onto an 8Sα-globulin. The black broken line surrounds the site near the N-terminal amino acid peptide and the N-terminal extended loop region, which forms a different formation in each globulin. (From Itoh, T., et al., 2006. *Acta Crystallographica* D62: 824–32.)

7S Globulins: These globulins are classified into three types based on their physico-chemical properties: β-conglycinin, γ-conglycinin, and basic 7S globulin (Friedman and Brandon 2001). β-Conglycinin accounts for 30%–50% of the total seed protein. Basic 7S globulin has a molecular mass of 168 kDa with four subunits and is a glycoprotein. γ-Conglycinin is a trimer glycoprotein with a molecular mass of 170 kDa. β-Conglycinin is a trimer with a molecular mass of 150–200 kDa. It has four subunits, three major: α′, α, and β, and one minor subunit: γ. β-Conglycinins are glycoproteins. The major 7S globulin genes and cDNAs in legume seeds have been studied and their nucleotide sequences determined.

11S Globulins: Glycinin, the 11S globulin in soybeans, is a hexamer with a molecular mass of between 300 and 380 kDa. Each subunit is composed of an acidic polypeptide and a basic polypeptide connected by a disulfide bond. Based on cloning and sequencing studies, each subunit is classified into two groups: group I with $A_{1a}B_{1b}$, A_2B_{1a}, and $A_{1b}B_2$; and group II with $A_5A_4B_3$ and A_3B_4. Among the soybean cultivars, glycinin was found to exhibit polymorphism in the subunit composition. The amino acid sequences in the subunits are different among soybean cultivars (Utsumi,

Matsumura, and Mori 1997). Due to the polymorphism and the heterogeneity of glycinin, there is a possibility of an exchange between the acidic and basic subunits and a single subunit formation during the construction of the molecule. Maruyama et al. (2004) studied the structure–function relationships of glycinin with respect to amino acids and found that the solubility of the subunits decreases because of the presence of the acidic amino acids in group I subunits, A3B4 and A5A4B3.

Relationship between 7S and 11S globulins: Soybean seeds have a higher β-conglycinin-to-glycinin ratio during maturation. Researchers studying the soy protein molecular structure found that the genes of the 7S and 11S globulins have a common ancestry. The entire region of the 11S globulin could be mapped onto the 7S sequence (Utsumi, Matsumura, and Mori 1997).

Gelation: The globulins glycinin and β-conglycinin have different gel-forming properties, which are the most important properties in food preparation and processing. The soluble aggregates and gel network were visualized by transmission electron microscopy to understand the mechanism of network formation in the thermal gelation process of glycinin (Figure 3.5) (Mills et al. 2003).

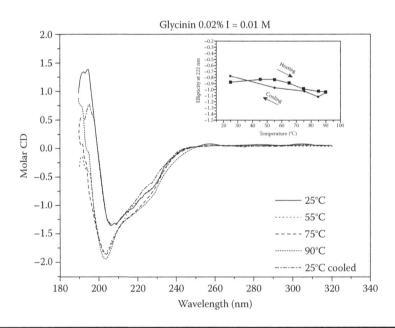

Figure 3.5 Thermal gelation phenomenon observed with glycinin. Far-UV CD spectra of dilute glycinin solution in situ during heating. Glycinin was at 0.02% (w/v) in 3.25 mM potassium phosphate, pH 7.6 buffer alone *I* = 0.01. Smoothed spectra are shown. Inset shows the change in molar ellipticity at 222 nm during heating and cooling. (From Mills, E. N., et al., 2003. *Biochimica Biophysica Acta* 1648: 105–14.)

The acidic and basic polypeptides disaggregate when heated at a low protein concentration, while gel formation is observed at a high protein concentration. Gelation is based on conditions such as high ionic strength, relatively high protein concentration, high heating temperature, and neutral pH. Under different heating conditions, the subunits may show association–dissociation gelation behavior. Hydrogen bonding and van der Waal forces were found to be the major forces in the gel-forming ability of soy proteins (Utsumi, Matsumura, and Mori 1997). Nakamura, Utsumi, and Mori (1984) studied the gelation property and found that disulfide exchange reactions may participate in network formation and gel formation. A study of the gelling properties of protein mixtures was conducted with soy protein and acid-induced casein gels in order to understand the influence of adding soy protein to milk protein. Whey protein, as the other component, when added to milk protein, contributed most to the gel-formation network in a soy–whey–casein system. Hence, it was concluded that the incorporation of soy proteins in a mixture of proteins is possible without a significant change in the structure of the casein gels (Roesch and Corredig 2006). The water-holding capacity of gels decreases with increasing salt concentration. Hydrophobic interactions at a pH of 2.75 with the addition of sodium chloride caused both 7S and 11S globulin subunits to participate in stabilizing the gels (Puppo and Añón 1998). Renkema et al. (2000) studied the influence of pH on the gel-forming properties of the SPI and purified glycinin in relation to denaturation and aggregation. Their study found the formation of more fine-stranded gels with a smooth, slightly turbid appearance at a pH of 7.6, against white coarse gels obtained with a high stiffness and a granulated surface at a pH of 3.8. Heat denaturation of β-conglycinin initiated the gelation process at a pH of 3.8, while its role seems to be minor at a pH of 7.6. The mechanism of gel formation was affected by the pH, wherein, on heating at a pH of 7.6, the disulfide bridge between the acidic and basic polypeptides of glycinin was broken, which was in contrast to the conditions at a pH of 3.8.

Emulsification: Soy protein is used as a macromolecular surfactant in food systems to stabilize oil-in-water emulsions. It was found to be a better emulsifier than whey protein or casein (Utsumi, Matsumura, and Mori 1997). β-Conglycinin especially showed better emulsifying activity than ovalbumin, which is a standard emulsifying protein used in studies. At varying pH values, β-conglycinin had the best ability to lower the interfacial tension between the two constituents of the emulsion in an experiment involving liquid paraffin/water. According to Yao, Tanteeratarm, and Wei (1983), a protein isolate prepared from mature seeds showed higher emulsion stability, which was due to a higher percentage of β-conglycinin. The differences in the emulsifying capabilities of β-conglycinin and glycinin were attributed to their chemical and structural features. Due to the differences in hydrophobic patches on the surface of the molecule, the absorption ability of an interface is more rapid in β-conglycinin than in glycinin (Utsumi, Matsumura, and Mori 1997).

These experiments conclude that β-conglycinin has greater emulsion stability and activity than glycinin.

Soy proteins have been used to prepare stable emulsions. Xu and Yao (2009) have successfully formed a stable oil-in-water emulsion using acid-soluble soy protein, dextran conjugates, and ultrasonic emulsification. The dextran molecules that conjugated to the protein were found to effectively enhance the hydrophilicity and steric repulsion of the oil droplets, making the emulsions stable against heat treatment, long-term storage, and changes in the pH and ionic strength.

The studies of Kiosseoglou and coworkers demonstrated that soy protein–polysaccharide conjugates can improve the emulsifying properties, especially in the reduction of oil droplet size and emulsion stabilization against creaming (Diftis and Kiosseoglou 2003, 2006a, 2006b; Diftis, Biliaderis, and Kiosseoglou 2005).

Foaming: The molecular flexibility of the protein film properties influences its foaming stability, while the nature of the protein environmental factors influences its foaming capacity. The foaming quality of HMW partly digested soy proteins was superior to that of native proteins. The foaming activity of proteins, which is the amount of interfacial area created, is related to their unique physicochemical properties under standard foaming conditions. The rheological properties of the lamella influenced by the environmental and physicochemical properties of the protein affect the stability of the foam.

SPIs have been modified to improve their foaming properties. A significant amount of variability in the foams formed from soy protein is due to the various degrees of denaturation during preparation. A relatively stable oligomeric structure of glycinin limits its functionality in foaming. Hence, molecular stability is needed when using glycinin as a foaming agent in food systems. Moderate heating up to 80°C and partial proteolysis improve the foaming properties of soy protein. Foam decay is dependent on environmental factors such as pH, temperature, ionic strength, and protein concentration. Foams formed with an unmodified SPI and soy 11S are more stable than foams made from the 7S fraction. This is due to the inability of the 7S fraction to retard gravitational drainage (Yu and Damodaran 1991).

Ruíz-Henestrosa et al. (2007a) studied the effect of limited enzymatic hydrolysis on the interfacial (dynamics of adsorption and surface dilatational properties) and foaming (foam formation and stabilization) characteristics of soy globulins (β-conglycinin, 7s fraction). They found that the interfacial characteristics of β-conglycinin improved with enzymatic treatment. Hydrolysates with a low degree of hydrolysis have improved functional properties (mainly foaming capacity and foam stability), especially at pH values close to the pI, because the protein is more difficult to convert into a film at fluid interfaces at a pH approximately equal to its pI.

Ruíz-Henestrosa, Sanchez, and Rodríguez Patino (2007b) also studied the effects of changes in the pH (5 or 7), ionic strength (at 0.05 and 0.5 M), addition of sucrose (at 1 M), and Tween 20 (at 1×10^{-4} M) on the interfacial characteristics (adsorption, structure, dynamics of adsorption, and surface dilatational properties) and the foam properties (foam capacity and stability) of soy globulins (7S and 11S). They observed that the rate of adsorption increases at a pH of 7, at a high ionic strength, and in the presence of sucrose. The surface dilatational properties reflect the fact that soy globulin adsorbed films exhibit viscoelastic behavior and improved foaming properties.

Structure–function relationships: The structural and functional properties of soy protein are due to their composition of heterogeneous protein subunits. The major proteins present in soy are glycinin and β-conglycinin. Heating and cooling of these proteins favor gelation with different gelling properties. β-Conglycinin denatures at a lower temperature than glycinin, while glycinin forms stronger gels (Renkema et al. 2001). The pH and the presence of ions regulate the gel strength and the formation of soy protein aggregates (Mills et al. 2001). Acidic pH allows all proteins to participate in the network, while a few acidic subunits of 11S take part in the network at a more neutral or basic pH (Renkema et al. 2002). A study comparing the interactions between 7S globulins, 11S globulins, and SPIs on the emulsifying activity and stability indexes at pHs of 7.5 and 6.5, at different concentrations (0.25%–0.75%) under high-pressure treatment (200–600 MPa), was performed. It was suggested that a pressure of 400 MPa dissociated the 7S fraction into partially or totally dena-tured monomers, which enhanced the surface activity, but at the same time, the unfolding of the 11S polypeptides within the hexamer led to aggregation, negatively affecting the surface hydrophobicity (Molina, Papadopoulou, and Ledward 2001).

Changes in the protein secondary and tertiary structures of the SPI on heating were investigated by Raman spectroscopy. The SPI was refrigerated and heated at 70°C, 100°C, and 200°C and then analyzed. Nonsignificant differences in the secondary structure (α-helix, β-sheet, unordered and turn) were observed in the SPI on heating at these temperatures. However, a comparison of the Raman spectra of the SPI refrigerated and heated at 70°C, 100°C, and 200°C revealed changes in the hydrophobic group environments. A decrease in the intensity of the 1340 cm^{-1} band attributable to tryptophan vibrations was observed, which involves an increase in the solvent exposure of tryptophan residues. The spectral results also showed an increasing intensity for the 1450 and 2935 cm^{-1} bands of the SPI on heating, which can be ascribed to breaking the hydrophobic contacts and the subsequent solvent exposure of the corresponding hydrophobic aliphatic groups. Finally, it was found that tertiary structural changes contributed significantly toward buried tyrosine residues on heating the SPI.

3.5 Egg Proteins and Peptides

Ovalbumin is the main protein found in egg white, making up 60%–65% of the total protein. The albumen forms foams and coagulates by processes such as whipping. The foams become stable by denaturation. The proteins denature and coagulate at the water–gas interface. Globulins orient their hydrophobic groups to the gas phase. The protein chains are stretched and linked by intermolecular forces. This causes an insoluble film to build up in the lamella, which stabilizes the foam. The foam volume of the egg white correlates with the globulin concentration. The stability of the foam is caused by ovomucin (Damodaran and Paraf 1997). The excellent foaming properties of egg white are due to electrostatic interactions. Lysozyme is positively charged and interacts with negatively charged proteins at the interface. A hen egg white lysozyme (HEWL) structure denoting charged domains is shown in Figure 3.6 (Ravichandran, Madura, and Talbot 2001). The flexibility of the protein molecule also dictates the foaming properties of egg white. In their natural state, ovalbumin and

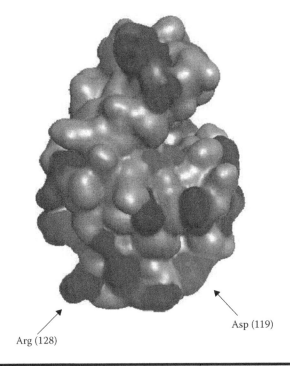

Arg (128)

Asp (119)

Figure 3.6 Structure of hen egg white lysozyme (HEWL). HEWL is shown in a molecular surface (GaussConnolly) representation. The colored surfaces represent the location of the positive, negative, and neutrally charged amino acids on HEWL. (From Ravichandran, S., Madura, J. S., and Talbot, J. 2001. *Journal of Physical Chemistry* B 105: 3610–13.)

lysozyme do not show much flexibility and foaming characteristics due to inter-molecular cross-links, but after heat-induced denaturation, their flexibility and foaming properties increase.

Ovalbumin is a globular protein. When it is denatured, it can produce different conformational states, each having different adsorption properties at a gas–liquid inter-face. Such changes in the adsorption can then affect the foaming behaviors of oval-bumin (Du et al. 2002). It has been observed that for ovalbumin, preheating (partial denaturation) improves the foaming properties (foam capacity and stability) by decreas-ing the surface tension. Du et al. (2002) studied foam stability by measuring foam decay relative to foam height at various denaturation temperatures. The foam decayed most rapidly (in the first 15 min) in the native ovalbumin (22°C) solution, as shown by the slopes of the curves in Figure 3.7. They concluded that foam stability could be improved by preheating the ovalbumin solution in order to denature that protein.

Gelation of egg proteins: Heat-induced coagulation occurs in a series of steps: native monomer-denatured monomer-soluble, aggregate-gel, and coagulum. The major change occurring during the thermal denaturation and aggregation of egg white and ovalbumin is the formation of stable intermolecular β-sheet structures. The β-sheets are formed in extensive regions antiparallel between the ovalbumin mol-ecules; however, an interaction with the β-sheets of ovotransferrin and lysozyme is also observed (Doi and Kitabatake 1997). Increased salt concentrations induce β-sheet formation. It has been observed that egg white proteins start to denature

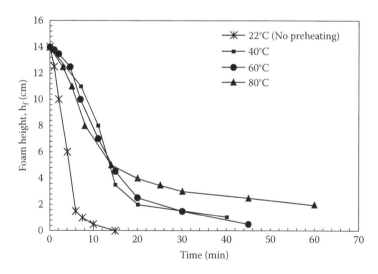

Figure 3.7   Denaturation effect of ovalbumin foam formation. Foam decays with time at different predenaturation temperatures for an initial bulk ovalbumin con-centration of 60 mg/L. (From Du, L., Prokop, A., and Tanner, R. D. 2002. *Journal of Colloid and Interface Science* **248: 487–92.)**

at a temperature of 70°C. Over 78°C, the egg white proteins are seen to coagulate. There is also a pH influence in this phenomenon; at low pH, the solubility decreases.

Observations on the thermal-induced denaturation of ovalbumin indicate that the conformation of ovalbumin at the secondary structure level does not vary much from its native state, but some of the hydrophobic areas that were buried in the native molecule are exposed after heating. This state is considered a "molten globule" state or a "compact denatured state."

A model to explain the heat denaturation and formation of aggregates of ovalubumin was proposed (Doi and Kitabatake 1989). When heated, the native ovalbumin attains a molten globule state. When heated at a pH far from its pI, aggregates form in a linear manner. At a high protein concentration, the aggregates are said to form a three-dimensional gel network. When heated at a pH near the pI, the denatured protein molecules aggregate randomly, giving a turbid gel suspension.

Using high-pressure technology, stiff gels from egg yolk and egg white are formed, although different thermal and pressure processing characteristics of the food protein structure, emulsifying properties, and interactions with food ingredients are observed. Novel functional oligophosphopeptides are prepared from the hen egg yolk protein, phosvitin, as a nutraceutical (Jiang and Mine 2000), which may also contribute to the enhanced functional characteristics.

Most food protein gels are formed during heating and are therefore referred to as heat-induced gels. For a relatively small number of proteins, gelation at ambient temperature is reported (Bryant and McClements 1998). One process to achieve this is the so-called cold gelation process. In the first step, a solution of native proteins is heated and soluble aggregates are formed by heating at a pH away from the pI and at a low ionic strength. After cooling, the aggregates remain soluble and no gelation occurs. In the second step, gelation is induced at an ambient temperature by a reduction of the electrostatic repulsion (Alting et al. 2002), either by adding salt or by changing the pH toward the pI of the proteins. Gelation caused by lowering the pH is called acid-induced gelation.

In contrast to heat-induced gelation, in which aggregation and gelation occur at the same time, in cold gelation, these two processes occur separately. The process of cold gelation of ovalbumin and the properties of the resulting cold-set gels were compared with those of the WPI (Alting et al. 2004). For both the ovalbumin and the WPI, repulsive aggregates were made consisting of disulfide cross-linked monomers and both possessed exposed thiol groups at their surface. The authors found that the differences in the size and shape of the protein aggregates were found to influence the small deformation properties of protein gels and that covalent bonds are natural determinants for gel hardness.

Choi, Lee, and Moon (2008) recently observed the influence of salt and glucose on acid-induced gelation of ovalbumin on heat denaturation. They found that to prepare cold-set gels, glucose did not have an effect on the rheological properties of the gel, whereas gels with 50 and 100 mM NaCl exhibited thixotropy during shearing at a constant shear rate. Perhaps in combination, these gels could have favorable properties for various food applications.

3.6 Proteins and Peptides of Animal Origin

The muscle proteins can be divided into contractile, regulatory, sarcoplasmic, and extracellular forms. The most important are the contractile proteins actin and myosin. Figure 3.8 shows the intermolecular associations in the actomyosin rigor complex (Milligan 1996). Among the regulatory proteins, troponin, tropomyosin, M-protein, β-actin, γ-actin, and C-protein are of great importance. Sarcolemma is also a protein component. Proteins such as myoglobin, myogen, myoalbumin, and γ-globulin are found in the sarcoplasm, and proteins such as elastin, collagen, and

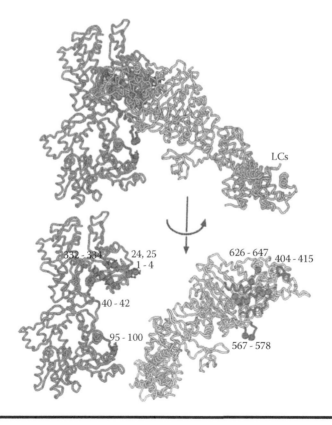

Figure 3.8 Intermolecular associations in the actomyosin rigor complex. Two long-pitch f-actin monomers interacting with the myosin head in the rigor conformation. (From Milligan, R. A. 1996. *Proceedings of the National Academy of Sciences of United States of America* 93: 21–26.)

reticulin are also found in the muscle. The remaining proteins include myofibrillar proteins of the Z-disc as well as small quantities of other proteins.

Various isoforms of myosin not only account for the difference in the physiological functions and biochemical activities of different fiber types of muscles, but also seem to differ in the functional properties of food systems. The functionality of various muscle proteins, especially myosin and actin, in the gelation process in model systems simulates structured meat products.

Several models, including broilers, salmon, Alaska pollock, surimi, and other aquatic foods, have been used to study the various morphological structural changes that occur in muscle proteins. For example, when broilers were exposed to heat, the oxidative stability of the broiler muscle protein was reduced, which has been found to be responsible for decreased protein functionalities such as gelation (Wang, Pan, and Peng 2009). Lefevre et al. (2007) studied the thermal denaturation and aggregation abilities of salmon myofibrils and myosin. The low gelation ability of the salmon muscle proteins was related to the limited extent of protein denaturation and aggregation on heating. These properties seem to be carried by the myosin molecules because a similar behavior was observed for both the myofibrils and myosin preparations. The higher thermal stability observed for red muscle proteins with higher transition temperatures in the rheological profiles was related to a shift to a higher temperature in the denaturation and aggregation processes. Visessanguan and An (2000) studied the effects of enzymatic proteolysis and gelation in the heat-induced gelation of fish myosin. It was observed that the addition of papain decreases the rate of heat-induced gelation; however, in the presence of a cysteine proteinase inhibitor, the myosin heavy chain was protected from degradation.

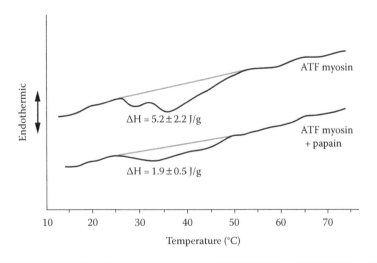

Figure 3.9 DSC endotherm profile of myosin. (From Visessanguan, W. and An, H. 2000. *Journal of Agricultural and Food Chemistry* 48: 1024–32.)

A DSC thermogram (Figure 3.9) and thermal denaturation and circular dichroism studies reveal that proteolysis affected the thermal properties and reactivity of myosin during heating, causing structural disruption and its low gelling ability.

3.7 Proteins and Peptides of Fish and Marine Invertebrates

Fish protein hydrolysate is obtained by the enzymatic hydrolysis of several raw materials of marine origin. As the proteins are hydrolyzed during production, they become increasingly soluble owing to the exposure of their amino and carboxyl groups that can interact with water. This improved solubility of fish protein hydrolysates provides for several applications in a variety of products. Several studies have demonstrated that fish protein hydrolysates can help retain water; hence, this property is being used in the food industry. This property arises due to the increased amount of terminal COOH and NH_3 groups on hydrolysis, which helps bind to the water in the food product. Moreover, the small peptides can be well distributed in the water phase and increase the osmotic pressure of the system, thereby preventing water loss. The peptides may also interact directly with the proteins, increasing their ability to hold more water. Studies have also shown that fish protein hydrolysates could protect proteins from denaturation (Kristinsson 2005). There is also evidence that surimi containing fish protein hydrolysates possesses better gelling ability than surimi without fish protein hydrolysates. Thus, fish protein hydrolysates can also aid in protein gelation.

Several studies on the emulsifying properties of fish protein hydrolysates revolved around the degree of hydrolysis that the fish protein underwent. For good emulsifying actions, it was suggested that the peptides generated after hydrolysis of the fish protein not be smaller than 20 amino acid residues. Kristinsson (2008) also supported this statement and showed in their experiment that as %DH (degree of hydrolysis) increases, the emulsifying properties of salmon hydrolysates decrease. It should also be taken into consideration that enzymes with narrow specificities could be used for hydrolysis.

Fish protein hydrolysates have also been found to exhibit good foaming properties. Such surface-active properties (interfacial properties) were found to be very sensitive to the peptide size. Shark and capelin fish protein hydrolysates were found to have good foaming ability, but substantially less than that of whey and egg white proteins (Kristinsson 2008).

Structural modifications can result in the formation of reordered structures within proteins and peptides. For example, heat treatment or thermal processing can reorient or expose certain groups, such as the carboxylic acid or the side chains of certain amino acids, which can add functionality. In this context, the following section discusses the formation of one such structure, the pyroglutamyl peptide, which has been studied for its potential functional role in the food industry.

3.8 Functional Role in the Food Industry: Pyroglutamyl Peptide in Food Protein Hydrolysates Prepared on Industrial Scale (Production, Structure, Digestibility, and Content)

3.8.1 Production of Pyroglutamyl Peptide in Protein Hydrolysates

Compared with proteins, peptides show higher solubility and adsorption rates and lower antigenicity. In some cases, peptides have better functionalities, such as emulsifying and foam-forming abilities, than intact proteins, as described previously. In addition, the enzymatic hydrolysate of proteins can be considered as a source of peptides that have health-promoting activities, which are reviewed in book II. Therefore, the enzymatic hydrolysates of food proteins have been used in various kinds of foods in order to improve their nutritional values and functionalities and decrease antigenicity.

It is well known that L-2-pyrrolidone-5-carboxylic acid (pyroglutamic acid) is readily induced by heat treatment of the glutamine solution. The amino-terminal glutaminyl residue of the peptide is also converted to a pyroglutamyl residue, which induces the formation of the pyroglutamyl peptide (Figure 3.10). The occurrences of pyroglutamyl peptides in food-grade enzymatic hydrolysates of wheat gluten, cow milk casein, and whey protein have been demonstrated (Sato et al. 1998; Suzuki, Motoi, and Sato 1999; Schlichtherle-Cerny and Amadò 2002).

The formation of pyroglutamyl peptides can be observed in a model food protein hydrolysate system. As shown in Figure 3.11, the pyroglutamyl peptide increases by heating the enzymatic hydrolysates of the food proteins in a boiling water bath and in an autoclave. These facts demonstrate that peptides with a glutaminyl residue at

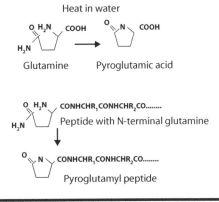

Figure 3.10 Formation of pyroglutamic acid and pyroglutamyl peptide from glutamine and glutaminyl peptide by heat treatment.

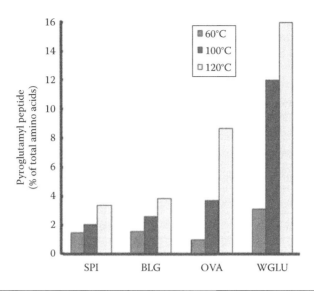

Figure 3.11 **Formation of pyroglutamic peptides by heating the enzymatic digests of soy protein isolate (SPI), β-lactoglobulin (BLG), ovalbumin (OVA), and wheat gluten (WGLU). Digestion was carried out in a laboratory at 1/43,000 scale of the industrial preparation. (From Sato, K., et al., 1998.** *Journal of Agricultural Food Chemistry* **46: 3403–405.) Pyroglutamyl peptide was determined by the method of Sato et al. (1998). Results are presented as mole percent of the constituting amino acids.**

the amino terminal, which can be induced by enzymatic digestion of food protein, are easily converted to pyroglutamyl peptides. To date, the protease, which does not cleave the peptide bond at the amino terminal side of the glutaminyl residue, has not been found. Therefore, it can be considered that most of the enzymatic hydrolysates of proteins that are prepared on an industrial scale, which suffer from heat treatment to terminate the enzyme reaction and sterilize, contain pyroglutamyl peptides in various amounts. On the other hand, a chemically or enzymatically deamidated peptide and protein may not induce a pyroglutamyl peptide due to the absence of glutaminyl residue.

3.8.2 Structure of Proglutamyl Peptides

The pyroglutamyl peptide was first found by exhaustive protease digestion of gluten hydrolysate (Sato et al. 1998). These peptides are resistant to in vitro protease digestion. The structure of the pyroglutamyl peptides after exhaustive protease digestion is summarized in Table 3.1 (Higaki-Sato et al. 2003). pyroGlu-Gln is the most abundant peptide in an exhaustive peptidase digest of wheat gluten. A polyglutamine motif up to octomers in size is distributed in the wheat storage proteins.

Table 3.1 Summary of the Structure of Pyroglutamyl Peptides after the Exhaustive Protease Digestion of Gluten Hydrolysate Prepared on an Industrial Scale

Sequence	Origin	Recovery (%)
pyroGlu-Asn-Pro-Gln	a	2
pyroGlu-Gln-Gln-Pro-Gln	e	1
pyroGlu-Gln-Pro-Gln	b,c,e	3
pyroGlu-Gln-Pro-Gly-Gln-Gly-Gln	d	5
pyroGlu-Gln	a–e	76
pyroGlu-Gln-Pro	a–e	5
pyroGlu-Ile-Pro-Gln	c	>1
pyroGlu-Ile-Pro	c	>1
pyroGlu-Gln-Pro-Leu	b	>1
pyroGlu-Gln-Phe-Pro-Gln	c	>1
pyroGlu-Ser-Phe-Pro-Gln	b	>1
pyroGlu-Phe-Pro-Gln	c	3
pyroGlu-Gln-Pro-Pro-Phe-Ser	e	3

Source: Higaki-Sato, N., et al. 2003. *Journal of Agricultural and Food Chemistry* 51: 8–13.

Note: a: α/β-gliadin; b: γ-gliadin; c: γ-gliadin B; d: glutenin high molecular weight subunit; e: glutenin low molecular weight subunit.

However, a pyroglutamyl peptide derived from a polyglutamine motif longer than a dimer could not be detected in the digest. Therefore, it can be assumed that the proteases (pronase E, leucine amino peptidase, carboxypeptidases A and W, and prolidase) can cleave the peptide bond between the glutaminyl residues in the pyroglutamyl peptides, but cannot cleave the peptide bond between the pyroglutamyl and glutaminyl residues. On the other hand, indigestible pyroglutamyl peptides longer than pyroGln-Gln contain prolyl and/or glycyl residues, which may also contribute to the indigestibility of the pyroglutamyl peptides. This information would be useful to design the peptide resistant to digestion.

3.8.3 Digestibility of Pyroglutamyl Peptide

As described above, pyroglutamyl peptides in the gluten hydrolysate resist exhaustive in vitro peptidase digestion. As shown in Figure 3.12, an increase in the free

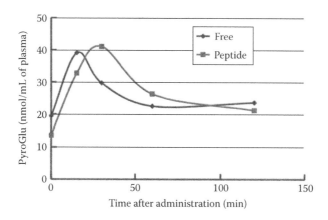

Figure 3.12 **Contents of the free and peptide forms of pyroglutamic acid in the rat portal plasma after ingestion of gluten hydrolysate. The rat was given 2 mL of 10% (w/v) acidic peptide fraction of food-grade wheat gluten hydrolysate. See the paper by Higaki-Sato, N., et al., 2006.** *Journal of Agricultural and Food Chemistry* **54: 6984–88, for the experimental details.**

and peptide forms of pyroglutamic acid in the plasma from rat portal blood was observed after the ingestion of gluten hydrolysates (Higaki-Sato et al. 2006). Then, some pyroglutamyl peptides can be degraded to free pyroglutamic acid and amino acids in the rat digestive tract and/or blood. However, some pyroglutamyl peptides in the wheat gluten hydrolysate are absorbed in peptide form into the circulatory system. Based on these facts, some pyroglutamyl peptides resist not only in vitro protease digestion but also in vivo digestion to some extent. The resistance of the pyroglutamyl peptide to protease digestion may contribute to the biological activity of pyroglutamyl peptide by ingestion.

3.8.4 Content of Pyroglutamyl Peptide

The pyroglutamyl peptide content can be estimated by amino acid analysis of the nonabsorbed fraction from the strong cation exchange column, as it loses the α-amino group. Using this technique, the pyroglutamyl peptide content (mol/mol of constituting amino acid) in the model enzymatic hydrolysates is shown in Figure 3.11. The content depends on the source of the food proteins, possibly due to the differences in the frequencies of the glutaminyl residue. The wheat gluten hydrolysate prepared on a laboratory scale contains up to 15% pyroglutamyl peptide, which coincides with that prepared on an industrial scale (Sato et al. 1998; Suzuki, Motoi, and Sato 1999).

The pyroglutamyl peptide content in protein hydrolysates and food can also be estimated by determining the amount of pyroglutamic acid after pyroglutamate

Table 3.2   Content of the Peptide Forms of Pyroglutamic Acid in Peptide-Based Foods

Category of Food	Form	Source of Peptide	Peptide Content (%)	Pyroglutamyl Peptide	
				µmol/g of peptide	µmol/g of product
Clinical nutrition	Liquid	Milk proteins	4.0[b]	12.84	0.51
Soup (FOSHU[a])	Powder	Sardine meat proteins	52.9	2.14	1.13
Sports drink	Powder	Soy proteins	68.7	29.0	19.94
Infant formulae	Powder	Milk proteins	12.6	34.94	4.40

[a] Food for special health use.
[b] Protein in liquid (w/v).

amino peptidase digestion, which liberates the pyroglutamic acid from the pyroglutamyl peptide. The content of the pyroglutamic acid in peptide form in some foods is shown in Table 3.2. The content depends on the source of the protein and also on the peptide content in the food. It is worth noting that peptide-based clinical nutrition formulas and infant milk contain significant amounts of pyroglutamyl peptide, on which patients and infants rely to obtain most of their nutrients. The potential effect of the pyroglutamyl peptide in these products should be evaluated.

It has been demonstrated that some endogenous pyroglutamyl peptides have significant biological activities (Cockle et al. 1989; Laerum et al. 1990; Khan et al. 1992). To the best of our knowledge, there are, however, few reports on the biological activity of food-derived pyroglutamyl peptides. Some pyroglutamyl peptides (pyroGlu-Pro-Ser, pyroGlu-Pro, pyroGlu-Pro-Glu, and pyroGlu-Pro-Gln) were suggested to have a glutamate-like taste (Schlichtherle-Cerny and Amadò 2002). Recently, pyroGlu-Leu was demonstrated to moderate galactosamine-induced rat hepatitis. There is a possibility that pyroglutamyl peptides in the protein hydrolysates play a significant role in exerting biological activity.

Most pyroglutamyl peptides do not have a positive charge and therefore have acidic properties. The presence of an acidic pyroglytamyl peptide might have an effect on the physicochemical functions of the protein hydrolysate. However, there are no available data on the effects of pyroglutamyl peptides, induced during food processing, on the physicochemical functions. Recently, a large-scale preparative isoelectric electrophoresis based on the amphoteric nature of the peptide has been developed (Hashimoto et al. 2006). This technique enables the fractionation of peptides on the basis of pI without using a chemically synthesized carrier ampholyte (see Chapter 15). Using this technique, sufficient amounts of pyroglutamyl peptide fraction can be prepared to evaluate the functional properties (Higaki-Sato et al. 2006), which would elucidate the functional properties of the pyroglutamyl peptide in protein hydrolysate.

3.9 Conclusion

Several models have been proposed to understand the properties of structures in relation to their functions. Since the isolated proteins and peptides for functionality can have varied structural arrangements, it is essential to elucidate the structural features of each protein or peptide to be able to relate to their properties and hence their functional characteristics. In addition to characterizing the structure–function relationships of food proteins and peptides, modifications to peptide structures can help stabilize the structures that are prone to changes due to processing. Finally, understanding the structural features, and their associated properties and modifications, if any, can be useful in selecting tools that can isolate these compounds on a large scale for commercial applications.

References

Achouri, A., Boye, J. I., Yaylayan, V. A., and Yeboah, F. K. 2005. Functional properties of glycated soy 11S glycinin. *Journal of Food Science* 70(4): C269–74.

Ahmedna, M., Prinyawiwatkul, W., and Rao, R. M. 1999. Solubilized wheat protein isolate: Functional properties and potential food applications. *Journal of Agricultural and Food Chemistry* 47(4): 1340–5.

Alting, A. C., de Jongh, H. J. J., Visschers, R. W., and Simons, J. F. A. 2002. Physical and chemical interactions in pH-induced aggregation and gelation of food proteins. *Journal of Agricultural and Food Chemistry* 50: 4674–81.

Alting, A. C., Weijers, M., de Hoog, E. H. A., van de Pijpekamp, A. M., Stuart, M. A. C., Hamer, R. J., de Kruif, C. G., and Visschers, R. W. 2004. Acid-induced cold gelation of globular proteins: Effects of protein aggregate characteristics and disulfide bonding on rheological properties. *Journal of Agricultural and Food Chemistry* 52(3): 623–31.

Aminlari, M., Ramezani, R., and Jadidi, F. 2005. Effect of Maillard-based conjugation with dextran on the functional properties of lysozyme and casein. *Journal of the Science of Food and Agriculture* 85(15): 2617–24.

Anjum, F. M., Khan, M. R., Din, A., Saeed, M., Pasha, I., and Arshad, M. U. 2007. Wheat gluten: High molecular weight glutenin subunits-structure, genetics and relation to dough elasticity. *Journal of Food Science* 72: R56–61.

Arrese, E. L., Sorgentini, D. A., Wagner, J. R., and Anon, M. C. 1991. Electrophoretic, solubility and functional properties of commercial soy protein isolates. *Journal of Agricultural Food Chemistry* 39(6): 1029–32.

Azuaga, A. I., Galisteo, M. L., Mayorga, O. L., Cortijo, M., and Mateo, P. L. 1992. Heat and cold denaturation of beta-lactoglobulin B. *FEBS Letters* 309: 258–60.

Bamforth, C. W. and Milani, C. 2004. The foaming of mixtures of albumin and hordein protein hydrolysates in model systems. *Journal of Sci. Food Agric.*, 84: 1001–4.

Banoglu, E. and Karata, S. 2001. High pressure effect on foaming behaviour of whey protein isolate. *Journal of Food Engineering* 47: 31–36.

Bigelow, C. C. 1967. On the average hydrophobicity of proteins and the relation between it and proteins structure. *Journal of Theoretical Biology* 16: 187–211.

Bryant, C. M. and McClements, D. J. 1998. Molecular basis of protein functionality with special consideration of cold-set gels derived from heat-induced whey. *Trends in Food Science and Technology* 9:143–51.

Cagampang, G. B., Cruz, L. J., Espiritu, S. G., Santiago, R. G., and Juliano, B. O. 1966. Studies on the extraction and composition of rice proteins. *Cereal Chemistry* 43: 145–55.

Cagampang, G. B., Perdon, A. A., and Juliano, B. O. 1976. Changes in salt-soluble proteins of rice during grain development. *Phytochemistry* 15(10): 1425–9.

Cayot, P. and Lorient, D. 1997. Structure–function relationships of whey proteins. In *Food Proteins and Their Applications*, eds. A. Damodaran and A. Paraf, Chap. 7. New York: Marcel Decker.

Chardot, T., Benetti, P. H., Canonge, M., Kim, S. I., Chaillot, D., Fouques, D., and Meunier, J. C. 2004. Enzymatic phosphorylation of food proteins by purified and recombinant protein kinase CK2T. *Molecular Nutrition & Food Research* 42: 145–47.

Cheftel, J. C., Cuq, J., and Lorient, D. 1985. Amino acids, peptides and proteins. In *Food Chemistry*, ed. R. Fennema, 2nd ed., p. 301. New York: Marcel Dekker.

Chevalier, F., Chobert, J. M., Popineau, Y., Nicolas, M. G., and Haertlé, T. 2001. Improvement of functional properties of β-lactoglobulin glycated through the Maillard reaction is related to the nature of the sugar. *International Dairy Journal* 11(3): 145–52.

Choi, S. J., Lee, S. E., and Moon, T. W. 2008. Influence of sodium chloride and glucose on acid-induced gelation of heat-denatured ovalbumin. *Journal of Food Sciences* 73: C313–22.

Cockle, S. M., Aitken, A., Beg, F., Morrell, J. M., and Smyth, D. G. 1989. The TRH-related peptide pyroglutamylglutamylprolinamide is present in human semen. *FEBS Letters* 252: 113–17.

Courthaudon, J. L., Colas, B., and Lorient, D. 1989. Covalent binding of glycosyl residues to bovine casein: Effects on solubility and viscosity. *Journal of Agricultural and Food Chemistry* 37(1): 32–6.

Culbertson, J. 2006. Food protein functionality. In *Handbook of Food Science, Technology and Engineering*, ed. Y. H. Hui, Vol. 1, pp. 7:1–7:12. Boca Raton, FL: CRC Press.

D'Agostina, A., Antonioni, C., Resta, D., Arnoldi, A., Bez, J., Knauf, U., and Wäsche, A. 2006. Optimization of a pilot-scale process for producing lupin protein isolates with valuable technological properties and minimum thermal damage. *Journal of Agricultural and Food Chemistry* 54(1): 92–8.

Dalgleish, D. 1997. Structure–function relationships of caesins. In *Food Proteins and Their Applications*, eds. A. Damodaran and A. Paraf, p. 199–223. New York: Marcel Decker.

Damodaran, S. 1996. Functional properties. In *Food Proteins: Properties and Characterization*, eds. S. Nakai and H. W. Modler, 168–220. New York: Wily-VCH, 168–220.

Damodaran, S. 2005. Protein stabilization of emulsions and foams. *Journal of Food Science* 70(3): R54–66.

Damodaran, S. and Paraf, A. 1997. *Food Proteins and Their Applications*, pp. 443–472. New York: Marcel Decker.

Das, K. P. and Kinsella, J. E. 1990. Stability of food emulsions: Physicochemical role of protein and nonprotein emulsifiers. *Advances in Food and Nutrition Research* 34: 81–201.

Deak, N. A. and Johnson, L. A. 2007. Effects of extraction temperature and preservation method on functionality of soy protein. *Journal of American Oil Chemists' Society* 84: 259–68.

Diftis, N. G., Biliaderis, C. G., and Kiosseoglou, V. D. 2005. Rheological properties and stability of model salad dressing emulsions prepared with a dry-heated soybean protein isolate–dextran mixture. *Food Hydrocolloids* 19: 1025–31.

Diftis, N. and Kiosseoglou, V. 2003. Improvement of emulsifying properties of soybean protein isolate by conjugation with carboxymethyl cellulose. *Food Chemistry* 81: 1–6.

Diftis, N. and Kiosseoglou, V. 2006a. Physicochemical properties of dry-heated soy protein isolate–dextran mixtures. *Food Chemistry* 96: 228–33.

Diftis, N. and Kiosseoglou, V. 2006b. Stability against heat-induced aggregation of emulsions prepared with a dry-heated soy protein isolate–dextran mixture. *Food Hydrocolloids* 20: 787–92.

Doi, E. and Kitabatake, N. 1989. Structure of glycinin and ovalbumen gels. *Food Hydrocolloids* 3: 327.

Doi, E. and Kitabatake, N. 1997. Structure–function relationship of egg proteins. In *Food Proteins and Their Applications*, eds. A. Damodaran and A. Paraf, pp. 325–40. New York: Marcel Decker.

Du, L., Prokop, A., and Tanner, R. D. 2002. Effect of denaturation by preheating on the foam fractionation behavior of ovalbumin. *Journal of Colloid and Interface Science* 248: 487–92.

Elfgam, A. A. and Wheelock, J. W. 1978. Heat interactions between lactalbumin and lactoglobulin and casein in bovine milk. *Journal of Dairy Science* 61: 159.

Euston, S. R., Finnigan, S. R., and Hirst, R. L. 2000. Aggregation kinetics of heated whey protein-stabilised emulsions. *Food Hydrocolloids* 14: 155–61.

Farrell, H. M. Jr., Qi, P. X., Brown, E. M., Cooke, P. H., Tunick, M. H., Wickham, E. D., and Unruh, J. J. 2002. Molten globule structures in milk proteins: Implications for potential new structure–function relationships. *Journal of Dairy Science* 85(3): 459–71.

Fox, P. F. and McSweeney, P. L. H. 1998. Milk proteins. In *Dairy Chemistry and Biochemistry*, 146. New York, NY: Kluwer Academic/Plenum publishers.

Friedman, M. and Brandon, D. L. 2001. Nutritional and health benefits of soy proteins. Review. *Journal of Agricultural and Food Chemistry* 49: 1069–86.

Galazka, V. B., Dickinson, E., and Ledward, D. A. 2000. Effect of high-pressure on surface behaviour of adsorbed films formed from mixtures of sulfated polysaccharides with various proteins. *Innovative Food Science and Emerging Technologies* 1: 177–85.

Galili, G. 1997. The prolamin storage proteins of wheat and its relatives. In *Cellular and Molecular Biology of Plant Seed Development*, eds. B. A. Larkins and I. K. Vasil, pp. 221–56. Dordrecht: Kluwer.

Gilbert, S. M., Wellner, N., Belton, P. S., Greenfield, J. A., Siligardi, G., Shewry, P. R., and Tatham, A. S. 2000. Expression and characterization of a highly repetitive peptide derived from a wheat seed storage protein. *Biochimica et Biophysica Acta* 1479: 135–46.

Goddik, L. 2003. High pressure leads to denatured proteins with optimal surface activity. *Emerging Food R&D*, Report 52(2): 193–202.

Haard, N. F. 2001. Enzymic modification of proteins in food systems. In *Chemical and Functional Properties of Food Proteins*. ed. Z. E. Sikorski. Lancaster, PA: Technomic Publishing.

Halling, P. J. 1981. Protein-stabilized foams and emulsions. *Critical Reviews in Food Science and Nutrition* 21: 155–203.

Harwalker, V. R. and Ma, C. Y. 1989. Effects of medium composition, preheating, and chemical modification upon thermal behaviour of oat globulin and β-lactoglobulin. In *Food Proteins*, eds. J. F. Kinsella and W. G. Soucie, pp. 210–51. Champaign, IL: The American Oil Chemists' Society.

Harwalker, V. R. and Ma, C. Y. 1996. Thermal analysis: Principles and applications. In *Food Proteins: Properties and Characterization*, eds. S. Nakai and H. W. Modler, pp. 405–27. New York: John Wiley & sons.

Hashimoto, K., Sato, K., Nakamura, Y., and Ohtsuki, K. 2006. Development of continuous type apparatus for ampholyte-free isoelectric focusing (autofocusing) of peptides in protein hydrolysates. *Journal of Agricultural and Food Chemistry* 54: 650–55.

Hettiarachcy, N. S., Griffin, V. K., and Gnanasambandam, R. 1999. Preparation and functional properties of a protein isolate from defatted wheat germ. *Cereal Chemistry* 73(3): 363–7.

Higaki-Sato, N., Sato, K., Esumi, Y., Okumura, T., Yoshikawa, H., Tanaka-Kuwajima, C., et al. 2003. Isolation and identification of indigestible pyroglutamyl peptides in an enzymatic hydrolysate of wheat gluten prepared on an industrial scale. *Journal of Agricultural and Food Chemistry* 51: 8–13.

Higaki-Sato, N., Sato, K., Inoue, N., Nawa, Y., Kido, Y., Nakabou, Y., et al. 2006. Occurrence of the free and peptide forms of pyroglutamic acid in plasma from the portal blood of rats that had ingested a wheat gluten hydrolysate containing pyroglutamyl peptides. *Journal of Agricultural and Food Chemistry* 54: 6984–88.

Holt, C. 1992. Structure and stability of bovine casein micelles. *Advanced Protein Chemistry* 43: 63–151.

Holt, C. and Horne, D. S. 1996. The hairy caseinmicelle: Evolution of the concept and its implications for dairy technology. *Netherland Milk and Dairy Journal* 50: 85–111.

Holt, C. and Roginski, H. 2001. Milk proteins: Biological and food aspects of structure and function. In *Chemical and Functional Properties of Food Proteins*, ed. Z. Sikorski, pp. 271–334. Boca Raton, FL: CRC Press.

Horax, R., Hettiarachchy, N. S., Chen, P., and Jalaluddin, M. 2004a. Preparation and characterization of protein isolate from cowpea (*Vigna unguiculata* L. Walp.). *Journal of Food Science* 69(2): fct114–18.

Horax, R., Hettiarachchy, N. S., Chen, P., and Jalaluddin, M. 2004b. Functional properties of protein isolate from cowpea (*Vigna unguiculata* L. Walp.). *Journal of Food Science* 69(2): fct119–21.

Horne, D. S. 2006. Casein micelle structure: Models and muddles. *Current Opinion in Colloid and Interface Science* 11: 148–53.

Horng, J. C., Demarest, S. J., and Raleigh, D. P. 2003. pH-dependent stability of the human alpha-lactalbumin molten globule state: Contrasting roles of the 6–120 disulfide and the beta-subdomain at low and neutral pH. *Proteins* 52(2): 193–202.

Houston, D. F., Iawasaki, T., Mohammad, A., and Chen, L. 1968. Radial distribution of protein by solubility classes in the milled rice kernel. *Journal of Agricultural and Food Chemistry* 16: 720.

Huebner, F. R., Bietz, J. A., and Juliano, B. O. 1990. Rice cultivar identification by high-performance liquid chromatography of endosperm proteins. *Cereal Chemistry* 67: 129–35.

Huppertz, T., Fox, P. F., and Kelly, A. L. 2004. High pressure-induced denaturation of alpha-lactalbumin and beta-lactoglobulin in bovine milk and whey: A possible mechanism. *Journal of Dairy Research* 7: 489–95.

Itoh, T., Garcia, N. R., Adachi, M., Maruyama, Y., Tecson-Mendoza, E. M., Mikami, B., and Utsumi, S. 2006. Structure of 8Sα globulin, the major seed storage protein of mung bean. *Acta Crystallographica* D62: 824–32.

Iwasaki, T., Shibuya, N., and Chikubu, S. 1972. Studies on rice protein. I. Albumin and globulin of rice endosperm; their heterogeneity and the difference in subunit composition among stored, heated, and fresh rice. *Journal of Food Science and Technology* 19: 70–5.

Iwasaki, T., Shibuya, N., Suzuki, T., and Chikubu, S. 1975. Studies on rice protein. 11. Albumins and globulins of regular rice and waxy rice. *Journal of Food Science and Technology* 22: 113–18.

Iwasaki, T., Shibuya, N., Suzuki, T., and Chikubu, S. 1982. Gel-filtration and electrophoresis of soluble rice proteins extracted from long, medium, and short grain varieties. *Cereal Chemistry* 59: 192–5.

Jiang, B., and Mine, Y. 2000. Preparation of novel functional oligophosphopeptides from hen egg yolk phosvitin. *Journal of Agricultural and Food Chemistry* 48: 990–94.

Jiménez-Castaño, L., López-Fandiño, R., Olano, A., and Villamiel, M. 2005. Study on β-lactoglobulin glycosylation with dextran: Effect on solubility and heat stability. *Food Chemistry* 93(4): 689–95.

Jiménez-Castaño, L., Villamiel, M., and López-Fandiño, R. 2007. Glycosylation of individual whey proteins by Maillard reaction using dextran of different molecular mass. *Food Hydrocolloids* 21(3): 433–43.

Juliano, B. O. 1985. Criterion and tests for rice grain qualities. In *Rice: Chemistry and Technology*, ed. B. O. Juliano, 2nd edn, 443–524. St. Paul, MN: American Association of Cereal Chemists.

Juliano, B. O. and Boulter, D. 1976. Extraction and composition of rice endosperm glutelin. *Phytochemistry* 15(11): 1601–6.

Jung, S., Lamsal, B. P., Stepien, V., Johnson, L. A., and Murphy, P. A. 2006. Functionality of soy protein produced by enzyme-assisted extraction. *Journal of the American Oil Chemists' Society* 83(1): 71–8.

Kato, A. 2002. Industrial applications of Maillard-type protein–polysaccharide conjugates. *Food Science and Technology Research* 8(3): 193–9.

Khan, Z., Aitken, A., Garcia, J. del R., and Smyth, D. G. 1992. Isolation and identification of two neutral thyrotoropin releasing hormone-like peptides, pyroglutamyl-phenylalanineproline amide and pyroglutamylglutamineproline amide, from human seminal fluid. *Journal of Biological Chemistry* 267: 7464–69.

Kinsella, J. E. 1976. Functional properties of food proteins: A review. *Critical Reviews in Food Science and Nutrition* 7: 219.

Kinsella, J. E. and Whitehead, D. M. 1988. Emulsifying and foaming properties of chemically modified protein. In *Advances in Food Emulsions and Foams*, eds. E. Dickinson and G. Stainsby, pp. 163–88. Amsterdam: Elsevier.

Krishnan, H. B. and White, J. A. 1990. Morphometric analysis of rice seed protein bodies (implication for a significant contribution of prolamine to the total protein content of rice endosperm). *Plant Physiology* 109(4): 1491–5.

Kristinsson, H. G. 2005. The production, properties and utilization of fish protein hydroly-sates. In *Food Biotechnology*, eds. K. Shetty, G. Paliyath, A. Pometto, and R. E. Levin, 1109–33. New York: CRC Press.

Kuraishi, C., Yamazaki, K., and Susa, Y. 2001. Transglutaminase: its utilization in the food industry. *Food Reviews International* 17: 221–46.

Laerum, O. D., Frostad, S., Tøn, H. I., and Kamp, D. 1990. The sequence of the hemoregu-latory peptide is present G1a proteins. *FEBS Letters* 269: 11–14.

Landstrom, K., Alsins, J., and Bergenståhl, B. 2000. Competitive protein adsorption between bovine serum albumin and -lactoglobulin during spray drying. *Food Hydrocolloids* 14: 75–82.

Lee, H. C., Htoon, A. K., Uthayakumaran, S., Paterson, J. L. 2007. Chemical and func-tional quality of protein isolated from alkaline extraction of Australian lentil cultivars: Matilda and Digger. *Food Chemistry* 102(4): 1199–207.

Lefevre, F., Fauconneau, B., Thompson, J. W., and Gill, T. A. 2007. Thermal denaturation and aggregation properties of Atlantic salmon myofibrils and myosin from white and red muscles. *Journal of Agricultural and Food Chemistry* 55: 4761–70.

López-Fandiño, R. 2006. Functional improvement of milk whey proteins induced by high hydrostatic pressure. *Critical Reviews in Food Science and Nutrition* 46(4): 351–63.

Losso, J. N. 2008. The biochemical and functional food properties of the Bowman–Birk inhibitor. *Critical Reviews in Food Science and Nutrition* 48: 94–118.

MacRitchie, F. and Lafiandra, D. 1997. Structure–function relationships of wheat proteins. In *Food Proteins and Their Applications*, eds. A. Damodaran and A. Paraf, pp. 293–323. New York: Marcel Decker.

Makri, E. A. and Doxastakis, G. I. 2006. Emulsifying and foaming properties of *Phaseolus vulgaris* and *coccineus* proteins. *Food Chemistry* 98(3): 558–68.

Maruyama, N., Prak, K., Motoyama, S., Choi, S.-K., Yagasaki, K., Ishimoto, M., and Utsumi, S. 2004. Structure-physicochemical function relationships of soybean gly-cinin at subunit levels assessed by using mutant lines. *Journal of Agricultural and Food Chemistry* 52(26): 8197–201.

Meade, S. J., Reid, E. A., and Gerrard, J. A. 2005. The impact of processing on the nutri-tional quality of food proteins. *Journal of AOAC International* 88(3): 904–22.

Milligan, R. A. 1996. Protein–protein interactions in the rigor actomyosin complex. *Proceedings of the National Academy of Sciences of United States of America* 93: 21–26.

Millqvist-Fureby, A., Elofsson, U., and Bergenståhl, B. 2001. Surface composition of spray-dried milk protein-stabilized emulsions in relation to pre-heat treatment of proteins. *Colloids and Surfaces B* 21: 47–58.

Mills, E. N., Marigheto, N. A., Wellner, N., Fairhurst, S. A., Jenkins, J. A., Mann, R., and Belton, P. S. 2003. Thermally induced structural changes in glycinin, the 11S globulin of soya bean (Glycine max) – an in situ spectroscopic study. *Biochimica Biophysica Acta* 1648: 105–14.

Mita, T., Nikai, K., Hiraoka, T., Matsuo S., and Matsumoto, H. 1977. Physicochemical studies on wheat protein foams. *Journal of Colloid and Interface Science* 59(1): 172–8.

Moatsou, G., Bakopanos, C., Katharios, D., Katsaros, G., Kandarakis, I., Taoukis, P., and Politis, I. 2008. Effect of high-pressure treatment at various temperatures on indigenous proteolytic enzymes and whey protein denaturation in bovine milk. *Journal of Dairy Research* 75: 262–69.

Molina, E. A., Papadopoulou, A., and Ledward, D. A. 2001. Emulsifying properties of high pressure treated soy protein isolate and 7S and 11S globulin. *Food Hydrocolloids* 15: 263–69.

Nakai, S. and Modler, H. W. 1996. *Food Proteins: Properties and Characterization*, p. 405–427. New York: VCH.

Nakamura, T., Utsumi, S., and Mori, T. 1983. Network structure formation in thermally induced gelation of glycinin. *Journal of Agricultural and Food Chemistry* 31: 1270.

Narsimhan, G. 1992. Maximum disjoining pressure in protein stabilized concentrated oil-in-water emulsions. *Colloids and Surfaces* 62(1–2): 41–55.

Olaofe, O., Adeyemi, F. O., and Adediran, G. O. 1994. Amino acid and mineral compositions and functional properties of some oilseeds. *Journal of Agricultural and Food Chemistry* 42(4): 878–81.

Osborne, T. B. 1924. *The Vegetable Proteins*. 2nd edn. London: Longmans, Green & Co.

Padhye, V. W. and Salunkhe, D. K. 1979. Extraction and characterization of rice proteins. *Cereal Chemistry* 56: 389–93.

Paraman, I., Hettiarachchy, N. S., Schaefer, C., and Beck, M. I. 2006. Physicochemical properties of rice endosperm proteins extracted by chemical and enzymatic methods. *Cereal Chemistry* 83(6): 663–7.

Paraman, I., Hettiarachchy, N. S., Schaefer, C., and Beck, M. I. 2007. Hydrophobicity, solubility, and emulsifying properties of enzyme-modified rice endosperm protein. *Cereal Chemistry* 84(4): 343–9.

Peltonen-Shallaby, R. and Mangino, M. E. 1986. Composition factors that affect the emulsifying and foaming properties of whey protein concentrates. *Journal of Food Science* 51: 91.

Phadungath, C. 2005. Casein micelle structure: A concise review. *Journal of Science and Technology* 27(1): 201–21.

Pierce, K. N. and Kinsella, J. E. 1978. Emulsifying properties of proteins: Evaluation of a turbidimetric technique. *Journal of Agricultural and Food Chemistry* 26: 716–23.

Pittia, P., Wilde, P. J., Husband, F. A., and Clark, D. C. 1996. Functional and structural properties of β-lactoglobulin as affected by high pressure treatment. *Journal of Food Science* 61: 1123–28.

Puppo, M. C. and Añón, M. C. 1998. Structural properties of heat-induced soy protein gels as affected by ionic strength and pH. *Journal of Agricultural Food Chemistry* 46: 3583–89.

Ravichandran, S., Madura, J. S., and Talbot, J. 2001. A Brownian dynamics study of the initial stages of hen egg-white lysozyme adsorption at a solid interface. *Journal of Physical Chemistry B* 105: 3610–13.

Renkema, J. M. S., Knabben, J. H. M., and van Vliet, T. 2001. Gel formation by β-conglycinin and glycinin and their mixtures. Food Hydrocolloids 15(4–6): 407–14.

Renkema, J. M. S., Lakemond, C. M. M., De Jongh, H. H. J., Gruppen, H., and van Vliet, T. 2000. The effect of pH on heat denaturation and gel forming properties of soy proteins. *Journal of Biotechnology* 79: 223–30.

Roesch, R. R. and Corredig, M. 2006. Study of the effect of soy proteins on the acid-induced gelation of casein micelles. *Journal of Agricultural and Food Chemistry* 54(21): 8236–43.

Ruíz-Henestrosa, V. P., Carrera Sánchez, C., Yust Mdel, M., Pedroche, J., Millan, F., and Rodríguez Patino, J. M. 2007a. Limited enzymatic hydrolysis can improve the interfacial and foaming characteristics of beta-conglycinin. *Journal of Agricultural and Food Chemistry* 55: 1536–45.

Ruíz-Henestrosa, V. P., Sanchez, C. C., and Rodríguez Patino, J. M. 2007b. Formulation engineering can improve the interfacial and foaming properties of soy globulins. *Journal of Agricultural and Food Chemistry* 55: 6339–48.

Sato, K., Nishimura, R., Suzuki, Y., Motoi, H., Nakamura, Y., Ohtsuki, K., and Kawabata, M. 1998. Occurrence of indigestible pyroglutamyl peptides in an enzymatic hydro-lysate of wheat gluten prepared on an industrial scale. *Journal of Agricultural Food Chemistry* 46: 3403–405.

Schlichtherle-Cerny, H. and Amadò, R. 2002. Analysis of taste-active compounds in an enzymatic hydrolysate of deamidated wheat gluten. *Journal of Agricultural and Food Chemistry* 50: 1515–22.

Schwenke, K. D. 1997. Enzyme and chemical modification of proteins. In *Food Proteins and Their Applications*, eds. S. Damodaran and A. Paraf. New York: Marcel Dekker.

Shewry, P. R., Tatham, A. S., Forde, J., Kreis, M., and Miflin, B. J. 1986. The classification and nomenclature of wheat gluten proteins: A reassessment. *Journal of Cereal Science* 4: 97–106.

Shih, F. F. and Daigle, K. W. 2000. Preparation and characterization of rice protein isolates. *Journal of the American Oil Chemists' Society* 77(8): 885–9.

Sliwinski, E. L., Lavrijsen, B. W. M., Vollenbroek, J. M., van der Stege, H. J., Van Boekel, M. A. J. S., and Wouters, J. T. M. 2003. Effects of spray drying on physico-chemical properties of milk protein-stabilised emulsions. *Colloid and Surface B* 31: 219–29.

Suzuki, Y., Motoi, H., and Sato, K. 1999. Quantitative analysis of pyroglutamic acid in pep-tides. *Journal of Agricultural and Food Chemistry* 47: 3248–51.

Tang, S., Hettiarachchy, N. S., Horax, R., and Eswaranandam, S. 2003. Physicochemical properties and functionality of rice bran protein hydrolyzate prepared from heat-stabilized defatted rice bran with the aid of enzymes. *Journal of Food Science* 68(1): 152–7.

Tecson, E. M., Esmama, B. V., Lontok, L. P., and Juliano, B. O. 1971. Studies on the extrac-tion and composition of rice endosperm glutelin and prolamin. *Cereal Chemistry* 48: 169–81.

Tilley, K. A., Benjamin, R. E., Bagorogoza, K. E., Okot-Kotber, B. M., Prakash, O., and Kwen, H. 2001. Tyrosine cross-links: Molecular basis of gluten structure and function. *Journal of Agricultural and Food Chemistry* 49: 2627–32.

Utsumi, S., Matsumura, Y., and Mori, T. 1997. Structure–function relationships of soy proteins. In *Food Proteins and Their Applications*, eds. A. Damodaran and A. Paraf, pp. 257–91. New York: Marcel Decker.

Van Camp, J. and Dierckx, S. 2003. Proteins. In *Handbook of Food Analysis, Physical Characterization and Nutrient Analysis*, ed. L. M. L. Nollet, 2nd edn, pp. 167–202. New York: Marcel Dekker.

Vega, C. and Roos, Y. H. 2006. Invited review: Spray-dried dairy and dairy-like emulsions—compositional considerations. *Journal of Dairy Science* 89: 383–401.

Visessanguan, W. and An, H. 2000. Effects of proteolysis and mechanism of gel weakening in heat-induced gelation of fish myosin. *Journal of Agricultural and Food Chemistry* 48: 1024–32.

Walstra, P. 1999. Caesin sub-micelles: Do they exist? *Internal Dairy Journal* 9: 189–92.

Wang, R. R., Pan, X. J., and Peng, Z. Q. 2009. Effects of heat exposure on muscle oxida-tion and protein functionalities of pectoralis majors in broilers. *Pollution Science* 88: 1078–84.

Wen, T. and Luthe, S. D. 1985. Biochemical characterization of rice glutelin. *Plant Physiology* 78: 172–7.

Wong, N. P. 1988. *Fundamental of Dairy Chemistry*, 3rd edn, 481–492. New York: Van Nostrand Reinhold.

Xu, K. and Yao, P. 2009. Stable oil-in-water emulsions prepared from soy protein–dextran conjugates. *Langmuir* 25(17): 9714–20.

Yao, J. J., Tanteeratarm, K., and Wei, L. S. 1983. Effects of maturation and storage stability of soybeans. *Journal of American Oil Chemistry Society* 60: 1245–49.

Yeboah, F. K., Alli, I., and Yaylayan, V. A. 1999. Reactivities of d-glucose and d-fructose during glycation of bovine serum albumin. *Journal of Agricultural Food Chem*istry 47(8): 3164–72.

Yeboah, F. K., Alli, I., Yaylayan, V. A., Konishi, Y., and Stefanowicz, P. 2000. Monitoring glycation of lysozyme by electrospray ionization mass spectrometry. *Journal of Agricultural Food Chemistry* 48(7): 2766–74.

Yu, M.-A. and Damodaran, S. 1991. Kinetics of destabilization of soy protein foams. *Journal of Agricultural and Food Chemistry* 39: 1563–7.

Zayas, J. F. 1997. *Functionality of Proteins in Food*. Berlin: Springer-Verlag.

Zhang T., Jiang, B., Mu, W., and Wang, Z. 2009. Emulsifying properties of chickpea protein isolates: Influence of pH and NaCl. *Food Hydrocolloids* 23(1): 146–52.

Chapter 4

Protein Solubility and Functionality

Shridhar K. Sathe

Contents

4.1 Introduction

It is generally accepted that when the molecular mass of a polypeptide is less than 10,000 Da, the molecule is typically referred to as a polypeptide, while those with molecular masses greater than 10,000 Da are considered to be proteins. The molecular masses of the food proteins may range from low, 10–30 kDa (e.g., milk proteins and several catalytic enzymes), to high, greater than 1 million kDa (e.g., wheat gluten proteins). Proteins perform a variety of functions in food systems (Table 4.1). Depending on the food systems and the proteins involved, such functions may be broadly categorized as desirable (e.g., egg proteins as foaming agents in ice creams) or undesirable (e.g., enzymatic browning of fruits and vegetables). Certain proteins

Table 4.1 Summary of Protein Functional Properties in Food Systems

Funtional Property	*Mode of Action*	*Food System*
Solubility	Protein solvation	Beverages, soups, feeding formulae
Water absorption and binding	Hydrogen bonding	Meats, bakery products
Viscosity	Thickening, hydrogen bonding	Soups, gravies, salad dressings
Gelation	Protein strands forming a matrix	Meats, curds, gelatin gels
Adhesion	Protein acts as an adhesive	Meats, baked goods, pasta products, breadings
Elasticity	Deformation (e.g., –S–S– bonds)	Baked goods, meats
Plasticity	Deformation (e.g., breaking H– bonds)	Wheat flour dough (e.g., pizza)
Emulsification	Reduction at fat–water interface	Sausages, ice cream, cakes, cookies
Fat adsorption	Binding free fat	Meats, sausages, donuts, crackers
Flavor binding	Adsorption, release, entrapment	Simulated meats, bakery products, micrencapsulated flavors, dry mixes
Foaming	Film formation at the gas-liquid interface, gas entrapment	Whipped toppings, angel cake, ice cream, protein shakes

Table 4.1 (Continued)   Summary of Protein Functional Properties in Food Systems

Funtional Property	Mode of Action	Food System
Catalysis	Make (synthesis) or break (degradation) chemical bonds	Amino acid derived flavors (e.g., cheese), cheese curd (e.g., κ-casein hydrolysis)
Fiber formation	Formation of fibers upon protein denaturation	Simulated meats (e.g., tofu turkey), formed/extruded cereals (e.g., shredded wheat)
Color	Maillard reaction (non-enzymatic browning)	Baked goods, roasted and fried foods, pan-seared foods
Flavor	Strecker degradation	Artificial flavors (e.g., chocolate flavor)

have unique properties that make them indispensable in the manufacture of some foods. For example, wheat gluten is unique in simultaneously providing elasticity and plasticity to the wheat flour dough. The former is thought to be the result of the presence of several disulfide bonds, while the latter is attributed to hydrogen bonds. This unique ability of gluten to provide for unique dough properties makes gluten indispensable in the manufacture of several wheat-based products, such as pasta, breads, cookies, and cakes. Disulfide and hydrogen bonds are found in several other food proteins; however, it remains unclear as to how these same chemical bonds manage to impart unique elasticity and plasticity properties in wheat and not in other cereals. Perhaps the neighboring amino acid residue side chains and the three-dimensional environment of the disulfide and hydrogen bonds in gluten may be contributing factors toward the elasticity and plasticity of the wheat flour dough.

Regardless of the function that a protein may serve in a food, protein solubility, or insolubility, is critical in understanding how the protein may perform in the food system of interest. The preparation of edible spun fibers from soybean proteins illustrates the importance of protein solubility as well as protein insolubility. In order to prepare the protein fibers, the soybean proteins are typically solubilized at an alkali pH (pH > 8.5) and are subsequently extruded under high shear force in an acid bath. Certain minerals, such as Ca^{2+}, may be added to the acid bath to aid protein insolubilization and subsequent protein fiber formation.

4.2 Solubility

Understanding protein solubility is essential in achieving targeted protein functionality. Protein solubility is dependent on several intrinsic (e.g., protein amino acid composition and protein amino acid sequence) and extrinsic (e.g., pH, ionic strength, and temperature) parameters. The first step in protein solubilization is protein wetting (solvation). This step is driven by several factors, including the protein amino acid composition, protein surface charge, protein folding, hydrophilic/hydrophobic balance on the protein surface, solvent properties, pH, ionic strength of the solvent, temperature, and accessibility of the amino acids to the solvent used for protein solubilization. Proteins with a large surface area, a high surface charge, and an open molecular structure are expected to be easily wetted in aqueous media and thus effectively solubilized. Protein solvation is also dependent on the presence of osmolytes (Auton, Wayne Bolen, and Rösgen 2008), such as sucrose, free amino acids, and peptides. The presence of osmolytes typically enhances protein solvation and, therefore, protein solubility. The role of hydrophobic amino acid residues in determining protein insolubility in water has been recently reviewed (Song 2009).

4.2.1 Protein Classification

Many attempts have been made to "classify" proteins in an effort to organize (or group) the ever-increasing list of well-characterized proteins. Typically, such attempts involve the use of a common property or a set of defined properties, either molecular or functional, to "group" several proteins together. However, since each protein is unique, such attempts at protein classification often yield mixed results. The earliest attempt to classify proteins was by Osborne and colleagues (Sathe [2002] and several references therein), where proteins were grouped based on their solubility in different solvents. Osborne described water, dilute salt, aqueous ethanol, and dilute acid/alkali soluble proteins as albumins (e.g., chicken ovalbumin, serum albumin, and α-lactalbumin), globulins (e.g., soybean glycinin, dry bean storage protein phaseolin, and oat globulin), prolamins (e.g., corn zein, wheat gliadin, and rice oryzin), and glutelins (e.g., wheat glutelin and rye secalin), respectively. While useful, one of the limitations encountered when using this classification includes the fact that certain proteins exhibit a solubility behavior of more than one class. For example, phaseolin, the major storage protein in dry beans, is partly soluble in water and partly soluble in dilute salt solutions. As a consequence, dry bean phaseolin is neither a true albumin nor a true globulin and is sometimes referred to as a euglobulin.

Another way to classify the food proteins would be to group them based on their functions. For example, many naturally occurring proteins in plants are found in different edible parts of the plants, such as the roots, stems, barks, flowers, fruits, leaves, tubers, and seeds. Many of these proteins may be categorized as metabolic, as they are responsible for a specific metabolic activity (e.g., lipoxygenase-catalyzing oxidation of polyunsaturated fatty acids), while many others may serve as storage proteins during the "dormant" state (e.g., many seed storage globulins). Depending on the protein and the tissue in which it is located, the protein may not always perform the same function.

The food proteins may be grouped based on their "functions" in the food systems. For example, proteins that are responsible for the foaming and emulsion properties may be considered as "surface active" proteins. However, when the same protein is responsible for more than one function, such attempts provide limited utility. Egg proteins are one such example. In many foods, egg proteins are added to enhance or control different functions, including foaming, emulsification, moisture holding, viscosity, flavor, color, and binding. Often, such an addition serves more than one purpose, as illustrated by the use of egg proteins in cake formulations, where they are responsible for foaming, emulsion, moisture holding, flavor, and color. While it is appreciated that egg white (mainly, the albumen protein) is mainly responsible for foaming (and not emulsification), the same protein does contribute to the flavor and color (through Maillard browning). Similarly, egg yolk proteins not only help in emulsion formation but also contribute to moisture holding and flavor.

As is evident from the preceding discussion, classification based on a single property or attribute of a food protein is often difficult.

4.3 Factors Affecting Protein Solubility

4.3.1 Protein Structure

Protein structure is an important determinant of protein functionality. For example, the role of controlled denaturation of egg proteins in the preparation of several food products (e.g., soufflés, ice creams, cakes, and cookies) is well-known. Another example is the removal of the C-terminal κ-casein macropeptide by chymosin, which causes destabilization of the milk casein micellar structure, permitting protein curd formation—an important step in cheese production from cow's milk. These two examples illustrate the role of protein structure where the proteins are outside a cell system. Examples of intracellular proteins include muscle proteins (e.g., actins, myosins, tropomyosins, and troponins) and their role in the meat structure and metabolic enzymes and in the quality of fruits and vegetables. A protein structure is mainly determined by its amino acid sequence. However, environmental factors, such as the pH, ionic strength, and presence/absence of nonprotein food components, also influence the protein structure. In addition, interaction(s) of the protein with its environment further influences the protein structure. For these reasons, assessing the influence of the three-dimensional structure of a protein on the protein functionality without consideration of the environmental factors is often difficult, especially in a food system. Investigating the molecular species of a protein, for instance, in the absence of other proteins and nonprotein components, offers several advantages. The availability of modern molecular biology and biochemical techniques has now made it possible to obtain homogeneous protein preparations that permit the assessment of well-defined molecular species. Significant advances in analytical methodologies have further enhanced the investigators' ability to examine targeted protein molecules or a specific motif in a protein molecule with unprecedented sensitivity, specificity, precision, and accuracy. An improved understanding of the protein structure–function relationships is therefore anticipated.

4.3.2 Amino Acid Properties

Proteins are composed of 20 common amino acids. Naturally occurring amino acids typically possess an L-configuration, although the occurrence of D-amino acids in nature is known. The L- and D-configurations are in relation to the L- and D-glyceraldehyde configurations and do not refer to the optical rotation properties. Over 900 nonprotein amino acids are known to naturally occur in many plants, especially Leguminosae. Although not part of proteins, nonprotein amino acids are considered to be important in nitrogen storage and as potent antimetabolites and antinutrients. The role of the nonprotein amino acids in metabolism with special reference to diet and diseases has been reviewed (Rubenstein 2000). All amino acids have a common structure, as shown in Figure 4.1.

Each amino acid has one free amino (–NH$_2$) group and one free carboxyl (–COOH) group attached to the α-carbon atom, with one hydrogen and one –R group fulfilling the remaining two valencies of the carbon atom. Proline, the side chain –R in the ring formation, possesses an imino rather than an amino group. The variation in the side chain –R distinguishes different amino acids and strongly influences the properties of those amino acids. At neutral to slightly acidic pH range, amino acids have free –NH$_3^+$ and –COO$^-$ and thus exist in Zwitterionic form when dissolved in water at neutral pH. Typically, an amino acid (or a protein) has a net zero charge at its isoelectric point/pH (pI) and, therefore, usually registers minimum solubility. Above its pI, the amino acid possesses a net negative charge and below its pI, it has a net positive charge.

$$\overset{K_1}{} \qquad \overset{K_2}{}$$
$$H_3^+N-CH_2-COOH \Omega H_3^+N-CH_2-COO^- \ \Xi \ H_2N-CH_2-COO^- \qquad (4.1)$$
$$+H^+ \qquad\qquad -H^+.$$

$$
\begin{array}{cc}
\text{COOH} & \text{COOH} \\
| & | \\
H-C_\alpha-NH_2 & H_2N-C_\alpha-H \\
| & | \\
R & R \\
\text{D-Amino acid} & \text{L-Amino acid}
\end{array}
$$

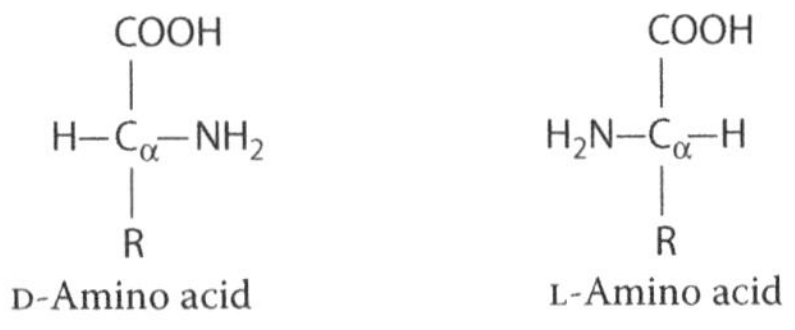

Figure 4.1   Structure of an amino acid. R = variable side chain.

When titrated with an acid (Equation 4.1), the $-COO^-$ group of an amino acid is protonated and becomes $-COOH$. The pH at which the $-COO^-$ and $-COOH$ are equal is $pK_1 = -log_{10} K_1$ (K_1 = dissociation constant). Similarly, on titration with an alkali, the $-NH_3^+$ group is deprotonated and the pH at which the concentration $-NH_3^+ = NH_2$ may be defined as pK_2, where K_2 is the dissociation constant. For an amino acid with no charged side chain, the pI is equal to $(pK_1 + pK_2)/2$. The pK values for the α-carboxyl, α-amino, and side chain $-R$ for several amino acids are summarized in Table 4.2. Among the common amino acids, glycine has the highest while aspartic acid and tyrosine have the least solubility in water (Table 4.2). It is recognized that amino acid solubility is not a linear function of the electrical charges on the free amino and carboxyl groups. As one would anticipate, the side chain $-R$ of an amino acid contributes to the amino acid properties including its solubility. Amino acids with aliphatic (Ala, Ile, Leu, Met, Pro, and Valine) and aromatic (Phe, Trp, and Tyr) $-R$ groups that are hydrophobic are less soluble in water as compared with amino acids with charged (Arg, Asp, Glu, His, and Lys) or uncharged/polar (Ser, Thr, Asn, Gln, and Cys) $-R$ groups. Arg and Lys by virtue of the guanidino and ε-amino groups, respectively, are positively charged (basic) amino acids. The imidazole group of His is weakly basic at neutral pH, while Asp and Glu contain a $-COOH$ group on the side chain, thereby rendering these amino acids acidic. Basic and acidic amino acids are hydrophilic while the remaining amino acids, depending on the $-R$ group, exhibit different hydrophobicities. For these and several other reasons, a number of attempts to compute the amount of energy associated with the effect of a side chain on amino acid solubility have been made (Whitney and Tanford 1962; Nozaki and Tanford 1963, 1965, 1970). The "hydrophobicity" values for the $-R$ side chains are shown in Table 4.2.

4.3.3 Amino Acid Composition

Theoretically, as the number of charged amino acids in a protein increases, the protein solubility is expected to increase. The interaction of a free amino acid with water can be different when compared with the same amino acid present in a polypeptide or protein, as the neighboring amino acids influence the properties of the concerned amino acid. In addition, the three-dimensional structure of the polypeptide or protein may also affect the interaction of water molecules with individual amino acids. Protein solubility depends on a number of additional parameters, including the protein surface area, protein surface charge, protein structure, and environmental conditions (pH and ionic strength). The amino acid sequence in a protein, the primary structure of a protein, is determined by the gene from which the protein is made. As the polypeptide chain becomes longer, the free energy associated with the polypeptide increases and, as a consequence, the polypeptide becomes thermodynamically unstable. To minimize the

Table 4.2 Common Acids and their Properties at 25°C

Amino Acid	Three Letters	One Letter	Molecular Mass	R Group Polarity	Genetic Code	pK' α-COOH	pK' NH₃⁺	pK' of R group	pI[a]	Solubility g/100 g Water[b]	pK[c]	Helix Forming Propensity[d]	Tastes of L-Amino Acids[e]	Δf_i	Hydration[g]
Alanine	Ala	A	89	Nonpolar	GC(N)	2.34	9.69		6	16.65		0	Sweet	+730	1.5
Arginine	Arg	R	174	Positively Charged	AGA, AGG, CG(N)	2.17	0.04	12.48	10.8			0.21	Bitter		3
Aspartic Acid	Asp	D	133	Negatively Charged	GAU, GAC	2.09	9.82	3.86	2.8	0.5	3.5 ± 1.2 (139)	0.69	Neutral	+540	6.0
Asparagine	Asn	N	132	Polar Uncharged	AAU, AAC				5.41			0.65	Neutral	−10	2.0
Cysteine	Cys	C	121	Polar Uncharged	UGU, UGC	1.71	10.78	8.33	5		6.8 ± 2.7 (125)	0.68	Neutral		1
Glutamic Acid	Glu	E	147	Negatively Charged	GAA, GAG	2.19	9.67	4.25	3.2	0.864	4.2 ± 0.9 (153)	0.4	Meat broth like	550	7.5
Glutamine	Gln	Q	146	Polar Uncharged	CAA, CAG	2.17	9.13		5.7			0.39	Neutral	−100	2
Glycine	Gly	G	75	Nonpolar	GG(N)	2.34	9.6		6	24.99		1	Sweet	+730	1
Histidine	His	H	155	Positively Charged	CAU, CAC	1.82	9.17	6	7.5	4.19	6.6 ± 1.0 (131)	0.61	Bitter		4
Isoleucine	Ile	I	131	Nonpolar	AUU, AUC, AUA				5.9	4.117		0.41	Bitter	+2970	1
Leucine	Leu	L	131	Nonpolar	UUA, UUG, CU(N)	2.36	9.6		6	2.426		0.21	Bitter	+2420	1

Lysine	Lys	K	146	Positively Charged	AAA, AAG	2.18	8.95	10.53	9.6		10.5 ± 1.1 (35)	0.26	Sweet, Bitter		4.5
Methionine	Met	M	149	Nonpolar	AUG				5.7			0.24	Sulfurous	+1300	1
Phenylalanine	Phe	F	165	Aromatic	UUU, UUC				5.5	2.965		0.54	Bitter	+2650	0
Proline	Pro	P	115	Polar Uncharged	CC(N)				6.3	16.23			Sweet, Bitter	+2600	3.0
Serine	Ser	S	105	Polar Uncharged	AGU, AGC	2.21	9.15		5.7			0.5	Sweet	+40	2
Threonine	Thr	T	119	Polar Uncharged	AC(N)	2.63	10.43		5.7			0.66	Sweet	+440	2
Tryptophan	Trp	W	204	Aromatic	UGG				5.9	1.136		0.49	Bitter		2
Tyrosine	Tyr	Y		Aromatic	AUA, UAC	2.2	9.11	10.07	5.7	0.453	10.3 ± 1.2 (20)	0.53	Bitter	+2870	3
Valine	Val	V	117	Nonpolar	GU(N)				6	8.85		0.61		+1690	1
C-terminal											3.3 ± 0.8 (22)				
N-terminal											7.7 ± 0.5				

Amino acids mentioned in bold are considered to be essential amino acids for 2–5 years and older individuals; Histidine is essential for infants.

a   From Belitz, H.-D. Amino acids, peptides, and proteins In *Food Chemistry*, p. 11.
b   From *Handbook of Chemistry and Physics*, 63rd Ed. CRC Press, Boca Raton, 1982–83, p. c-758.
c   From Grimsley et al. 2009. Numbers inside the parenthesis indicate the number of measurements.
d   From Pace and Scholtz (1998). The values are expressed in kcal/mol in relation to Alanine (value set at 0).
e   From Belitz, H.-D. Amino acids, peptides, and proteins In *Food Chemistry*, p. 28.
f   Free energy change in calories per mole for transfer from ethanol to water at 25°C (side chain contribution). From Tanford (1962).
g   From Kuntz 1971. Values are expressed as moles of water per mole of amino acid at −35°C and pH 6–8.

free energy associated with a molecule, the polypeptide chain folds and assumes higher order structures. The native protein molecule therefore has the lowest free energy and is the most stable. The primary structure largely determines the higher order structures—secondary, tertiary, and quaternary. The secondary structure refers to the protein structure along the polypeptide backbone. Depending on the amino acid in the primary sequence, the –R group varies. These variations in the –R group coupled with the angle of rotation in a peptide bond largely determine the nature of the secondary structure. Common secondary structures include α-helices, β-sheets, and random coils. The helix structure is considered to be the stable one. Average globular proteins contain ~30% α-helix (Pace and Schlotz 1998). Certain amino acids have a higher propensity for α-helix formation (Table 4.2). These propensity values (ΔG) are in relation to an assigned value of ΔG = 0 for alanine, as alanine has the highest propensity for helix formation. The tertiary structure refers to polypeptide folding, while the quaternary structure is the spatial arrangement of the individual polypeptides of a protein. The contribution by individual amino acids differs. For example, Pace (2001) reported that burying the R groups containing amide groups made a more favorable contribution to protein folding than burying the aliphatic R groups. Depending on the protein structure, the R groups of different amino acids in a protein may be buried or exposed to water, which, in turn, may influence the protein solubility. A recent study illustrates this point. Trevino, Scholtz, and Pace (2007) investigated the contribution of 20 amino acids to the solubility of ribonuclease A (RNAse Sa). The RNAse Sa residue #76 (Thr) is normally exposed to water in the native protein. The investigators substituted the desired amino acid in place of the residue #76 (Thr) in the RNAse Sa, one at a time. The solubility of the RNAse Sa, both the native and the substituted forms, was then measured to determine which variants were highly soluble. The findings indicated that Asp, Glu, and Ser contributed more favorably than the other hydrophilic amino acids to the solubility of the protein. Although the study was done using only RNAse Sa and therefore may not be globally applicable, it does provide a model to assess the relative contribution of the amino acid to protein solubility in a defined environment. Another important observation made by these researchers was that the contribution of polar amino acids to protein solubility appeared to be mainly determined by their water-binding ability (i.e., hydration) rather than their hydrophobicity. The word "hydration" in the preceding sentence refers to unfreezable water when an aqueous macromolecular solution is rapidly frozen and then equilibrated at –20°C to –60°C. The water hydration values for several amino acids (Kuntz 1971) are summarized in Table 4.2. As stated by the author, these values are assigned based on the assumption that bovine albumin has a hydration value of 0.40 ± 0.04 g/g protein (i.e., error is ~10%). The amino acid hydration values, in Table 4.2, for Asp and Glu are assigned assuming that these amino acids are

Twenty standard amino acids

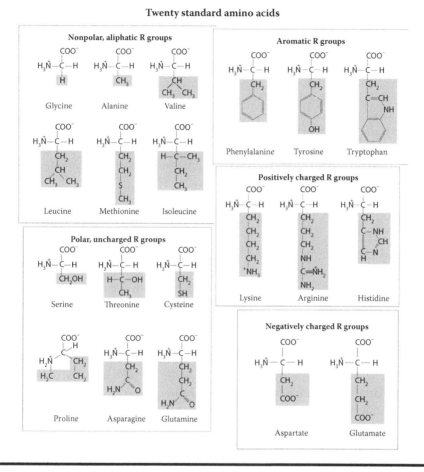

Figure 4.2   Structures of 20 amino acids. The side chain –R is shaded.

at a pH of 4, while the Arg and Lys values for the amino acids are at a pH of 10 and 10–11, respectively.

Typically, as the net charge on a free amino acid increases, so does its solubility. Theoretically, proteins are expected to behave the same way (Figure 4.2). Since proteins contain several amino acids, each in varying proportions, the net charge on the protein is not always equal to the algebraic sum of all the electrical charges present on the amino acids in the protein. This is because the charge on the amino acid side chains is partly governed by the neighboring side chains, the three-dimensional protein folding, the microenvironment of the amino acid residue, and the protein environment (the medium, pH of the medium, ionic strength of the medium, and protein solubility/aggregation).

4.3.4 Salts

Salts, inorganic or organic, affect protein solubility. The ionic strength of a salt solution is expressed as

$$\mu = 0.5\Sigma C_i Z_i^2, \tag{4.2}$$

where μ is the ionic strength, C_i is the molar concentration of the ion, and Z_i is the valence of the ion.

Salts, being water-soluble, influence the surface charge on the protein. The protein solubility (S) is related to μ by

$$\text{Log}\,S = K\mu, \quad \text{where K is a constant.} \tag{4.3}$$

At low salt concentrations, neutral salts increase the protein solubility, the "salting in" effect, while at high salt concentrations, neutral salts cause a decrease in the protein solubility, the "salting out" effect. Mathematically, salting out is expressed as

$$\text{Log}\,S = \log S_0 - K\mu, \tag{4.4}$$

where S = protein solubility and S_0 = protein solubility at $\mu = 0$.

At low ionic strengths ($\mu < 0.5$), the protein surface charge is neutralized. Such a charge neutralization may either decrease or increase the protein solubility, depending on the polar and nonpolar patches on the protein surface. At low ionic strength, the neutral salt suppresses the protein–protein interactions, thereby typically improving the protein solubility. In instances where the presence of salt increases the hydrophobic interactions within or between the polypeptides, the protein solubility is likely to decrease, whereas when such hydrophobic interactions are discouraged, the protein solubility is likely to increase. At higher ionic strengths ($\mu > 1.0$), salt-specific effects are observed. At constant $\mu = 1.0$, typically, one observes that the salts follow the Hofmeister series where the anions promote solubility in the order $SO_4^{2-} < F^- < Cl^- < Br^- < I^- < ClO_4^- < SCN^-$, while cations decrease solubility in the order $NH_4^+ < K^+ < Na^+ < Li^{2+} < Mg^{2+} < Ca^{2+}$.

4.3.5 Nonprotein Components

The nonprotein nitrogen (NPN) includes all nitrogen present in the sample that is not a part of the protein molecules. Such nonprotein molecules may include small peptides (molecular mass <10,000 Da), free amino acids, free bases (e.g., pyrimidines and pyridines), and nucleic acids. Traditionally, the solubility of proteins in trichloroacetic acid (TCA) has been used to define the NPN. The operational definition of the NPN is the nitrogen fraction soluble in 10% (w/v) TCA. Although useful, this operational definition of the NPN is limited in its utility, as not all proteins register minimum protein solubility at 10% TCA. As with other salts, the TCA protein solubility is known to be dependent on the salt concentration (Wolf 1994, 1995). To what extent the NPN affects the protein solubility is largely unknown. Recent research on the influence of specific amino acids (e.g., Arg) in helping to improve the protein solubility (insulin and yeast alcohol dehydrogenase [ADH]) is important in this regard (Baynes, Wang, and Trout 2005; Lyutova, Kasakov, and Gurvits 2007). The investigators found that Arg at 1–10 mM (i.e., the physiological concentration range) suppressed intermolecular aggregation by shifting the hydrodynamic radii of large nanoparticles to the smaller hydrodynamic radii. The mechanism of conversion of a high molecular mass species (by Arg) to a lower molecular mass species remains to be elucidated.

4.3.6 Accessible Interfacial Area

Compared with a linear polypeptide, a folded polypeptide is expected to have a reduced exposed area. In the case of proteins composed of several polypeptides, such a reduction in the exposed area may even be greater. Depending on the protein polypeptide composition and the three-dimensional structure of the protein, the accessible surface area and the accessible three-dimensional space (accessible interfacial area) are variable (Miller et al. 1987a, 1987b). These investigators indicated that the accessible interfacial area (A_s) of a protein is related to its molecular weight (M) by the equation

$$A_s = 6.3 M^{0.73}. \tag{4.5}$$

The total accessible interfacial area (A_t) of a protein (linear molecule) is also correlated with its M as follows:

$$A_t = 1.48 M + 21. \tag{4.6}$$

Thus, knowing M, one can calculate the accessible interfacial area using Equations 4.5 and 4.6. The authors estimated an accessible interfacial area of several dimeric, tetrameric, hexameric, and octameric proteins with various degrees of deviations from the expected values indicating contributions from higher order structures that are important in determining the accessible interfacial area of polypeptides and proteins. Whether these methods can be applied to proteins containing an odd number of monomeric polypeptides (e.g., trimers, pentamers, and heptamers) remains to be determined. As the accessible interfacial area increases, the protein solubility is expected to increase, provided that the predominant amino acids in such area are not hydrophobic.

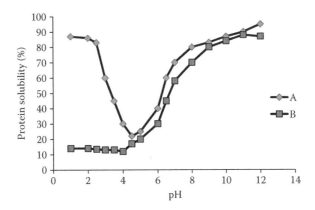

Figure 4.3 Theoretical protein solubility curves. (A) Typical bell-shaped curve. (B) Solubility curve for a protein that aggregates and insolubilizes at its pI.

4.3.7 pH

The pH of the medium strongly influences the protein solubility as the electrical charge on a protein is greatly influenced (Figure 4.3). Typically, at its pI, the protein has a minimum charge and therefore a minimum solubility. On either side of the pI, the protein is expected to have a net charge, a positive charge below and a negative charge above the pI, resulting in the typical bell-shaped curve (Figure 4.3A). If the protein molecule irreversibly aggregates at pI, it may lose its solubility, resulting in minimum solubility at pH ≤ pI (Figure 4.3B) in the medium used to solubilize the protein.

4.3.8 Protein Partitioning

Often, the food proteins exhibit partial solubility in different solvents. For example, prolamins are soluble in aqueous alcohol solutions. Such solubility behavior is mainly due to the surface charge distribution on the protein. Determining the solubility in mixed solvents has therefore been of interest, as understanding the behavior of the proteins in such systems has practical applications in downstream processing (e.g., the removal of unwanted residual proteins in fermentations). It has been reported that protein partitioning is linearly related to protein hydrophobicity; therefore, measuring the protein hydrophobicity may provide important information about protein partitioning. The hydrophobicity values have been shown to correlate well with the partition coefficients, K, in a polyethylene glycol (PEG)/salt system at high NaCl concentrations (r = 0.92–0.93) (Andrews, Schmidt, and Asenjo 2005):

$$\log K = R \log P - R \log P_0, \tag{4.7}$$

Table 4.3   Hydrophilicity (m*) and Hydrophobicity (P) of Select Proteins
Using Ammonium Sulfate Precipitation[a]

Protein	Molecular Mass (Da)	m* (mol/kg)	P (mol/kg)$^{-1}$
α-Amylase	50,000	2.07	0.484
Amyloglucosidase	97,000	3.11	0.321
Bovine serum albumin	67,000	3.21	0.312
α-Chymotrypsinogen A	23,600	1.84	0.544
Conalbumin	77,000	2.90	0.345
Invertase	270,000	2.95	0.339
α-Lactalbumin	14,400	2.83	0.353
Lysozyme	14,400	2.39	0.418
Ovalbumin	45,000	2.80	0.358
Subtilisin	27,500	2.16	0.462
Thaumatin	22,200	2.11	0.473
Trypsin inhibitor	24,500	1.99	0.503

[a] Adapted from Andrews et al. (2005). Phase systems were prepared using polyethylene glycol and aqueous salt solutions.

where $P = (1/m*)$ is the protein hydrophobicity in solution measured by precipitation, and $\log P_0$ represents the intrinsic hydrophobicity of the given partitioning system. The salt concentration at which the protein begins to precipitate in ammonium sulfate at a given protein concentration is referred to as m*. When $K = 1$, $\log P_0 = \log P$. R represents the hydrophobic resolution, which is the ability of the system to discriminate between the proteins with different hydrophobicities (Andrews Schmidt and Asenjo 2005). Andrews, Schmidt, and Asenjo (2005) reported hydrophobicity values for several proteins (Table 4.3).

4.4 Methods to Test Protein Solubility and Function

Depending on the protein and the function of that protein in the food system, a suitable method is needed to assess the protein solubility and protein function. For example, in bread manufacturing, the bread loaf volume is one of the most important parameters considered to be indicative of the bread quality. Since wheat gluten is mainly responsible for the bread volume, wheat gluten functionality is often assessed by measuring the bread volume. The American Association of Cereal Chemists (AACC) has an approved mustard seed displacement method for this purpose. The test employs AACC-approved standardized baking procedures and seed displacement equipment to determine the bread volume and thus the gluten performance. Another example is the percent overrun (i.e., the percent increase in volume) that is often used as an index to determine the ice-cream quality. The increase in the volume of ice cream is mainly due to the foaming agents (e.g., chicken hen eggs or other foaming agents) used in the ice-cream mix. The foaming agents enable the mechanical incorporation of small air bubbles into the ice-cream mix. Egg proteins, the typical foaming agents in the ice-cream mix, form an elastic skin around the air bubbles, providing a stable, light, fluffy product. The functionality of a protein may be assessed for a food protein in isolated form, in a food system, or both. Isolated proteins are often assessed for their potential as functional ingredients in foods prior to determining their performance in a test food system. The advantages of investigating isolated proteins include defined protein preparation, the ability to eliminate the influence of the nontargeted proteins and nonprotein components, the precise control of the protein environment (e.g., pH, ionic strength, and temperature), and the ability to systematically evaluate the influence of extrinsic factors on the protein function. As protein solubility has a significant influence on protein functionality, accurate protein solubility determination is important. The first step in such an assessment is to select the solvent. As water is the most commonly used medium in food processing, determining the protein solubility in water is desirable. Often, the investigators use water and several aqueous buffers/solvents to compare the protein solubilities in the selected solvents to learn the influence of the food components on protein solubility. For the proteins that are insoluble in water or aqueous solvents, the use of mixed solvents such as aqueous alcohol (e.g., to solubilize prolamins) is necessary. Depending on the solvent used for protein solubilization, different methods for protein determination may be used.

4.4.1 Protein Solubility

A soluble protein may be determined by a variety of methods, depending on the protein source and the presence or absence of the substances that may interfere with the method utilized to determine the protein content of the food sample. Commonly used protein determination methods have been reviewed (Petersen 1979; Stoscheck 1990; Baret et al. 1997; Chang 1998; Kolakowski 2005). Depending on the protein and its end use, the food protein is quantified using one of several methods that are summarized in Table 4.4.

The Kjeldahl method is by far the most widely accepted and officially recognized method used for food protein quantification. In principle, the Kjeldahl method considers the nitrogen content of the food proteins to be typically 16% by weight and therefore utilizes a factor of 6.25 (100 g protein/16 g nitrogen = 6.25).

Table 4.4 Summary of Protein Determination Methods

Method	Principle	Advantages	Limitations	Range	References
Kjeldahl/ Micro-Keldahl	N converted to NH_3 which is collected in excess acid to titrimetrically determine N. N × Factor = % Protein	Official Method, Accurate and Direct nitrogen determination, universally applicable for soluble and insoluble proteins	Non-protein nitrogen has to be determined separately, expensive, time consuming, uses highly corrosive chemicals, requires special fume hoods to handle acid fumes	mg-g/ μg-mg	AOAC (1995)
Biuret	Cupric ions, under alkaline conditions, interact with peptide bonds to produce purplish colored complex read at ~540 nm	Rapid, inexpensive,	Interference from: protein precipitating agents, reducing agents, fiber, clouding agents, certain additives and preservatives, certain pigments, some minerals	μg-mg	Robinson and Hodgen (1940a,b)
Bicinchoninic acid (BCA)	Reduction of cupric to cuprous ions (Cu^{1+}) by proteins under alkali conditions to produce purplish color complex read at ~560 nm. BCA acts as the Cu^{1+} capture reagent instead of Folin-Ciocalteau reagent.	Rapid, relatively inexpensive, sensitive, nonionic detergents, buffer salts, 4M guanidine-HCl and 3 M urea typically do not interfer, low protein to protein variation in color yield	Reducing agents interfere, color instability with time	μg	Smith et al. (1985)

(continued)

Table 4.4 (Continued) Summary of Protein Determination Methods

Method	Principle	Advantages	Limitations	Range	References
Lowry	Alkali copper reacts with peptide bonds + reduction of Folin-Ciocalteau phenol (phosphomolybdic-tungstic acid) by aromatic amino acids tryptophan and tyrosine, the bluish-purple color complex is read at wavelengths 750 nm (05–25 µg/ml range) at 500 nm (≥25 µg/ml)	Rapid, sensitive (0.2 µg for micro-procedure), inexpensive, less interference by turbidity	Protein must be soluble at pH 10, protein dependent color yield, reducing agents and many commonly used biochemicals may interfere, requires quick (≤ 10 seconds) mixing of Folin-Ciocalteau reagent with sample to obtain reproducible color yield, color yield is not strictly proportional to protein concentration	0.2–200 µg	Lowry et al (1951), Seevaratnam et al. (2009)
Bradford	For an acidic solution of Coomassie Brilliant Blue G-250 dye absorbance maximum shifts from 465 nm to 595 nm when the dye binds to protein	Rapid, sensitive, several salts and biochemicals that interfere in Lowry procedure do not interfere in Bradford	Strong alkali buffers and SDS (> ~0.1%) interfere	1–20 µg range for micro-procedure	Bradford (1976)
Ninhydrin	Amino acids, ammonia, and primary amino groups of protein react with ninhydrin in pH 5.5	Simpler than Kjeldahl, can be used to quantify amino acids	Ammonia, free amino acids, and non-protein primary amines may interfere, color varies		Spackman et al. (1958)

	Description	Advantages	Disadvantages	Range	References
	buffer when boiled. The resultant Ruhemann purple color is measured at 570 nm		with different amino acids (e.g. proline absorbtion maximum is at 440 nm), low precision	μg range	Layne (1957), Iwata and Nishizake (1979), Wieser (2000)
Turbidimetric	Low (3%–10%) concentrations of trichloroacetic acid, sulfosalycylic acid, and potassium ferricyanide in acetic acid used to precipitate extracted protein to form turbid suspension. Turbidity can be measured from reduction in transmission of the radiant energy (typically measured at 540 nm)	Simple and rapid	Nucleic acids interfere, turbidity varies with protein as well as the acid reagents, corrosive chemicals		
Dumas	Sample is combusted at high temperatures (700°C–800°C) and the released nitrogen is quantified (gas chromatography) using thermal detector	No hazardous chemicals, with use of automated intstruments the method is rapid. Results are comparable to Kjeldahl	Requires large (mg) quantity of the sample analyzed, expensive instrumentation, determines total nitrogen (protein + nonprotein nitrogen)	mg–g/ μg–mg	Simonne et al. (1997), Miller et al. (2007)

(continued)

Table 4.4 (Continued) Summary of Protein Determination Methods

Method	Principle	Advantages	Limitations	Range	References
Ultraviolet (UV)	Strong absorption at 280 nm by tryptophan and tyrosine present in proteins. Beer's law is used to quantify protein. $A = \varepsilon l c$ where ε = molar absorptivity, l = light path length (cm), and c = analyte concentration (moles/liter)	Rapid, sensitive, inexpensive, nondestructive, free of interference from ammonium sulfate and several buffer salts	Molar extinction coefficient for each protein (ε280 nm) must be determined as each protein has unique number of tryptophan and tyrosine residues, interference from non-protein UV absorbing materials such as phenols, nucleic acid, and others	μg–mg	Whitaker and Granum (1980), Peterson (1983), Manchester (1996), Kolakowski (2005)
Colloidal Gold	Protein binds to colloidal gold under acidic conditions to cause a shift in its absorbance proportional to the amount of protein. Absorbance is measured at 560 nm	Simple, rapid, sensitive	pH of solution must be acidic for most proteins to carry a positive charge to enable its interaction with negatively charge on colloidal gold. Expensive reagent, proteins stick to glass and plastic surfaces, overloading erroneously produces low protein values	ng	Stoschek (1990), Sanchez-Martinez et al. (2009)

Infrared	Absorption of radiation energy by food proteins in the near (3300–3500 nm, 2080–2220 nm, 1560–1670 nm) or mid infra-red (6.47 µm) regions	Rapid	Expensive equipment		Wesley et al (2001), Kim et al. (2007), Prieto et al. (2009)
Fluorescence	Select reagents produce fluorescent derivatives (extrinsic fluorescence)- e.g. opthalaldehyde (OPA) reacts with primary amines to form fluorescent product with an excitation maximum at 340 nm and a broad emission maximum with maximum at ~450 nm. With proteins, OPA reacts with NH2-terminal amino acid and ε-amino group of lysine residues. In protein hydrolyzates all free primary amines from α-amino amino acids react.	Rapid, sensitive, stable	Expensive equipment, non-protein fluorogenic substances interfere, stability of fluorescent products, more accurate for hydrolyzed, compared to in-tact proteins	pg–ng	Benson and Hare (1975), Petersen (1983), Nielsen et al. (2001), Held (2006)

(continued)

Table 4.4 (Continued)   Summary of Protein Determination Methods

Method	Principle	Advantages	Limitations	Range	References
Immunoassays	Specific primary antibodies (poly- or monoclonal) raised against targeted protein bind to the targeted protein. Antibodies against primary antibodies, the secondary antibodies, bind to the primary antibodies. Suitable signal amplication (radiation, visible color, or fluorescence) system quantifies secondary antibody binding to primary antibody binding indirectly measuring protein	Sensitive, specific, rapid	Requires standardized antibodies, expensive assay euipment, and specialized training. Must be standardized and validated for each target protein	pg–ng	Sasakura et al. (2006), Tessler et al. (2009), Sanchez-Martinez et al. (2009)

While useful, this factor is not the same for all proteins, as the nitrogen content of proteins is known to vary considerably. The primary source of variation in the protein nitrogen is the nitrogen residues in the side chains of the amino acids in the protein (notably, asparagine and glutamine residues). For this reason, the nitrogen conversion factors for the proteins from wheat (5.33), almonds (5.18), peanuts (5.46), and several tree nuts (5.3) have already been approved in the AOAC procedures. Salo-Väänänen and Koivistoinen (1996) investigated the true protein (NP) values (sums of amino acid residues) and compared them with the corresponding crude protein (PA) values for a number of foods, including milk and milk products, human milk, eggs, infant formulae (28 items), meats and meat products including offal (32 items), fish and fish products including shrimps and roe (28 items), cereals and bakery products (19 items), fruits, vegetables, root crops, berries/berry products (24 items), and miscellaneous processed foods—mainly ready-to-eat products from mixed raw materials (17 items). The investigators found that the traditional PA values (N × 6.25) significantly deviated from the corresponding NP values and they obtained the following conversion factors:

1. Milk and milk products: 5.94
2. Meat and meat products: 5.17
3. Fish and fish products: 4.94
4. Cereals and bakery products: 5.40
5. Vegetables, fruits, and berries: 5.36
6. Miscellaneous processed foods: 5.51

Therefore, an accurate determination of the nitrogen conversion factors for the Kjeldahl protein contents of many commonly consumed foods and food ingredients is warranted.

All spectrophotometric methods are susceptible to interferences by the suspended matter in the sample, and great care must be taken to remove the suspended matter during sample preparation. One collaborative study (Morr et al. 1985) designed to develop a "rapid, simple and reliable procedure for determining the solubility of food protein products, e.g., spray-dried whey protein concentrate, sodium caseinate, egg white protein and soy protein isolate," found that the micro-Kjeldahl and biuret procedures were generally useful (standard deviations were 0.83–4.12) for the protein samples investigated—milk caseinates registered a standard deviation of 13.95. The investigators also reported that the biuret method exhibited considerable error and variability for several proteins. The study also highlighted the need to develop standardized methodologies for assessing protein solubility. A recent study (Lozzi et al. 2008) compared seven different spectrophotometric methods—Bradford (standard, micro, and 590/450 nm ratio), Lowry, bicinchoninic acid (BCA), UV spectrophotometry at 280 nm, and Quant-iT fluorescence-based determination—for interference by suspended clay (particle size <2 μm; particles not easily visible to the eye) on bovine serum albumin (BSA) quantification and found that the suspended clay did interfere in all but the Lowry method. Although the primary goal of this study was to assess the utility of the methods for application in the analysis of soil samples, the study illustrates the importance of preparing protein samples free of suspended matter. In the case of milk powder samples (skim milk powder, whole milk powder, whey protein powder, and buttermilk powder), a comparison of the Kjeldahl method and several spectrophotometric methods (UV 280 and 220 nm, biuret 340 and 550 nm, Bradford, Lowry, and p-chloranil) revealed that the Bradford method could be used for skim milk and whole milk powders, without removing the lipids in the powders, instead of the Kjeldahl method (Kamizake et al. 2003). The investigation also reported on the least variation of specific absorbance for cow milk casein and BSA, suggesting that both proteins could be used as standard proteins in the Lowry procedure. Khodabux et al. (2007) found the near infrared (NIR) spectroscopic method to be accurate in determining the protein contents of two species of Atlantic tuna: skipjack (*Katsuwonus pelamis*) and yellow fin (*Thunnus albacares*). Similarly, Adamopoulos, Goula, and Petropakis (2001) reported the NIR method to be accurate and satisfactory for protein determination in feta cheese production. A brief overview of the NIR methodology, instrumentation, and applications (Blanco and Villarroya 2002) provides a succinct coverage of NIR. Occasionally, methods are needed to quantify the proteins adsorbed on solid surfaces (e.g., protein film formation on solid surfaces). The currently available methods do not readily permit protein determination in such situations, as protein solubilization is often necessary prior to protein determination. Protein solubilization, as discussed earlier, often requires tedious and systematic sample preparation. Orschel et al. (1998) investigated several methods and found most spectroscopic methods to be unsuitable for quantification of the proteins adsorbed on silica and gold surfaces. The investigators did find enzyme-linked immunosorbent assay (ELISA) as well as fluorescence-based methods to be suitable for such analyses.

While analytical techniques suitable for food protein quantification are being constantly improved, advances in the protein quantification of clinical specimens are worthy of note. Protein analysis of clinical specimens often requires not only speed and accuracy, but also sensitivity, specificity, and robustness. Recent advances using miniaturized systems coupled with the use of specific and sensitive reagents for targeted proteins are expected to help the routine development of protein quantification at the picogram level (Kartalov et al. 2008). Such methods may potentially be adapted for the analysis of food specimens. The adaptation of several established assay procedures, for example, the Bradford method, in microformats (e.g., the ELISA plate format) coupled with sensitive signal detection systems (e.g., fluorescence and radiation) is expected to help develop robust, accurate, and sensitive methods that may also afford throughput of a large number of samples.

Since food matrixes are often chemically and physically diverse and complex, the method of choice for protein determination is often complex and difficult and must be done judiciously. For these reasons, a single method for food protein quantification may not always be easy to achieve.

4.4.2 *Protein Function*

The food proteins perform a variety of functions in food systems, including hydration, water holding, foaming, emulsion, gelation, catalysis, structure, and flavor production as a result of the Strecker degradation and browning due to the Maillard reactions. Depending on the functional property to be evaluated, different methods are used. Since proteins and water are polar, they may interact, resulting in water immobilization. The amount of water that a food protein may hold depends on the protein, the amount of water available, the presence or absence of other polar molecules in the system that may compete for water binding, the biochemical and physicochemical properties of the food system, and the environmental conditions of the food system. There is no official definition of the water-holding capacity (WHC) of the food proteins. However, the amount of water held by a protein, under defined conditions, is referred to as the WHC. Since the WHC influences many of the functional properties of the food proteins, WHC determination is an important parameter in the characterization of the functional properties of the food proteins. For example, as the WHC is critical in meat texture (juiciness), the WHC of meat and meat products has been extensively investigated (Huff-Lonergan and Lonergan 2005). Similarly, the importance of the WHC in the functionality of milk and milk products (Kneifel et al. 1991; Haque 1993; Morr and Ha 1993; Panyam and Kilara 1996; Chobert 2003; Augustin and Udabage 2007), collagen (Ashgar and Henrickson 1982), fish proteins (Kristinsson and Rasco 2000), oilseed proteins (Kinsella 1979; Moure et al. 2006; González-Pérez and Vereijken 2007), and dry bean proteins (Sathe 2002) has been recognized.

Although WHC is an important functional property of the food proteins, there is no universal method to quantify this property. Part of the difficulty is that the WHC of a protein, depending on its solubility, may be viewed as a bimodal system. For example, a partly soluble protein in the chosen solvent leaves the insoluble portion of the protein in a dispersed state. Both the soluble and dispersed protein portions contribute to the WHC, and the proportion of their contribution to the total WHC may differ depending on the degree of protein solubility and the polarity of the soluble and insoluble protein fractions. For operational simplicity, one may determine the WHC by suspending a known amount of protein in an excess (but fixed) volume of water under specified conditions (e.g., time, pH, temperature, mechanical stirring, and ionic strength), separating the unbound water using a defined force (e.g., filtration under gravity or a fixed centrifugal force), and quantifying the amount of water withheld by the protein. The amount of soluble protein may be quantified using an appropriate method to account for the mass balance considerations before determining the WHC of the investigated protein. While useful, the WHC determination of an isolated protein has its limitations, as one cannot accurately determine the influence of the nonprotein components on the WHC. As a consequence, ideally, the WHC should be determined by using the protein in the food system for which it was prepared.

4.5 Conclusion

Although the molecular properties of common amino acids and several proteins are well documented, a gap remains in our understanding of these important molecules. Despite the significant advances in the molecular tools available, several areas warrant further research, including understanding (a) the role of the amino acid properties on the molecular properties of proteins, such as protein hydration and protein folding; (b) the role of the protein surface properties on the food functional properties; (c) the mechanisms of protein denaturation/aggregation; and (d) developing novel methods to control protein solubility, leading to improved functionality of the protein in the desired food.

References

Adamopoulos, K., Goula, A. M., and Petropakis, H. J. 2001. Quality control during processing of feta cheese-NIR application. *Journal of Food Composition and Analysis* 14: 431–40.

Andrews, B. A., Schmidt, A. S., and Asenjo, J. A. 2005. Correlation for the partition behavior of proteins in aqueous two-phase systems: Effect of surface hydrophobicity and charge. *Biotechnology and Bioengineering* 90: 380–90.

AOAC. 1995. *Official Methods of Analysis*, 16th ed. Arlington, VA: Association of Official Analytical Chemists.

Ashgar, A. and Henrickson, R. L. 1982. Chemical, biochemical, functional, and nutritional characteristics of collagen in food systems. *Advances in Food Research* 28: 231–372.

Augustin, M. A. and Udabage, P. 2007. Influence of processing on functionality of milk and dairy proteins. *Advances in Food Research* 53: 1–38.

Auton, M., Wayne Bolen, D., and Rösgen, J. 2008. Structural thermodynamics of protein preferential solvation: Osmolyte solvation of proteins, aminoacids, and peptides. *Proteins* 73: 802–13.

Baret, P., Angeloff, A., Rouch, C., Pabion, M., and Cadet, F. 1997. Microquantification of proteins by spectrophotometry. Part I. From 190–1100 nm, selection of wavelengths. *Spectroscopy Letters* 30: 1067–88.

Baynes, B. M., Wang, D. I. C., and Trout. B. L. 2005. Role of arginine in the stabilization of proteins against aggregation. *Biochemistry* 44: 4919–25.

Belitz, H. D. and Grosch, W. 1987. *Food Chemistry*. Berlin: Springer Verlag.

Benson, J. R. and Hare, P. E. 1975. O-pthalaldehyde: Fluorogenic detection of primary amines in the epicomole range. Comparison with fluorescamine and ninhydrin. *Proceedings of the National Academy of Sciences of the United States of America* 72: 619–22.

Blanco, M. and Villarroya, I. 2002. NIR spectroscopy: A rapid-response analytical tool. *Trends in Analytical Chemistry* 21: 240–50.

Bradford, M. M. 1976. A rapid and sensitive method for the quantitation of microgram quantities of protein utilizing the principle of protein-dye binding. *Analytical Biochemistry* 72: 248–54.

Chang, S. K. C. 1998. Protein analysis. In *Food Analysis*, ed. S. S. Nielsen, 2nd ed., pp. 237–49. Gaithersburg, MD: Aspen Publishers.

Chobert, J.-M. 2003. Milk protein modification to improve functional and biological properties. *Advances in Food Research* 47: 1–71.

González-Pérez, S. and Vereijken, J. M. 2007. Sunflower proteins: Overview of their physicochemical, structural and functional properties. *Journal of the Science of Food and Agriculture* 87: 2173–91.

Grimsley, G. R., Scholtz, J. M., and Pace, C. N. 2009. For the record: A summary of the measured pK values of the ionizable groups in folded proteins. *Protein Science* 18: 247–51.

Haque, Z. 1993. Influence of milk peptides in determining the functionality of milk proteins – A review. *Journal of Dairy Science* 76: 311–20.

Held, P. G. 2006. Quantitation of total protein using OPA. BioTek Instruments, Inc., Winooski, VT. Application note: http://www.biotek.com/resources/docs/FL600_Quantitation_Total_Protein_Using_OPA.pdf (Accessed July 25, 2009).

Huff-Lonergan, E. and Lonergan, S. M. 2005. Mechanisms of water-holding capacity of meat: The role of postmortem biochemical and structural changes. *Meat Science* 71: 194–204.

Iwata, J. and Nishikaze, O. 1979. A new micro-turbidimetric method for determination of protein in cerebrospinal-fluid and urine. *Clinical Chemistry* 25: 1317–19.

Kamizake, N. K. K., Gonçalves, M. M., Zaia, C. T. B. V., and Zaia, D. A. M. 2003. Determination of total proteins in cow milk powder samples: A comparative study between the Kjeldahl method and spectrophotometric methods. *Journal of Food Composition and Analysis* 16: 507–16.

Kartalov, E. P., Lin, D. H., Lee, D. T., Anderson, W. F., Taylor, C. R., and Scherer, A. 2008. Internally calibrated quantification of protein analytes in human serum by fluorescence immunoassays in disposable elastomeric microfluidic devices. *Electrophoresis* 28: 5010–16.

Khodabux, K., L'Omelette, M. S. S., Jhaumeer-Laulloo, S., Ramasami, P., and Rondeau, P. 2007. Chemical and near-infrared determination of moisture, fat and protein in tuna fishes. *Food Chemistry* 102: 669–75.

Kim, Y., Singh, M., and Kays, S. E. 2007. Near-infrared spectroscopic analysis of macronutrients and energy in homogenized meals. *Food Chemistry* 105: 1248–55.

Kinsella, J. E. 1979. Functional properties of soy proteins. *Journal of the American Oil Chemists Society* 56: 242–58.

Kneifel, W., Paquin, P., Abert, T., and Richard, J. P. 1991. Water-holding capacity of proteins with special regard to milk proteins and methodological aspects. A review. *Journal of Dairy Science* 74: 2027–41.

Kolakowski, E. 2005. Analysis of proteins, peptides, and amino acids in foods. In *Methods of Analysis of Food Components and Additives*, ed. S. Ötleş, pp. 59–96. Boca Raton, FL: CRC Press Taylor & Francis Group.

Kristinsson, H. G. and Rasco, B. A. 2000. Fish protein hydrolyzates: Production, biochemical, and functional properties. *CRC Critical Reviews in Food Science & Nutrition* 40: 43–81.

Kuntz, I. D. 1971. Hydration of macromolecules. III. Hydration of polypeptides. *Journal of the American Chemical Society* 93: 514–16.

Layne, E. 1957. Spectrophotometric and turbidimetric methods for measuring proteins. *Methods in Enzymology* 3: 447–54.

Lowry, O. H., Rosebrough, N. J., Farr, A. L., and Randall, R. J. 1951. Protein measurement with the Folin phenol reagent. *Journal of Biological Chemistry* 193: 265–75.

Lozzi, I., Pucci, A., Pantani, O. L., D'Acqui, L. P., and Calamai, L. 2008. Interferences of suspended clay fraction in protein quantitation by several determination methods. *Analytical Biochemistry* 376: 108–14.

Lyutova, E. M., Kasakov, A. S., and Gurvits, B. Y. 2007. Effects of arginine on kinetics of protein aggregation studied by dynamic laser light scattering and tubidimetry techniques. *Biotechnology Progress* 23: 1411–16.

Manchester, K. L. 1996. Use of UV methods for measurement of protein and nucleic acid concentrations. *Biotechniques* 20: 968–70.

Miller, E. L., Bimbo, A. P., Barlow, S. M., and Sheridan, B. 2007. Repeatability and reproducibility of determination of the nitrogen content of fishmeal by the combustion (Dumas) method and comparison with the Kjeldahl method: Interlaboratory study. *Journal of AOAC International* 90: 6–20.

Miller, S., Lesk, A. M., Janin, J., and Chotia, C. 1987a. Interior and surface area of monomeric proteins. *Journal of Molecular Biology* 196: 641–56.

Miller, S., Lesk, A. M., Janin, J., and Chotia, C. 1987b. The accessible surface area and stability of oligomeric proteins. *Nature* 328: 834–36.

Morr, C. V., German, B., Kinsella, J. E., Regenstein, J. M., Van Buren, J. P., Kilara, A., et al. 1985. A collaborative study to develop a standardized food protein solubility procedure. *Journal of Food Science* 50: 1715–18.

Morr, C. V. and Ha, E. Y. W. 1993. Whey protein concentrates and isolates – Processing and functional properties. *CRC Critical Reviews in Food Science & Nutrition* 33: 431–76.

Moure, A., Sineiro, J., Domínguez, H., and Parajó, J. C. 2006. Functionality of oilseed protein products: A review. *Food Research International* 39: 946–63.

Nielsen, P. M., Petersen, D., and Dambmann, C. 2001. Improved method for determining food protein degree of hydrolysis. *Journal of Food Science* 66: 642–46.

Nozaki, Y. and Tanford, C. 1963. The solubility of amino acids and related compounds in aqueous urea solutions. *Journal of Biological Chemistry* 238: 4074–81.

Nozaki, Y. and Tanford, C. 1965. The solubility of amino acids and related compounds in aqueous ethylene glycol solutions. *Journal of Biological Chemistry* 240: 3568–73.

Nozaki, Y. and Tanford, C. 1970. The solubility of amino acids, diglycine, and triglycine in aqueous guanidine hydrochloride solutions. *Journal of Biological Chemistry* 245: 1648–52.

Orschel, M., Katerkamp, A., Meusel, M., and Cammann, K. 1998. Evaluation of several methods to quantify immobilized proteins on gold and silica surfaces. *Colloids and Surfaces B: Biointerfaces* 10: 273–79.

Pace, C. N. 2001. Polar group burial contributes more to protein stability than non-polar group burial. *Biochemistry* 40: 310–13.

Pace, C. N. and Scholtz, J. M. 1998. A helix propensity scale based on experimental studies of peptides and proteins. *Biophysical Journal* 75: 422–27.

Panyam, D. and Kilara, A. 1996. Enhancing the functionality of food proteins by enzymatic modification. *Trends in Food Science & Technology* 7: 120–25.

Peterson, G. L. 1979. Review of the Folin phenol protein quantitation method of Lowry, Rosebrough, Farr and Randall. *Analytical Biochemistry* 100: 201–20.

Peterson, G. L. 1983. Determination of total protein. *Methods in Enzymology* 91: 95–119.

Prieto, N., Roehe, R., Lavin, P., Batten, G., and Andres, S. 2009. Application of near infra-red reflectance spectroscopy to predict meat and meat product quality: A review. *Meat Science* 83: 175–86.

Robinson, H. W. and Hogden, C. G. 1940a. The biuret reaction in the determination of serum proteins I. A study of the conditions necessary for the production of a stable color which bears a quantitative relationship to the protein concentration. *Journal of Biological Chemistry* 135: 707–25.

Robinson, H. W. and Hogden, C. G. 1940b. The biuret reaction in the determination of serum proteins II. Measurements made by a Duboscq colorimeter compared with values obtained by the Kjeldahl procedure. *Journal of Biological Chemistry* 135: 727–31.

Rubenstein, E. 2000. Biologic effects and clinical disorders caused by nonprotein amino acids. *Medicine* 79: 80–89.

Salo-Väänänen, P. P. and Koivistoinen, P. E. 1996. Determination of protein in foods: Comparison of net protein and crude protein (N × 6.25) values. *Food Chemistry* 57: 27–31.

Sánchez-Martínez, M. L., Aguilar-Caballos, M. P., and Gómez-Hens, A. 2009. Homogeneous immunoassay for soy protein determination in food samples using gold nanoparticles as labels and light scattering detection. *Analytica Chimica Acta* 636: 58–62.

Sasakura, Y., Kanda, K., and Fukuzono, S. 2006. Microarray techniques for more rapid protein quantification: Use of single spot multiplex analysis and a vibration reaction unit. *Analytica Chimica Acta* 564: 53–58.

Sathe, S. K. 2002. Dry bean protein functionality. *CRC Critical Reviews in Biotechnology* 22: 175–223.

Seevaratnam, R., Patel, B. P., and Hamadeh, M. J. 2009. Comparison of total protein concentration in skeletal muscle as measured by Bradford and Lowry assays. *Biochemical Journal* 145: 691–97.

Simonne, A. H., Simonne, E. H., Eitenmiller, R. R., Mills, H. A., and Cresman III, C. P. 1997. Could the Dumas method replace the Kjeldahl digestion for nitrogen and crude protein determinations in foods? *Journal of the Science of Food and Agriculture* 73: 39–45.

Smith, P. K., Krohn, R. I., Hermanson, G. T., Mallia, A. K., Gartner, F. H., Provensano, M. D., et al. 1985. Measurement of protein using bicinchoninic acid. *Analytical Biochemistry* 150: 76–85.

Song, J. 2009. Insight into "insoluble proteins" with pure water. *FEBS Letters* 583: 952–59.

Spackman, D. H., Stein, W. H., and Moore, S. 1958. Automatic recording apparatus for use in the chromatography of amino acids. *Analytical Chemistry* 30: 1191–1206.

Stoscheck, C. M. 1990. Quantitation of protein. *Methods in Enzymology* 182: 50–68.

Tanford, C. 1962. Contribution of hydrophobic interactions to the stability of the globular conformation of proteins. *Journal of the American Chemical Society* 84: 4240–47.

Tessler, L. A., Reifenberger, J. G., and Mitra, R. D. 2009. Protein quantification in complex mixtures by solid phase single-molecule counting. *Analytical Chemistry* doi: 10.1021/ac901068x (Accessed July 25, 2009).

Trevino, S. R., Scholtz, J. M., and Pace, C. N. 2007. Amino acid contribution to protein solubility: Asp, Glu, and Ser contribute more favorably than the other hydrophilic amino acids in RNAse Sa. *Journal of Molecular Biology* 366: 449–60.

Wesley, I. J., Larroque, O., Osborne, B. G., Azudin, N., Allen, H., and Skerritt, J. H. 2001. Measurement of gliadin and glutenin content of flour by NIR spectroscopy. *Journal of Cereal Science* 34: 125–33.

Whitaker, J. R. and Granum, P. E. 1980. An absolute method for protein determination based on difference in absorbance at 235 and 280 nm. *Analytical Biochemistry* 109: 156–59.

Whitney, P. L. and Tanford, C. 1962. Solubility of amino acids in aqueous urea solutions and its implications for the denaturation of proteins by urea. *Journal of Biological Chemistry* 327: 1735–37.

Wieser, H. 2000. Simple determination of gluten protein types in wheat flour by turbidimetry. *Cereal Chemistry* 77: 48–52.

Wolf, W. J. 1995. Gel electrophoresis and amino acid analysis of the nonprotein nitrogen fractions of defatted soybean and almond meals. *Cereal Chemistry* 72: 115–21.

Wolf, W. J., Schaer, M. L., and Abbott, T. P. 1994. Nonprotein nitrogen content of defatted jojoba meals. *Journal of the Science of Food and Agriculture* 65: 277–88.

Chapter 5

Proteins–Peptides as Emulsifying Agents

Yasuki Matsumura and Kentaro Matsumiya

Contents

5.1 Introduction: Features of Proteins as Emulsifying Agents

Proteins and low molecular weight surfactants are the two major components included in foods as emulsifying agents. Protein is a macromolecular, linear polypeptide consisting of hydrophilic and hydrophobic amino acids. The amphiphilic nature of protein, which is essential for emulsifying agents, originates from the well-balanced distribution of the hydrophobic and hydrophilic regions in the protein molecule. The emulsifying activity of both the protein and the surfactant is mainly based on this amphiphilic nature, and these components

125

contribute to the process of emulsification and emulsion stabilization in different ways.

Table 5.1 summarizes the key factors affecting the emulsification and emulsion stability. Such key factors contribute to the emulsification process and emulsion stability differently depending on the emulsifying agents used for the preparation (Dickinson 1992a; McClements 2005a). Surface activity, that is, the ability to decrease the surface tension or the interfacial tension, is of great importance for both the surfactants and the proteins, but this factor is more crucial for producing fine oil droplets in the case of surfactants. During emulsification, the fine oil droplets that are produced will coalesce again without the adsorption of the emulsifying agents onto the newly exposed surface area, indicating the importance of the rapid and efficient adsorption ability of the emulsifying agents. Since small surfactant molecules can usually adsorb more rapidly and efficiently than the macromolecular proteins, the use of surfactants as emulsifiers normally leads to the production of fine emulsified oil droplets as compared to the use of proteins (McClements 2005b).

The Gibbs–Marangoni effect is a phenomenon in which the surfactant molecules flow toward the region of low surfactant concentration according to the surface tension gradient at the oil–water interface, dragging some of the liquid in the surrounding continuous phase (McClements 2005c). The accumulation of the surfactants and the motion of the aqueous phase liquid prevent the close approach and coalescence of the oil droplets, thereby increasing their stability. This effect is most important for the surfactants that are relatively mobile at the oil–water interface rather than the surface-active biopolymers such as proteins.

Table 5.1 Key Factors Affecting the Emulsification and Emulsion Stability (Comparison of the Cases of Proteins and Low Molecular Weight Surfactants)

Factors	Proteins	Low Molecular Weight Surfactants
Surface activity	2	1
Gibbs–Marangoni effect	4	1
Electrostatic repulsion	1	2 (ionic surfactant: 1)
Steric hindrance	1	3
Hydration of adsorbed layer	2 (conjugate with the polysaccharide: 1)	2
Viscoelastic properties of the adsorbed film	1	4

The importance of each factor in the emulsification and emulsion stability is shown as numbers: 1, generally important; 2, often important; 3, sometimes important; and 4, not important.

Electrostatic repulsion does not play an important role when using surfactants in food emulsions because the surfactants normally used in food emulsions are nonionic (Krog and Sparso 2004). However, one should consider the importance of electrostatic repulsion among the surfactants in the emulsification process, when some organic esters, fatty acids, and lecithins are used in the production of food emulsions. Proteins include amino acid residues with positive and negative charge groups. After adsorption, some of the charged amino acid residues in the adsorbed protein molecule are exposed to the aqueous phase surrounding the oil droplets and can prevent the close contact of the oil droplets via electrostatic repulsion (McClements 2005c). The addition of high concentrations of electrolytes very often destabilizes the emulsion because of the electrostatic screening effects.

The steric hindrance is very important for proteins. When two oil droplets coated with adsorbed protein layers approach each other very closely, the polypeptide chain in the overlapped layer causes a repulsive force between the oil droplets via the mixing (osmotic) effect or the elastic effect (Dickinson 1992a; McClements 2005c). Since the surfactant possesses a small-sized hydrophilic moiety as a polar head group, there is little contribution from the steric hindrance. The hydration of the polar head groups in the surfactants and the hydrophilic amino acid residues should also play important roles in the emulsion stability (Dickinson 1992a; McClements 2005d). Charged groups and the alcoholic OH group have strong affinity to water, causing the structuring of the water molecules around these groups. Hydration interactions arise from the structured water. The greater the degree of hydration, the more repulsive and long range is the interaction. Another biopolymer, polysaccharides, has numerous OH groups and shows strong hydration. Peptide-bound polysaccharides, such as gum arabic (Williams and Phillips 2000) and soluble soybean polysaccharides (Maeda 2000), therefore function as superior emulsifying agents by having both the surface activity originating from the peptide moiety and the hydrated protective layer originating from the polysaccharide chain.

The final key factor is the viscoelastic properties of the protein films adsorbed at the oil–water interfaces. In the long term, these properties are closely relevant to the stability of the food emulsions. That is, the protein film at the oil–water interface with high viscoelastic properties or high mechanical strength is resistant to rupture by the collision or the close contact among oil droplets, thereby preventing the coalescence of the droplets (Dickinson 1992b, 2001). The viscoelastic properties of the adsorbed films are measured by surface rheological techniques. The surface modulus and viscosity are extremely low for the adsorbed film of surfactants compared with the film of proteins, indicating that surfactants cannot form a rigid adsorbed film resistant to the mechanical stress (Murray and Dickinson 1996).

The phenomenon most relevant to the real food systems is the competitive adsorption among the protein molecules or between the proteins and the surfactants. Examples of competitive adsorption among protein molecules at the interface and its relationship to the food systems will be described in Sections 5.2 through 5.4 for each protein. The competitive adsorption between the proteins and the emulsifiers has been studied extensively by several groups (Courthaudon, Dickinson, and Dalgleish 1991; Euston et al. 1995; Fang and Dalgleish 1993; Mackie et al. 2001; Wilde et al. 2004). As described above, surfactants with higher surface activity adsorb the oil–water interface more rapidly than proteins. Therefore, a reduction in the amount of adsorbed proteins can be observed if surfactants are present when homogenizing the protein solutions and oils to make emulsions. The addition of surfactants after the emulsification can also induce a decrease in the amount of adsorbed proteins, indicating the displacement of the proteins at the oil–water interface by the surfactants. It is strange but interesting that although both the proteins and the emulsifiers can stabilize emulsions alone, their individual mechanisms of stabilization are different, often causing dramatic destabilization when both species are present at the oil–water interface. The use of both proteins and surfactants is inevitable in many food emulsions to meet various required applications. For example, bacteriostatic emulsifiers are used with the aim of providing bacteriostatic effects on heat-resistant sporeformers in milk-based emulsion products, such as canned coffee or tea. We have recently found that these surfactants destabilize the emulsions by displacing the milk proteins at the oil–water interface (Matsumiya et al. 2010).

Of the food proteins possessing emulsifying activity, we focus on three proteins in this chapter: milk proteins, egg yolk proteins, and legume proteins. The former two classes of proteins have been chosen because of their large production and wide application (Walstra, Wouters, and Guerts 2006a; Doi and Kitabatake 1997). To understand the features of these proteins, it is helpful to explore their new applications and improve or design the functional properties of other proteins based on the knowledge gained from milk and egg yolk proteins. The latter class of proteins, legume proteins, are chosen as the representative proteins from plants. Although the amino acid balance of the plant proteins is usually inferior to that of the proteins from animals, legume proteins, especially soybean proteins, are well known to have not only a high nutritional value but also excellent physiological and health-promoting effects (Schaafsma 2000; Potter 1995). In Western countries, most soybeans are used for extracting oils, and the components, including soy proteins, in the resultant residues are not fully utilized. The use of soy proteins as emulsifying agents has great advantages from the viewpoints of ecology, food production, and the health promotion of humans.

Another reason for choosing these three proteins is their different structural features. Whereas the major proteins of milk, caseins, are flexible proteins with a disordered secondary structure, the legume proteins are composed of subunit polypeptides with a compact globular conformation. The strong emulsifying activity of egg yolk is attributable to lipoproteins, that is, a complex of protein and lipid. By comparing the emulsification properties of these three proteins with different structural features, we try to show the diverse possibility of food proteins as emulsifying agents. In Section 5.5, we introduce techniques for improving the emulsifying properties of the food proteins. The enzyme-catalyzed hydrolysis of proteins is one of the most popular methods for improving the emulsifying properties. In describing the emulsification properties of the resultant hydrolyzed peptides, we discuss the features that make these peptides good emulsifying agents.

5.2 Milk Proteins

Milk proteins consist of approximately 80% caseins and 20% whey proteins (Walstra, Wouters, and Guerts 2006b). The casein fraction is a composite of four different proteins, αs1-, αs2-, β-, and κ-casein in weight proportions of approximately 4:1:4:1 (Dalgleish 1997). The whey protein fraction contains a variety of proteins, but β-lactoglobulin (about 50% in the whey protein fraction) and α-lactalbumin (25%) are the major components (Cayot and Lorient 1997). Four caseins associate to form a "micelle" in milk. The structure of a casein micelle is described in many books and reviews (Dalgleish 1997; Walstra, Wouters, and Guerts 2006c; Fox 1989). Although the casein micellar structure seems to be retained in some milk products, such as skimmed milk and milk protein concentrates (MPCs), the four caseins dissociate to individual molecules in the more processed products, such as sodium caseinate, which is used as a common ingredient in a wide range of emulsions. Therefore, the studies on the structures and functions of individual caseins are of great importance from both fundamental and practical viewpoints.

Caseins are flexible proteins without a specific secondary structure, and their adsorption behavior is different from that of the globular proteins with a compact rigid structure (Dalgleish 1997; Matsumura 2001). A schematic representation of the adsorbed structure of β-casein is shown in Figure 5.1a. The conformation of the adsorbed β-casein can be explained by the "train-loop-tail" model, which is devised for the synthetic flexible polymers. About three quarters of the β-casein molecule is close to the interface in trains and loops, whereas the rest, consisting of N-terminal 40–50 residues, is in a tail extending into the aqueous phase. The tail length, corresponding to the thickness of the adsorbed layer of β-casein, has been shown to be in the range of 10–15 nm. Whereas the region of trains and loops is rich in hydrophobic amino acid residues, the N-terminal tail region is highly charged (Dalgleish 1997). The high charge at the N-terminal tail is due to the high content of phosphorylserine residue. It is thought that the conformation of αs1- and αs2-caseins is similar to that of β-casein except that they have a long loop instead of a long tail at the interface (Dickinson 2001), because the hydrophilic regions of

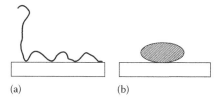

(a)   (b)

Figure 5.1   Schematic representation of the adsorbed configurations of (a) β-casein and (b) β-lactoglobulin. (From Matsumura, Y., *Protein-based Surfactants,* **Marcel Dekker, New York, 2001.)**

the former caseins are located near the center of the molecules. Such an amphiphilic structure of the whole molecule may be the molecular basis for the excellent emulsifying properties of the $\alpha s1$-, $\alpha s2$-, and β-caseins.

Conformations of the α-lactalbumin and β-lactoglobulin at the interface are quite different from those of the caseins. The structure of the β-lactoglobulin adsorbed at the interface is shown in Figure 5.1b (Matsumura 2001). Some experimental evidence suggests that the rigid compact conformation of the native β-lactoglobulin is not unfolded extensively, that is, much of the secondary structure is retained after adsorption. The situation is probably the same for α-lactalbumin. Therefore, it is appropriate to regard an adsorbed monolayer of globular whey protein molecules as a two-dimensional system of interacting deformable particles that are not very different in structure from the native protein molecules in solution (Dickinson 2001).

As described in Section 5.1, with respect to the emulsification process, fine oil droplets are produced by using more surface-active agents. The same tendency was observed when comparing the emulsifying properties of each milk protein. Caseins, especially β-casein, exhibited high surface activity (estimated by the rate of surface tension decrease) and higher emulsifying activity (estimated by the ability to produce fine emulsified oil droplets) than β-lactoglobulin and α-lactalbumin (Kinsella 1984). This result shows that the conformation of the adsorbed β-casein, that is, the high flexibility and an optimum distribution of hydrophilic and hydrophobic regions, is more advantageous for the emulsification process as compared with the compact rigid conformation at the interface of the whey proteins. Generally, the rate of adsorption of the globular proteins to the oil–water interface is slower than that of the flexible caseins. This is partly because of the need to undergo a conformational change and the recent exposure of the hydrophobic amino acid residues from the molecular interior to the oil–water interface (Magdassi and Kamyshny 1996). Nevertheless, whey proteins still have excellent emulsifying properties as compared with other food proteins.

As also described in Section 5.1, the viscoelastic properties of an adsorbed film of proteins are closely related to the resistance of the coalescent emulsified oil droplets, measured by surface rheological techniques. At the hydrocarbon–water interface, the surface shear viscosity of the adsorbed film is 10^3–10^4 times larger for β-lactoglobulin than for β-casein (Murray and Dickinson 1996). The highly viscoelastic character of the adsorbed β-lactoglobulin is attributed to a high two-dimensional packing density and strong protein–protein interactions, as compared with the loose packing and weak protein–protein interactions of the casein monolayers. This result suggests that the oil droplets coated with β-lactoglobulin (and other whey proteins) exhibit more resistance to coalescence.

In oil-in-water emulsions stabilized by milk proteins, some competitive adsorption of the milk proteins at the interface occurs during the emulsification, which may cause differences in the compositions of the protein layers. Dickinson and Stainsby (1988) demonstrated that β-casein was adsorbed in preference to $\alpha s1$-casein in the emulsions stabilized by a mixture of the two caseins. However, others observed no preference for αs1-casein and β-casein in sodium caseinate–stabilized emulsions (Hunt and Dalgleish 1994). Such a contradiction may originate from the concentration of the proteins used for the emulsification. Several reports demonstrated that β-casein was preferentially adsorbed to the oil–water interface from sodium caseinate during homogenization at low protein concentrations. However, at high protein concentrations, the amount of αs-casein adsorbed at the interface was larger than that of β-casein (Euston et al. 1996; Srinivasan, Singh, and Munro 1996, 2000).

In the case of whey proteins, contradictory results have also been reported. Although several papers reported that there was no preferential adsorption between β-lactoglobulin and α-lactalbumin (Euston et al. 1996; Hunt and Dalgleish 1994), Cayot and Lorient (1997) demonstrated that β-lactoglobulin was adsorbed in preference to α-lactalbumin in the emulsions formed with whey proteins or with a mixture of β-lactoglobulin and α-lactalbumin. For the emulsions stabilized by a mixture of caseinate and whey proteins, there was no preferential adsorption between the caseins and the whey proteins at low concentrations, but the caseins were preferentially adsorbed at high concentrations.

Recently, Ye (2008) carried out a systematic approach to the composition of the adsorbed proteins at the oil–water interface in the emulsions made with mixtures of commercial sodium caseinate and whey protein concentrate, especially emphasizing the effects of protein concentration. First, the emulsion was prepared by mixing soya oil with sodium caseinate solution or whey protein solution, individually. Figure 5.2 shows the total surface protein concentration and the surface concentrations of the individual protein components. In Figure 5.2a, the total surface protein concentration increased gradually with an increase in the protein concentration. At caseinate concentrations of less than 2%, β-casein was preferentially adsorbed at the oil droplet surface, but αs-casein was adsorbed most preferentially at concentrations larger than 2%. As described above, β-casein is more hydrophobic than αs-casein. At low concentrations, the hydrophobic nature of β-casein is advantageous for efficient adsorption, but it may lose its competitive adsorption ability because of its self-aggregation tendency at high concentrations. κ-Casein is a glycosylated protein and seems to provide sufficient hydrophilic character to be adsorbed in preference to β-casein at higher protein concentrations. Figure 5.2b shows the results of whey protein concentrate. For the composition of α-lactalbumin and β-lactoglobulin, the relative proportions of α-lactalbumin (about 18%) at the interface were slightly lower and the relative amounts of β-lactoglobulin (about 82%) were slightly higher, than those in the original whey protein concentrate (about 23% and 77%). This result indicates the preference of β-lactoglobulin to α-lactalbumin.

Ye (2008) also investigated the emulsion system produced by homogenizing oil with a mixture of sodium caseinate and whey protein concentrate. Figure 5.3b shows that the total surface protein concentration increased with an increase of the protein concentration in solution. The surface concentration of whey proteins

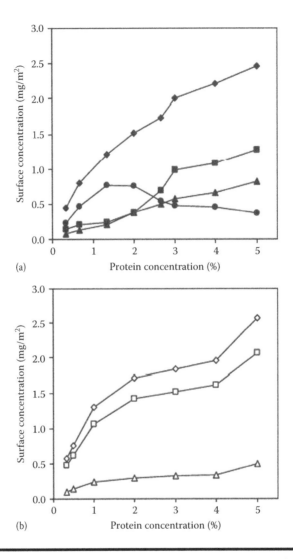

Figure 5.2 Surface protein concentration of αs-casein (■), β-casein (●), κ-casein (▲), β-lactoglobulin (□), and α-lactalbumin (△), in the cream phase of the emulsions made with various concentrations of sodium caseinate (a) and whey protein concentrate (b), in 30% soya oil, pH 7.0. Total surface protein concentrations (◆, ◇). (From Ye, A., *Food Chemistry*, 110, 946–52, 2008.)

was larger than that of sodium caseinate at a lower protein concentration (less than 3%), but higher at 4% and 6%. This tendency is clearly depicted by the proportion of proteins at the surface (Figure 5.3a), in which both curves crossed around 3.5% of the protein concentration. For the whey proteins adsorbed at the oil droplet surface, the proportion of β-lactogobulin was especially high (data not shown).

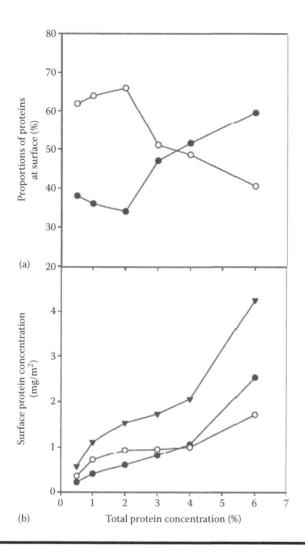

Figure 5.3   Changes in the relative proportions (a) and surface concentrations (b) of the casein (●) and whey proteins (○), as well as the total surface protein concentration (▼) at the droplet surface, in the emulsions made with a binary mixture of sodium caseinate and whey protein concentrate (1:1 by weight), in 30% soya oil, pH 7.0, as a function of the protein concentration in the mixture. (From Ye, A., *Food Chemistry,* **110, 946–52, 2008.)**

β-Lactoglobulin, the small compact globular protein (MW: 18,362), might be adsorbed to the oil droplet surface in preference to the casein molecules, which tend to associate, forming a micelle or aggregate at lower protein concentrations (less than 3%). β-Casein has the potential of superior emulsifying properties over β-lactoglobulin, as described previously in relation to the results of Kinsella (1984).

If one applies experimental conditions in which the casein molecules are well dispersed, one may get results that show that caseins are adsorbed to the oil–water interface in preference to whey proteins.

A sharp increase of surface protein concentration above 4% in Figure 5.3b was attributed to the formation of a secondary layer of adsorbed proteins around the emulsion droplet. In this second adsorbed layer, the caseins predominated because of their tendency to aggregate.

The results of Ye (2008) are very interesting and important, but more experimental evidence is necessary for understanding the competitive adsorption of milk proteins at the oil–water interface because the adsorption behavior changes according to the surrounding conditions for emulsification (i.e., the protein concentration, pH, and addition of ions, especially calcium ions) as well as the treatments on protein samples during the production process (i.e., heat and dry) (Ye 2011). The effects of heat treatment on the aggregation and emulsifying properties of milk proteins will be described in Section 5.5.

5.3 Egg Yolk Proteins

Hen egg yolk is an excellent emulsifying agent and is widely used in many food applications (Doi and Kitabatake 1997; Juneja and Kim 1997). Especially, since egg yolk is able to stabilize emulsions at low pH, it is used in the preparation of mayonnaise. Egg yolk proteins consist of 65% low-density lipoprotein (LDL), 16% high-density lipoprotein (HDL), 10% livetin, and 4% phosvitin (Juneja and Kim 1997). It has generally been accepted that the excellent emulsifying activity of egg yolk is closely related to the LDL (Le Denmat, Anton, and Beaumal 2000; Anton, Martinet, and Dalgalarrondo 2003).

The LDL is a spherical particle, approximately 35 nm in diameter, consisting of a core of triglycerides, cholesterol, and cholesterol esters, surrounded by a monolayer of phospholipids and proteins (Martin, Augustyniak, and Cook 1964; Matsumura 2001). Another lipoprotein, HDL, has a similar structure but is smaller in size. Whereas LDL is composed of 89% lipid and 11% protein, the contents of lipid and protein in HDL are 20% and 80%, respectively. Because of the high lipid content, the density of LDL is very low, 0.98.

The composition of LDL apoproteins analyzed by SDS-PAGE showed nine protein bands in the range of 8–150 kDa (Jolivet et al. 2006). The identification of these proteins was attempted by sequence analysis of their tryptic hydrolysates with liquid chromatography with tandem mass spectrometry (LC-MS/MS) (Jolivet et al. 2006, 2008). The results have demonstrated that the protein of 8 kDa corresponds to the apo-VLDL found in the VLDL of hen blood. This protein is the only apoprotein from blood lipoproteins to be transferred to the yolk in large amounts without any modification and is called apovitellenin I. The protein of 15 kDa was shown to be a homodimer of apovitellenin I linked via a disulfide bond. The other seven protein bands of the LDL (from 55 to 150 kDA) were shown to be peptide fragments resulting from apolipoprotein B (MW: 500 kDa) in blood VLDL by enzymatic processing. The same group further investigated the topology of the apoproteins in egg yolk LDL, that is, the location of the lipoproteins at the surface of LDL (Jolivet et al. 2008). LDL was treated with proteinase K, and the peptides resistant to the enzyme were shown to be embedded in the lipid layer. Such fragments were the peptide of 98 kDa and the peptide of 62 kDa, with both peptides originating from apolipoprotein B. These two domains were characterized with a very basic isoelectric point (pI = 9.14 and 9.32, respectively), and the hydropathy plots show the hydrophobic behavior. These domains could be an anchoring point in egg yolk LDL. Interestingly, these domains are enriched in β-sheets. Oleosin on the surface of oil bodies (organelles for oil storage) from many plant species also has a region rich in β-sheets and is embedded in the lipid layer (Matsumura 2001).

Nilsson et al. (2007) reported that the apolipoproteins of LDL are selectively and strongly adsorbed to the oil–water interface when the whole egg yolk is used for the preparation of oil-in-water emulsions, indicating the dominant role of LDL in the superior emulsification ability of egg yolk. The studies of competitive adsorption between egg yolk lipoproteins and milk proteins showed that LDL predominates at the oil–water interface and substantially displaces whey proteins and caseins (Aluko, Keeratiurai, and Mine 1998; Mine and Keeratiurai 2000). This means that the emulsifying properties of the egg yolk lipoproteins, especially LDL, are so high as to expel even the surface-active milk proteins from the interface.

Although LDL contains other surface-active components, such as phospholipids and cholesterols (Le Denmat, Anton, and Beaumal 2000), the emulsifying activity of LDL seems to be mainly due to the apoproteins rather than the polar lipids (Martinet et al. 2003). However, LDL apoproteins are insoluble in water or in aqueous buffer. Therefore, the complex structure of apoproteins and other lipids is essential for the excellent emulsifying properties of LDL. In other words, in the emulsifying process, this complex structure of LDL enables the transportation of the apoproteins and phospholipids in a soluble form in the vicinity of the interface and then the release of them at the interface (Martinet et al. 2003). In this adsorption process, it is hypothesized that apoproteins serve as the first anchorage of LDL at the interface, and the subsequent structural change of LDL particles leads to the release of the LDL components at the interface. The lipid core of LDL (triacylglycerol and cholesterol ester) may coalesce with the oil phase of the emulsions. However, controversies persist about the adsorption mechanism of LDL, that is, some authors still hypothesize that LDL particles adsorb and stabilize the oil droplet surface without breakage or coalescence at the interface (Mine 1998).

5.4 Legume Proteins (Soy Proteins)

The storage proteins of legumes are divided into two groups, 7S globulin and 11S globulin (Wright 1987). Soybeans, broad beans, and peas contain both types of globulins,

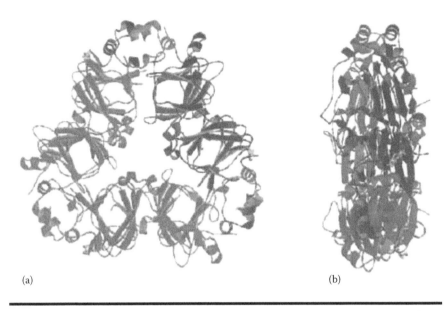

(a)   (b)

Figure 5.4   Three-dimensional structures of β-conglycinin; homotrimer of the β-subunit. The three subunits are shown. (a) Trimer seen along the threefold axis. (b) View after a 90° rotation around the vertical axis. (From Maruyama, N., et al., *European Journal of Biochemistry***, 268, 3595–604, 2001.)**

but the major storage globulin in haricots and red beans is the 7S globulin type. The 7S globulin is a trimer (MW: 150,000–200,000) composed of subunit peptides of molecular weights in the range 50,000–70,000. Figure 5.4 shows the three-dimensional structure of the 7S globulin of soybean, called β-conglycinin (Maruyama et al. 2001). In the case of β-conglycinin from soybean, there are three kinds of subunits, α, α′, and β (Utsumi 1992). Figure 5.4 illustrates a homotrimer of the β-subunit. Other trimers of β-conglycinin are naturally present in soybean, that is, a homotrimer of the α- and α′-subunits and heterotrimers consisting of different subunits (e.g., αα′β, αβ$_2$, α′β$_2$, α$_2$β, α$_2$′β). All the trimers are thought to show a similar three-dimensional structure. One or two carbohydrate moieties are attached to all the subunits.

The molecular weights of α, α′, and β are approximately 67, 71, and 50 kDa, respectively. The α- and α′-subunits have extension regions at the N-terminal (α, 125 residues; α′, 141 residues) and the core region (418 residues) (Utsumi, Matsumura, and Mori 1997; Utsumi 1992), while the β-subunit consists of only the core region (416 residues). Homology of the core region is very high among the three subunits, that is, 90.4%, 76.2%, and 75.5% between α and α′, between α and β, and between α′ and β, respectively. The extension regions of the α- and α′-subunits exhibit 57.3% sequence identity and are rich in acidic amino acids.

In order to test the emulsifying properties of the β-conglycinin subunits, homotrimers of each subunit protein expressed in *Escherichia coli* were prepared and used for the emulsification of soybean oil (Maruyama et al. 1999). The average particle size was calculated from the particle size distribution. At pH 7.6 and $\mu = 0.5$, the α-subunit homotrimer exhibited the best value (4.2 μm) among the three subunits, then α′ (16.4 μm) and β (52.9 μm). In this report, the authors also expressed α- and α′-subunits without extension regions, that is, consisting of only core regions. These truncated subunits exhibited poor values (33.3 and 46.2 μm for α- and α′-subunits, respectively), indicating the importance of extension regions in the emulsification of β-conglycinin.

Emulsification using the expressed β-conglycinin subunits was also carried out under different conditions (at pH 7.6 and $\mu = 0.08$), and the particle size distributions of the resultant emulsions are shown in Figure 5.5 (Maruyama et al.

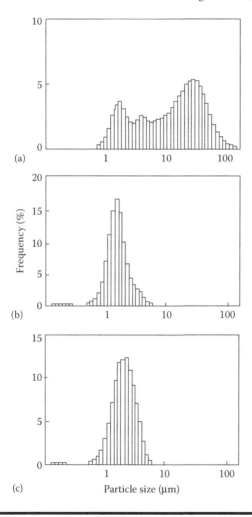

Figure 5.5 Typical particle size distributions of the native β-conglycinin (a), the recombinant α-subunit (b), and α′-subunit (c) at pH 7.6 and m=0.08. (From Maruyama, N., et al., *Journal of Agricultural and Food Chemistry*, 47, 5278–84, 1999.)

1999). Under this condition, β-subunits and truncated α- and α′-subunits could not be subjected to measurement because of their low solubility, and the data are not shown. The average particle sizes of emulsions of the α- and α′-subunits were 1.6 and 2.8 μm, respectively, indicating the production of fine emulsions (Figures 5.5b and c). Figure 5.5a indicates the particle size distribution of an emulsion prepared from the β-conglycinin separated from soybean. This emulsion gave the two major peaks (1.9 and 28.3 μm). Such a wide range of distribution could be due to the heterogeneity of natural β-conglycinin molecules including homotrimers and heterotrimers. In this study, the important role of carbohydrate moiety in the emulsification of β-conglycinin was also pointed out.

Maruyama et al. (2002) purified heterotrimers consisting of α- and β-subunits or α′- and β-subunits from mutant soybean cultivars lacking α′- and α-subunits, respectively. They demonstrated that the emulsifying abilities of the heterotrimers containing one β-subunit were similar to those of the α- and α′-homotrimers, whereas those of the heterotrimers containing two β-subunits were similar to those of the β-homotrimer. These results are very helpful in the evaluation and prediction of the emulsification properties of β-conglycinin from various soybean cultivars based on the data of the subunit composition.

Another major storage protein of soybean is the 11S type globulin, called glycinin. The nutritional value of glycinin is high because of its relatively high content of sulfur amino acids and tryptophans (Utsumi, Matsumura, and Mori 1997). Glycinin exhibits excellent gelling ability (Nakamura, Utsumi, and Mori 1984), which enables the production of soybean curd "Tofu". Like another 11S type globulin from other beans, the glycinin molecule is a hexamer (MW: 300,000–350,000) of subunits (MW: approximately 50,000). Each subunit is composed of an acidic polypeptide and a basic polypeptide, which are linked together by a disulfide bond. As a constituent subunit of glycinin, five kinds of subunit are identified: A1aB1b, A1bB2, A2B1a, A3B4, and A5A4B3. The structure of each subunit is reviewed in detail elsewhere (Utsumi, Matsumura, and Mori 1997; Adachi et al. 2003). It is known that the emulsifying ability of glycinin is inferior to that of β-conglycinin (Utsumi, Matsumura, and Mori 1997). Therefore, the emulsifying properties of glycinin are not discussed, although there are some papers about the relationship between the glycinin subunit structure and the emulsifying properties (Maruyama et al. 2004).

From the practical viewpoints, the emulsification properties of soy protein isolate (SPI) are of great importance. Commercial SPI is composed of approximately 40% glycinin, 20% conglycinin, and 40% lipoproteins (Samoto et al. 2007). It is generally accepted that the emulsifying properties of SPI are inferior to those of milk proteins. For example, coarse emulsions (average diameter: more than 10 μm) were obtained when mixing the SPI solution (protein concentration: 1%) with soybean oil at 1:9 (Roesch and Corredig 2002), although fine emulsions (average diameter: approximately 1 μm) could be produced by sodium caseinate solution under similar composition. More SPI concentrations were necessary to produce fine emulsions (Dalgleish 1997). Moreover, the coalescence stability of the oil droplets emulsified by SPI was inferior to that of the droplets stabilized by sodium caseinate (Palazolo, Sorgentini, and Wagner 2005).

The poor emulsifying ability of SPI could be due to the presence of glycinin. Glycinin is a gigantic protein with a complicated subunit structure described above, which may be a disadvantage for the rapid adsorption of the glycinin molecule to the newly exposed oil droplet surface during emulsification because of the slow diffusion and the slow dissociation to individual subunits. In addition, conformational rearrangement of the amino acid residues of glycinin at the oil–water interface may be prevented by intrapolypeptide and inter-polypeptide disulfide bonds (Utsumi, Matsumura, and Mori 1997) that are lacking in β-conglycinin. The absence of carbohydrate moiety may also be responsible for the inferior emulsifying properties of glycinin. It was reported that the ratio of β-conglycinin to glycinin was a key factor affecting the emulsifying properties of SPI (Yao, Tanteeratarm, and Wei 1990).

Although it had long been believed that the major fractions of SPI are only glycinin and β-conglycinin, Samoto et al. (2007) recently found that approximately 40% of another fraction was included in SPI. This fraction is called lipoprotein because of the coexistence of lipids. Little attention has been paid to lipoproteins because the polypeptides of this fraction are difficult to detect by electrophoresis (difficult to stain with dyes). The details and functional properties, including the emulsifying ability of the lipoproteins fraction, are not clear.

5.5 Techniques for Improving Emulsifying Properties of Proteins

Of the proteins reviewed to this point, the emulsifying properties of the legume (soybean) proteins are inferior to those of the milk proteins or egg yolk proteins. If the emulsifying properties of the legume proteins are improved, the possibility of their application to emulsions will increase. However, with respect to milk proteins and egg yolk proteins having high emulsifying properties, there are problems that need to be resolved for their use in food products. Generally, food proteins suffer several stresses, such as heating and drying, during the production process from materials, thereby affecting the emulsifying and other functional properties of the final food protein products via protein denaturation and the compositional change of other ingredients. For example, Ye (2011) demonstrated that the emulsifying properties of MPCs are closely related to the calcium content of MPCs.

There are two major methods for improving the emulsifying properties of proteins, that is, physical and enzymatic modifications. For the physical modifications, heat treatment is the most common and useful tool. In this section, the effectiveness of heat treatment and enzymatic modification on the emulsifying properties of the food proteins, mainly milk, egg yolk, and soy proteins, is described.

5.5.1 Heat Treatment

Heat treatment has both positive and negative effects on the emulsifying properties of proteins. Heat treatment of sodium caseinate near the pI enhanced its emulsifying ability, which was attributed to the heat-induced exposure of the hydrophobic domains (Jahaniaval et al. 2000). However, heating of whey proteins at 80°C and 90°C, which are high enough to denature most whey proteins, induced a decrease in their emulsifying ability (Millqvist-Fureby, Elofsoon, and Bergenstal 2001). After denaturation, whey protein molecules form large aggregates that are unable to efficiently cover the oil droplet surfaces, leading to a decrease in their emulsifying ability. Contrasting results about the heat treatment on sodium caseinate and whey proteins with respect to their emulsifying properties can be attributed to the interaction of denatured molecules. It is well known that whey proteins have an excellent gelling ability when heated (Zayes 1997), indicating the efficient intermolecular interaction between heat-denatured molecules to form the gel network structure, whereas heat treatment normally cannot induce the gelation of caseins. Instead of a rigid gel, only coagulants can be formed from caseins by treatment with rennet or acidification (Dalgleish 1997). The strong interaction of the denatured molecules by heating may be beneficial to gelation but disadvantageous for the emulsification.

The emulsion stabilizing effect of caseins under heating conditions was demonstrated in the case of a protein mixture system. When oil droplets emulsified by whey proteins were heated at 80°C or 90°C, the oil droplets flocculated dramatically. However, the replacement of just 5% of the whey proteins with caseinate in this emulsion almost completely suppressed the flocculation of the oil droplets (Dickinson and Parkinson 2004). Such a protective effect of caseins may be due to the steric hindrance by the long tangling tails of the casein molecules adsorbed at the oil droplet surfaces with the whey protein molecules.

The impact of heat treatment on the emulsifying properties of soy proteins has also been investigated. The emulsifying properties of SPI were improved by heating (Maneephan and Corredig 2009). However, such an improvement was not observed for the two major proteins, glycinin and β-conglycinin, when each protein was separated and tested individually. Whereas no significant change in the emulsifying properties was observed for β-conglycinin, heat treatment induced a decrease in the emulsifying activity of glycinin (Aoki 1980). The reason why the improvement of the emulsifying properties is observed only for SPI may be due to the interaction of the glycinin and β-conglycinin subunits during heating. When glycinin is subjected to heating above the denaturation temperature, the molecule becomes unfolded, leading to the exposure of the hydrophobic regions to the molecular surface. As described in Figures 5.4 and 5.5, glycinin is composed of six subunits in which the acidic and basic polypeptides are linked via disulfide bonds. Basic polypeptides are very hydrophobic and embedded within the molecule in the native conformation but are exposed to the molecular surface (Nakamura, Utsumi, and

Mori 1984). These hydrophobic regions of the basic subunits induce the aggregation of glycinin molecules, decreasing the emulsifying activity. According to the environmental conditions, basic polypeptides are also liberated from glycinin molecules via sulfhydryl–disulfide (SH–SS) interchange reactions and precipitate because of excess aggregation. However, it is known that the β-subunit of β-conglycinin associates with the basic polypeptides, when glycinin is heated in the presence of β-conglycinin (Utsumi, Damodaran, and Kinsella 1984). As a result, the basic polypeptides attached by the β-subunits of β-conglycinin can remain soluble or avoid excess aggregation, whereas having the proper hydrophobicity is beneficial to the emulsifying properties. Such a complicated process of subunit interaction between glycinin and β-conglycinin is believed to be involved in the improvement of the emulsifying properties of SPI by heating.

5.5.2 Enzymatic Modification

Of the various enzymatic modifications potentially applied to improve the functional properties of proteins, protease-catalyzed hydrolysis is the most widely used (Nielsen and Olsen 2002). Enzymatic hydrolysis of proteins produces peptides with smaller molecular sizes and less secondary structure than the original proteins. These peptides are expected to have increased solubility over a wide pH range (especially the isoelectric point), decreased viscosity, and significant changes in their gelling, foaming, and emulsifying properties (Schwenke 1997). In particular, the large molecular weight and rigid conformation of proteins are disadvantageous for the rapid diffusion and the following reorientation at the oil–water interface as compared with low molecular weight emulsifiers. Therefore, many trials for the improvement of the emulsifying properties of proteins by protease treatments have been reported. A summary of the important features of peptides with respect to their emulsifying properties was presented, especially focusing on the peptides produced by protease treatments mainly from milk proteins.

Generally, the hydrolysis of proteins with a limited degree of hydrolysis improved their emulsifying properties, while excess hydrolysis led to the decrease or loss of their emulsifying properties, indicating the importance of length or molecular weight on emulsification (Schwenke 1997). By digesting milk whey proteins with trypsin, Chobert, Bertrand-Harb, and Nicolas (1988) investigated the relationship between the molecular weight and the emulsifying properties of the generated peptides. Good emulsifying properties were observed for the peptides of molecular weights greater than 5000 Da. However, Singh and Dalgleish (1998) reported that a peptide length of only 500 Da is needed for emulsion stabilization. Rahali et al. (2000) prepared oil-in-water emulsions using the chemical and enzymatic hydrolysates of β-lactoglobulin and identified peptides adsorbed on the oil droplet surface. They found that the length of the peptides was not a limiting factor for anchorage at the oil–water interface. In order to understand the relationship of the emulsifying properties and the molecular sizes of peptides more clearly, van der Ven et al. (2001) hydrolyzed caseins and whey proteins using 11 different commercially available enzymes and investigated the emulsion-forming ability (emulsifying activity) and the emulsion stability of the digests under the same conditions. It was shown that emulsion-forming behavior is generally independent of the molecular weight distribution and the degree of hydrolysis. However, coalescence of the emulsion droplets (an index of the long-term stability of the emulsions) was correlated to the molecular weight distribution of the hydrolysates. That is, the hydrolysates with a high proportion of peptides with a molecular weight greater than 2000 Da formed stable emulsions toward coalescence.

In addition to the chain length and molecular weight, hydrophobicity and amphiphilicity of peptides are important for the interfacial and emulsifying properties of peptides. Turgeon, Gauthier, and Paquin (1992) showed that the hydrophobic peptides liberated from whey protein concentrate by tryptic digestion had better emulsifying properties than the chymotryptic-digested peptides in which the hydrophobic amino acid sequences might be cleaved by the enzyme. As described above, Rahali et al. (2000) analyzed the amino acid sequence of β-lactoglobulin hydrolysate at the oil–water interface. They pointed out that the amphiphilic character, that is, the hydrophilic or hydrophobic distribution of the amino acids in the sequence of the peptide fragments, was more relevant to the adsorption of the peptides than the peptide length. A computer program was applied to the known peptide sequences for the purpose of providing a reasonable explanation to the underlying mechanism of their emulsification (Nakai et al. 2004). It was shown that the hydrophobic periodicity, defined as the hydrophobic similarity density in the polar or apolar cycles within the sample sequences, plays the most important role in peptide emulsification. Hydrophobicity distribution in a peptide sequence thus seems to be the crucial criterion for forming stable oil-in-water emulsions. The importance of the amphiphilic structure of peptides for emulsification was discussed by us elsewhere (Matsumura and Kito 2001).

The combination of protease hydrolysis and physical means is sometimes useful. For example, peptide fractions with excellent emulsifying properties were produced from whey protein concentrate by the combined effects of (1) heat treatment under acidic conditions, (2) enzymatic hydrolysis, and (3) ultrafiltration separation of the hydrolysates (Turgeon, Gauthier, and Paquin 1992).

Recently, a novel enzyme, called protein-glutaminase, was purified from *Chryseobacterium proteolyticum* (Yamaguchi, Jeenes, and Archer 2001). This enzyme catalyzes the deamidation of glutamine residues in short peptides or proteins. Unlike peptide-glutaminase, which catalyzes the deamidation of glutamine in short peptide chains (Hamada 1992), protein-glutaminase prefers proteins to short peptides as substrates. It has also been shown that protein-glutaminase is different from transglutaminase and has no ability to catalyze the formation of ε-(γ-glutamyl) lysine isopeptide bonds and the incorporation of amines to glutamine residues in proteins.

Cereal storage proteins, such as wheat gluten and maize zein, are classified as prolamins or glutelins and are insoluble in aqueous solutions (Chung and Pomeranz 2000). Since solubility is closely related to other functional properties of proteins, such as foaming, emulsification, and gelling ability, the insolubility or low solubility of the cereal proteins is responsible for their limited applications in various types of foods. Cereal storage proteins are characterized by their high content of glutamine residues, which can be target sites of protein-glutaminase (Belitz, Grosch, and Schieverle 2004). Therefore, the effects of protein-glutaminase–catalyzed deamidation on the functional properties of zein and gluten were investigated (Yong et al. 2004; Yong, Yamaguchi, and Matsumura 2006). Figure 5.6a shows the appearance of gluten dispersions at neutral pH with and without protein-glutaminase treatment. It is obvious that protein-glutaminase treatment dramatically improves the dispersion state of gluten. The solubility of gluten increased from 20% to 70% when only 20% glutamine residues were deamidated by protein-glutaminase (data not shown). The untreated gluten has a low ability to emulsify corn oil, as shown in Figure 5.6b. After one day, the cream layer was clearly separated with some oil-off. However, a stable emulsion was observed for not just 1 day but 8 days of storage when the deamidated gluten was used for emulsification.

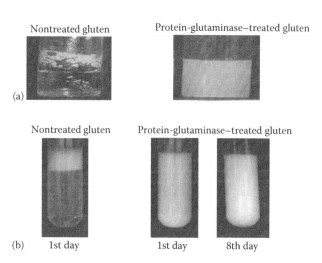

Figure 5.6 Appearance of gluten dispersions (a) and emulsions stabilized by gluten (b). Gluten (10 mg/mL in phosphate buffer) was treated with protein-glutaminase (0.13 U/mL) at 40°C for 30 h. (a) Nontreated and enzyme-treated glutens (1 mg) were dispersed in 1 mL phosphate buffer. (b) Gluten dispersion was mixed with corn oil and homogenized. The final oil and protein concentrations were 10% and 0.09% (W/W), respectively. The appearance of emulsions was observed on the first and eighth days of storage at room temperature. (From Yong, Y. H., Yamaguchi, S., and Matsumura, Y., *Journal of Agricultural and Food Chemistry*, 54, 6034–40, 2006.)

144 ■ *Food Proteins and Peptides*

The action of protein-glutaminase has been investigated for various proteins, including milk proteins (Yamaguchi, Jeenes, and Archer 2001; Gu et al. 2001). Globular whey proteins as well as caseins were shown to be good substrates for this enzyme. Deamidation reaction is believed to improve the interfacial and emulsifying properties of proteins by introducing negative charges, thereby loosening the rigid native conformation of the protein molecules. Therefore, protein-glutaminase–catalyzed deamidation could be a new promising tool for producing proteinous emulsifying agents from various protein resources, such as milk, soybean, and egg.

With respect to the improvement of the emulsifying properties of lipoproteins in egg yolk, the enzymes targeting phospholipids rather than the enzymes targeting protein molecules are useful. Industrial egg yolk has to be pasteurized to ensure microbiological safety. Nowadays, phospholipase A_2 is used in the egg industry to improve the heat stability of egg yolk and enhance its functionality as an emulsifying agent, for example, producing a stable emulsion toward freezing and thawing stresses. Mine (1997) reported the structural and functional changes of LDL in egg yolk by phospholipase A_2 (Dutilh and Groger 1981). The mechanism whereby phospholipase A_2 treatment improves the emulsifying properties of LDL and egg yolk remains unclear. Daimer and Kulozik (2008) demonstrated that the product of this enzymatic reaction, lisophospholipids, plays an important role in the improvement of the emulsifying properties of egg yolk. However, they also pointed out that the solubility and denaturation behavior of proteins were indirectly affected by the phospholipase A_2 reaction. Further studies are necessary for understanding the mechanism of the improved emulsifying properties of egg yolk by phospholipase A_2 reaction from the point of view of the interaction of the apoprotein molecules and lipid fractions in lipoproteins.

5.6 Conclusion

Proteins and low molecular weight surfactants are the major components included in foods as emulsifying agents. As described in Section 5.1, proteins and low molecular weight surfactants contribute to the emulsification process and emulsion stability in different ways, although both components show surface activity. Despite the fact that proteins and low molecular weight surfactants are used together most often in food production, a decrease in the use of low molecular weight surfactants is preferred by people because they are synthetic additives rather than natural compounds (proteins). Since proteins and peptides, which are derived from proteins by proteolysis, show a great variety, it is possible that they satisfy the diverse demands for better-quality emulsion food products. However, the emulsifying properties of many proteins and peptides are not fully understood. In this chapter, the structural features and emulsifying properties of milk proteins and egg yolk proteins have been described. Although a lot of fundamental and practical data on milk and egg yolk proteins are available, systematic approaches to the behavior of the proteins in more real food model systems are necessary. Similar to the results of Ye (2008, 2011) and Dickinson and Parkinson (2004), which are introduced in Sections 5.2 and 5.5, respectively, competitive adsorption or interaction among the protein components at the oil–water interface in emulsions should be further investigated using commercial protein products or protein mixture systems simulating real foods. In addition to the popular food proteins, such as milk and egg yolk proteins, the development of new protein sources for the use of emulsifying agents is required from various points of view, such as the economy, ecology, and health promotion. Plant proteins, particularly legume storage proteins, have a possibility to satisfy such requirements. In this chapter, the structure of and the relationship between the structure and the emulsifying properties of soybean 7S globulin, β-conglycinin, were reviewed. Further studies should be carried out to understand the relationship between the structure and the emulsifying properties of storage proteins from other legumes as well as soybeans.

References

Adachi, M., Kanamori, J., Masuda, T. et al. 2003. Crystal structure of soybean 11S globulin: Glycinin A3B4 homohexamer. *Proceedings of the National Academy of Science USA* 100: 7395–400.

Aluko, R. E., Keeratiurai, M., and Mine, Y. 1998. Competitive adsorption between egg yolk lipoprotein and whey protein on oil-in-water interfaces. *Colloids and Surfaces B: Biointerfaces* 10: 385–93.

Anton, M., Martinet, V., Dalgalarrondo, M. et al. 2003. Chemical and structural characterization of low-density lipoproteins purified from hen egg yolk. *Food Chemistry* 83: 175–83.

Aoki, H. 1980. Emulsifying properties of soy protein: Characteristics of 7S and 11S proteins. *Journal of Food Science* 45: 534–38.

Belitz, H.-D., Grosch, W., and Schieverle, P. 2004. Cereals and cereal products. In *Food Chemistry*, 3rd rev. ed., pp. 673–746. Berlin: Springer-Verlag.

Cayot, P. and Lorient, D. 1997. Structure–function relationships of whey proteins. In *Food Proteins and Their Applications*, eds. S. Damodaran and A. Paraf, pp. 225–56. New York: Marcel Dekker.

Chobert, J. M., Bertrand-Harb, C., and Nicolas, M. G. 1988. Solubility and emulsifying properties of caseins and whey proteins modified enzymatically by trypsin. *Journal of Agricultural and Food Chemistry* 36: 883–92.

Chung, O. K. and Pomeranz, Y. 2000. Cereal processing. In *Food Proteins: Processing Applications*, eds. S. Nakai and H. W. Modler, pp. 243–307. Toronto: Wiley-VCH.

Courthaudon, J. L., Dickinson, E., and Dalgleish, D. G. 1991. Competitive adsorption of β-casein and nonionic surfactants in oil-in-water emulsions. *Journal of Colloid and Interface Science* 145: 390–95.

Daimer, K. and Kulozik, U. 2008. Impact of a treatment with phospholipase A₂ on the physiocochemical properties of hen egg yolk. *Colloids and Surfaces B: Biointerfaces* 56: 4172–80.

Dalgleish, D. G. 1997. Structure–function relationships of caseins. In *Food Proteins and Their Applications*, eds. S. Damodaran and A. Paraf, pp. 199–223. New York: Marcel Dekker.

Dickinson, E. 1992a. Emulsions. In *An Introduction to Food Colloids*, pp. 79–122. Oxford University Press.

Dickinson, E. 1992b. Rheology. In *An Introduction to Food Colloids*, pp. 51–78. Oxford University Press.

Dickinson, E. 2001. Milk protein interfacial layers and the relationships to emulsion stability and rheology. *Colloids and Surfaces B: Biointerfaces* 20: 197–210.

Dickinson, E. and Parkinson, E. L. 2004. Heat-induced aggregation of milk protein-stabilized emulsions: Sensitivity to processing and composition. *International Dairy Journal* 14: 635–45.

Dickinson, E. and Stainsby, G. 1988. Emulsion stability. In *Advances in Food Emulsions and Foams*, eds. E. Dickinson and G. Stainsby, pp. 1–44. London: Elsevier Applied Science.

Doi, E. and Kitabatake, N. 1997. Structure and functionality of egg proteins. In *Food Proteins and Their Applications*, eds. S. Damodaran and A. Paraf, pp. 325–40. New York: Marcel Dekker.

Dutilh, C. E. and Goger, W. 1981. Improvement of product attributes of mayonnaise by enzymatic hydrolysis of egg yolk with phospholipase A_2. *Journal of Agricultural and Food Chemistry* 32: 451–58.

Euston, S. E., Singh, H., Munro, P. A., and Dalgleish, D. G. 1995. Competitive adsorption between sodium caseinate and oil-soluble and water-soluble surfactants in oil-in-water emulsions. *Journal of Food Science* 60: 1124–31.

Euston, S. E., Singh, H., Munro, P. A., and Dalgleish, D. G. 1996. Oil-in-water emulsions stabilized by sodium caseinate or whey protein isolate as influenced by glycerol mono-stearate. *Journal of Food Science* 61: 916–20.

Fang, Y. and Dalgleish, D. G. 1993. Casein adsorption on the surface of oil-in-water emulsions modified by lecithin. *Colloids and Surfaces B: Biointerfaces* 1: 357–64.

Fox, P. F. 1989. The milk protein systems. In *Developments in Dairy Chemistry-4*, ed. P. F. Fox, pp. 1–53. London: Elsevier.

Gu, Y., Matsumura, Y., Yamaguchi, S., and Mori, T. 2001. Action of protein-glutaminase on α-lactalbumin in the native and molten globule states. *Journal of Agricultural and Food Chemistry* 49: 5999–6005.

Hamada, J. S. 1992. Modification of food proteins by enzymatic methods. In *Biochemistry of Food Proteins*, ed. B. J. F. Hudson, pp. 249–70. London: Elsevier Applied Science.

Hunt, J. A. and Dalgleish, D. G. 1994. Adsorption behavior of whey protein isolate and caseinate in soya oil-water emulsions. *Food Hydrocolloids* 8: 175–87.

Jahaniaval, F., Kakuda, Y., Abraham, V., and Marcone, M. F. 2000. Soluble protein fractions from pH and heat treated sodium caseinate: Physicochemical and functional properties. *Food Research International* 33: 637–47.

Jolivet, P., Boulard, C., Beaumal, V., Chardot, T., and Anton, M. 2006. Protein components of low-density lipoproteins purified from hen egg yolk. *Journal of Agricultural and Food Chemistry* 54: 4424–29.

Jolivet, P., Boulard, C., Chardot, T., and Anton, M. 2008. New insights into the structure of apolipoprotein B from low-density lipoprotein and identification of a novel YGP-like protein in hen egg yolk. *Journal of Agricultural and Food Chemistry* 56: 5871–79.

Juneja, L. R. and Kim, M. 1997. Egg yolk protein. In *Hen Eggs: Their Basic and Applied Science*, eds. T. Yamamoto, L. R. Jeneja, H. Hatta, and M. Kim, pp. 57–71. Boca Raton: CRC Press.

Kinsella, J. E. 1984. Milk proteins: Physicochemical and functional properties. *CRC Critical Review of Food Science and Nutrition* 21: 197–262.

Krog, N. J. and Sparso, F. V. 2004. Food emulsifiers: Their chemical and physiological properties. In *Food Emulsions*, eds. S. E. Friberg, K. Larsoon, and J. Sjoblom, pp. 45–91. New York: Marcel Dekker.

Le Denmat, M., Anton, M., and Beaumal, V. 2000. Characterization of emulsion properties and of interface composition in oil-in-water emulsions prepared with hen egg yolk, plasma and granules. *Food Hydrocolloids* 14: 539–49.

Mackie, A. R., Gunning, A. P., Ridout, M. J., Wilde, P. J., and Morris, V. J. 2001. Orogenic displacement in mixed β-lactoglobulin/β-casein films at the air/water interface. *Langmuir* 17: 6593–98.

Maeda, H. 2000. Soluble soybean polysaccharide. In *Handbook of Hydrocolloids*, eds. G. O. Phillips and P. A. Williams, pp. 309–20. Cambridge: CRC Press.

Magdassi, S. and Kamyshny, A. 1996. Introduction: Surface activity and functional properties of proteins. In *Surface Activity of Proteins*, ed. S. Magdassi, pp. 1–38. New York: Marcel Dekker.

Maneephan, K. and Corredig, M. 2009. Heat-induced changes in oil-in-water emulsions stabilized with soy protein isolate. *Food Hydrocolloids* 23: 2141–48.

Martin, W. G., Augustyniak, J., and Cook, W. H. 1964. Fractionation and characterization of the low-density lipoprotein of hen's egg yolk. *Biocimica et Biophysica Acta* 84: 714–20.

Martinet, V., Saulnier, P., Beaumal, V., Courthaudon, J.-L., and Anton, M. 2003. Surface properties of hen egg yolk low-density lipoproteins spread at the air-water interface. *Colloids and Surfaces B: Biointerfaces* 31: 185–94.

Maruyama, N., Adachi, M., Takahashi, K., et al. 2001. Crystal structures of recombinant and native soybean β-conglycinin β homotrimers. *European Journal of Biochemistry* 268: 3595–604.

Maruyama, N., Prak, K., Motoyama, S., et al. 2004. Structure-physicochemical function relationships of soybean glycinin at subunit level assessed by using mutant lines. *Journal of Agricultural and Food Chemistry* 52: 8197–201.

Maruyama, N., Salleh, M. R. H., Takahashi, K., et al. 2002. Structure-physicochemical function relationships of soybean β-conglycinin heterotrimers. *Journal of Agricultural and Food Chemistry* 50: 4323–26.

Maruyama, N., Sato, R., Wada, Y., et al. 1999. Structure–physicochemical function relationships of soybean β-conglycinin constituent subunits. *Journal of Agricultural and Food Chemistry* 47: 5278–84.

Matsumiya, K., Takahashi, W., Inoue, T., and Matsumura, Y. 2010. Effects of bacteriostatic emulsifiers on stability of milk-based emulsions. *Journal of Food Engineering* 96: 185–91.

Matsumura, Y. 2001. Protein interaction at interfaces. In *Protein-based Surfactants*, eds. I. A. Nnanna and J. Xia, pp. 45–74. New York: Marcel Dekker.

Matsumura, Y. and Kito, M. 2001. Enzyme-catalyzed synthesis of protein-based surfactants. In *Protein-based Surfactants*, eds. I. A. Nnanna and J. Xia, pp. 123–46. New York: Marcel Dekker.

McClements, D. J. 2005a. Interfacial properties and their characterization. In *Food Emulsions*, 2nd ed., pp. 175–231. New York: CRC Press.

McClements, D. J. 2005b. Emulsion formation. In *Food Emulsions*, 2nd ed., pp. 233–68. New York: CRC Press.

McClements, D. J. 2005c. Colloid interactions. In *Food Emulsions*, 2nd ed., pp. 53–93. New York: CRC Press.

McClements, D. J. 2005d. Emulsion ingredients. In *Food Emulsions*, 2nd ed., pp. 95–174. New York: CRC Press.

Millqvist-Fureby, A. M., Elofsoon, U., and Bergenstal, B. 2001. Surface composition of spray-dried milk protein-stabilized emulsions in relation to pre-heat treatment of proteins. *Colloids and Surfaces B: Biointerfaces* 21: 47–58.

Mine, Y. 1997. Structural and functional changes of hen's egg yolk low density lipoproteins with phopholipase A_2. *Journal of Agricultural and Food Chemistry* 45: 4558–63.

Mine, Y. 1998. Adsorption behavior of egg yolk low-density lipoproteins in oil-in-water emulsions. *Journal of Agricultural and Food Chemistry* 46: 36–41.

Mine, Y. and Keeratiurai, M. 2000. Selective displacement of caseinate proteins by hens egg yolk lipoproteins at oil-in-water interfaces. *Colloids and Surfaces B: Biointerfaces* 18: 1–11.

Murray, B. S. and Dickinson, E. 1996. Interfacial rheology and the dynamic properties of adsorbed films of food proteins and surfactants. *Food Science and Technology International (Japan)* 2: 131–45.

Nakai, S., Alizadeh-Pasdar, N., Dou, J., et al. 2004. Pattern similarity analysis of amino acid sequences for peptide emulsification. *Journal of Agricultural and Food Chemistry* 52: 927–34.

Nakamura, T., Utsumi, S., and Mori, T. 1984. Network structure formation in thermally induced gelation of glycinin. *Journal of Agricultural and Food Chemistry* 32: 349–52.

Nielsen, P. M. and Olsen, H. S. 2002. Enzymic modification of food proteins. In *Enzymes in Food Technology*, eds. R. J. Whitehurst and B. A. Law, pp. 109–43. Sheffield: Sheffield Academic Press.

Nilsson, L., Osmark, P., Fernandez, C., and Bergenstahl, B. 2007. Competitive adsorption of protein from total hen egg yolk during emulsification. *Journal of Agricultural and Food Chemistry* 55: 6746–53.

Palazolo, G. G., Sorgentini, D. A., and Wagner, J. R. 2005. Coalescence and flocculation in O/W emulsions of native and denatured whey soy proteins in comparison with soy protein isolates. *Food Hydrocolloids* 19: 595–604.

Potter, S. M. 1995. Overview of proposed mechanism for the hypocholesterolemic effect of soy. *Journal of Nutrition* 125: 606S–11S.

Rahali, V., Chobert, J. M., Haertle, T., and Gueguen, J. 2000. Emulsification of chemical and enzymatic hydrolysates of β-lactoglobulin: Characterization of the peptide adsorbed at the interface. *Nahrung* 44: 89–95.

Roesch, R. R. and Corredig, M. 2002. Characterization of oil-in-water emulsions prepared with commercial soy protein concentrate. *Journal of Food Science* 67: 2837–42.

Samoto, M., Maebuchi, M., Miyazaki, C., et al. 2007. Abundant proteins associated with lecithin in soy protein isolate. *Food Chemistry* 102: 317–22.

Schaafsma, G. 2000. The protein digestibility-corrected amino acid score. *Journal of Nutrition* 130: 1865S–67S.

Schwenke, K. D. 1997. Enzymes and chemical modification of protein. In *Food Proteins and Their Applications*, eds. S. Damodaran and A. Paraf, 393–423. New York: Marcel Dekker.

Singh, A. M. and Dalgleish, D. G. 1998. The emulsifying properties of hydrolyzates of whey proteins. *Journal of Dairy Science* 81: 918–24.

Srinivasan, M., Singh, H., and Munro, P. A. 1996. Sodium caseinate-stabilized emulsions: Factors affecting coverage and composition of surface protein. *Journal of Agricultural and Food Chemistry* 44: 3807–11.

Srinivasan, M., Singh, H., and Munro, P. A. 2000. The effect of sodium chloride on the formation and stability of sodium caseinate emulsions. *Food Hydrocolloids* 14: 497–507.

Turgeon, S. L., Gauthier, S. F., and Paquin, P. 1992. Emulsifying property of whey peptide fractions as a function of pH and ionic strength. *Journal of Food Science* 57: 601–604.

Utsumi, S. 1992. Plant food protein engineering. In *Advances in Food Nutrition Research 36*, ed. J. E. Kinsella, pp. 89–208. San Diego: Academic Press.

Utsumi, S., Damodaran, S., and Kinsella, J. E. 1984. Heat-induced interactions between soybean proteins: Preferential association of 11S basic subunits and β subunit of 7S. *Journal of Agricultural and Food Chemistry* 32: 1406–12.

Utsumi, S., Matsumura, Y., and Mori, T. 1997. Structure-function relationships of soy proteins. In *Food Proteins and Their Applications*, eds. S. Damodaran. and A. Paraf, 257–91. New York: Marcel Dekker.

van der Ven, C., Gruppen, H., de Bont, D. B. A., and Voragen, A. G. 2001. Emulsion properties of casein and whey protein hydrolysates and the relation with other hydrolysate characteristics. *Journal of Agricultural and Food Chemistry* 49: 5005–12.

Walstra, P., Wouters, J. T. M., and Guerts, T. T. 2006a. Protein preparations. In *Dairy Science and Technology*, pp. 537–50. Boca Raton: Tailor & Francis.

Walstra, P., Wouters, J. T. M., and Guerts, T. T. 2006b. Milk components. In *Dairy Science and Technology*, pp. 17–108. Boca Raton: Tailor & Francis.

Walstra, P., Wouters, J. T. M., and Guerts, T. T. 2006c. Colloidal particles of milk. In *Dairy Science and Technology*, pp. 109–57. Boca Raton: Tailor & Francis.

Wilde, P., Mackie, A., Husband, F., Gunning, P., and Morris, V. 2004. Proteins and emulsifiers at liquid interfaces. *Advances in Colloid and Interface Science* 108–109: 63–71.

Williams, P. A. and Phillips, G. O. 2000. Gum Arabic. In *Handbook of Hydrocolloids*, eds. G. O. Phillips and P. A. Williams, pp. 155–168. Cambridge: CRC Press.

Wright, D. J. 1987. The seed globulins. In *Developments in Food Proteins-5*, ed. B. J. F. Hudson, pp. 81–157. London: Elsevier Applied Science.

Yamaguchi, S., Jeenes, D. J., and Archer, D. B. 2001. Protein-glutaminase from *Chryseobacterium proteolyticum*, an enzyme that demidates glutamyl residues in proteins. Purification, characterization and gene cloning. *European Journal of Biochemistry* 268: 1410–21.

Yao, J. J., Tanteeratarm, K., and Wei, L. S. 1990. Effects of maturation and storage solubility and gelation properties of isolated soy proteins. *Journal of the American Oil Chemists' Society* 67: 974–79.

Ye, A. 2008. Interfacial composition and stability of emulsions made with mixtures of commercial sodium caseinate and whey protein concentrate. *Food Chemistry* 110: 946–52.

Ye, A. 2011. Functional properties of milk protein concentrates: Emulsifying properties, adsorption and stability of emulsions. *International Dairy Journal* 21: 14–20.

Yong, Y. H., Yamaguchi, S., Gu, Y. S., Mori, T., and Matsumura, Y. 2004. Effects of enzymatic deamidation by protein-glutaminase on structure and functional properties of α-zein. *Journal of Agricultural and Food Chemistry* 52: 7094–7100.

Yong, Y. H., Yamaguchi, S., and Matsumura, Y. 2006. Effects of enzymatic deamidation by protein-glutaminase on structure and functional properties of wheat gluten. *Journal of Agricultural and Food Chemistry* 54: 6034–40.

Zayes, J. F. 1997. Gelling properties of proteins. In *Functionality of Proteins in Food*, pp. 310–66. New York: Springer.

Chapter 6

Proteins and Peptides as Foaming Agents

Arvind Kannan, Navam Hettiarachchy, and Maurice Marshall

Contents

6.1 Introduction

The most general definition of foam is a substance that is formed by trapping many gas bubbles in a liquid or a solid. Foams are normally an extremely complex system consisting of polydispersed gas bubbles separated by draining films (Lucassen 1981). Fine foam can be considered a type of colloid.

The foam makes a network of interconnected films called lamellae. Ideally, the lamellae are connected and radiate at 120° from the connection points, known as the Plateau borders. Several conditions are needed to produce foam: there must be mechanical work, surface-active components that reduce the surface tension, and the formation of foam must be faster than its breakdown.

To create foam, work (W) is needed to increase the surface area (ΔA):

$$W = \gamma \Delta A,$$

where γ is the surface tension.

The stabilization of foam is caused by the van der Waals forces between the molecules in the foam, the electrical double layers created by the dipolar surfactants, and the Marangoni effect, which acts as a restoring force to the lamellae. Several destabilizing effects can break the foam down. Gravitation causes drainage of liquid to the foam base; osmotic pressure causes drainage from the lamellae to the Plateau borders due to the internal concentration differences in the foam; and the Laplace pressure (the pressure difference between the inside and the outside of a bubble or a droplet) causes diffusion of the gas from small bubbles to large bubbles.

6.1.1 Formation of Protein Foams

In order to develop or produce foam, it is imperative to develop the protein film that surrounds a gas bubble. This foam should be able to build up into its overall structure by packing more gas bubbles. The spontaneous adsorption of the proteins from the solution to the air–aqueous interface is of central importance to their foaming performance. This phenomenon is thermodynamically favorable due to the simultaneous dehydration of the hydrophobic interfaces and the hydrophobic portions of the proteins (Dickinson 1986).

It has been observed that between the interfacial protein films and the foams, there is a lack of foam stability. For example, foams made with whey or egg white proteins tend to deform or undergo structural changes within minutes of forming, than model interfacial systems. Therefore, it has become challenging to observe or monitor the properties of foams that undergo spontaneous changes involving thermodynamic considerations.

The extent to which a protein can foam depends on the protein solution being able to accept the aeration processes (Wilde and Clark 1996). In general, under experimental conditions, the aeration process is usually whipping or sparging. The typical overrun values given for beaten egg white and meringue range from 500% to 800% (Campbell and Mougeot 1999). Whey and egg white protein foams have been shown to vary experimentally in overrun values from 500% to 1700%, depending on such factors as the pH, cosolutes, and protein concentration (Davis and Foegeding 2004; Davis, Foegeding, and Hansen 2004; Halling 1981; Luck, Bray, and Foegeding 2002; Pernell et al. 2002a; Phillips, Schulman, and Kinsella 1990). It has been observed that of the several proteins examined for their foaming capacities, the whey and egg white proteins show similar abilities to incorporate air into a foam structure.

The stability of an aerated system, as in foam, is influenced by three basic mechanisms: gravitational drainage, coalescence, and disproportionation. Creaming or vertical phase separation due to the upward movement of large bubbles is usually observed with spherical bubble foams (Dutta et al. 2002; Prins 1986). In a polyhedric foam, gravity acts directly on the draining film liquid, and also indirectly through suction of the Plateau borders, with the local concentration of the liquid increasing from the top to the bottom of the foam (Prins 1988). Furthermore, coalescence of the adjacent bubbles by rupture of the interbubble lamellae and disproportionation due to the polydispersity in the bubble-size distribution combine to produce a gradual coarsening of the gaseous dispersion. These structural changes also serve to accelerate the gravity-driven separation processes.

6.1.2 Ostwald Ripening

Ostwald ripening is the process whereby gas diffusion from small bubbles to large bubbles is driven by the differences in the Laplace pressure (Dickinson 1992). The phenomenon of Ostwald ripening results in larger bubbles (or particles, droplets, or crystals) growing in size at the expense of smaller ones. When applied to bubbles, the term "disproportionation" is often used instead of Ostwald ripening. Disproportionation can also mean the process of gas transport between the bubbles containing gases of different solubilities (like CO_2 and air). For a constant surface tension at the air–liquid interface, the process of disproportionation is self-accelerating until the smallest bubbles disappear completely (Prins 1986). When a viscoelastic protein layer is adsorbed at the surface of the gas bubbles, the rate of disproportionation is reduced. This can be attributed to the adsorbed protein layer having a finite surface dilatational modulus.

The destabilization of protein foams occurs due to several factors, including creaming, drainage (from the lamellae and the Plateau borders), bubble coalescence, and disproportionation (Dickinson 1992). Recently, Lau and Erickson (2005) studied the instability mechanisms in an aerated system consisting of egg albumen and invert sugar. They observed that by reducing the invert sugar concentration in the serum phase (from 82% total solids), a higher overrun of the foam was produced; however, this caused an increased rate of destabilization. Confocal image microscopy revealed dynamic changes observed during foam formation with this system. An interesting feature was the formation of a gel-like network composed of flocculated, uniform, small bubbles in the middle foam layer of the samples that were aged (Figure 6.1). This bubble network was observed to be permeated by mobile liquid pores, allowing larger bubbles to pass through the network under the influence of a gravity-driven hydrodynamic flow. On the other hand, the whey protein isolate (WPI) foams were found to be less stable against gravity-induced

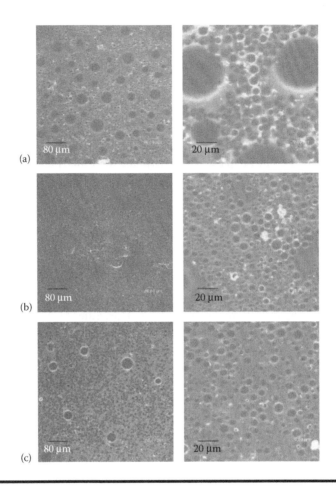

Figure 6.1 Confocal microscopy of an aerated system containing albumen. Confocal microscopy of different regions of an aerated system containing 2 wt% egg albumen and 50% total sugar solids. Samples of three different layers were observed after 2 h of standing: (a) upper layer, (b) middle layer, and (c) bottom layer. The left and right images represent low- and high-resolution micrographs, respectively. (From Lau and Dickinson 2005.)

drainage, as compared with the egg white foams at a concentration of 5% protein (Phillips et al. 1989).

In protein foams, it is most likely that several of the destabilization mechanisms occur simultaneously. This makes the detection of individual processes difficult and calls for highly sophisticated and sensitive methods, such as the use of confocal microscopy. All of these instability mechanisms act together, gradually resulting in a nonuniform microstructure and a reduced product quality, which is obviously undesirable.

6.2 Factors Determining Physical Properties of Protein Foams

6.2.1 Proteins at Air–Water Interfaces

The hydrophobic patches on a protein's surface initially drive the process of foam formation, and surface hydrophobicity has been correlated with improved foaming properties (Kato and Nakai 1980; Moro, Gatti, and Delorenzi 2001; Townsend and Nakai 1983). Once contacts are made with the interface, the natural flexibility within the molecules can expose previously buried hydrophobic portions to the interface, potentially leading to interfacial denaturation of the molecules (Graham and Phillips 1979).

The most obvious outcome of protein adsorption is a reduction in the interfacial (surface) tension. Note that the surface tension of water is 72 mN/m at room temperature and the equilibrium surface tensions of concentrated protein solutions are often around 45 mN/m (Prins et al. 1998). More important than the equilibrium surface tension values is the capacity of the protein to rapidly decrease the surface tension, as this has been correlated with an improved foam capacity (Wilde and Clark 1996). Indeed, the increased adsorption rate of the protein fragments generated during hydrolysis, due to their larger diffusion coefficients compared with the intact proteins, seems to be a primary cause behind the improved foaming properties of various protein hydrolysates (Caessens et al. 1999; Turgeon, Gauthier, and Paquin 1991; van der Ven et al. 2002).

The intermolecular interactions between the adsorbed proteins can lead to an interfacial film with measurable rheological properties. Interfacial rheology has long been recognized as a primary factor contributing to the bulk behavior of foams and emulsions. Two general types of measurements are commonly reported in the scientific literature: those based on shear and those based on dilatational deformations. Detailed descriptions of the principles behind these measurements, as well as their pros and cons, have been well documented (Murray 2002). The measurements of proteins at model interfaces have been reviewed by Bos and van Vliet (2001) and Murray (2002). A notable example that we do not cover is the large amount of work being applied to understanding the mixed protein/small molecular weight surfactant systems (Patino, Nino, and Sanchez 2003; Wilde et al. 2004).

The interfacial dilatational modulus (E) was originally defined by Gibbs and is the change in the interfacial tension (γ) upon a small change in the interfacial area (A) at a constant shape (Lucassen-Reynders 1993):

$$E = d\gamma/d\ln A.$$

This modulus can be thought of as the total resistance of a surfactant-covered interface to dilatational deformations, that is, stretching and compression. This modulus contains both an elastic (E') component and a viscous (E'') component corresponding to the energy recovered and lost during an interfacial deformation. Interfacial shear rheology involves measuring the forces encountered upon shearing a surfactant-covered interface parallel to the interfacial plane. Note that the interfacial area does not change during such deformations, but the shape does change. Both a shear elasticity (G) and a shear viscosity (η) can be defined, again corresponding to the energy recovered and lost during such deformations (Dickinson 1999). The primary difference between dilatational rheology and shear interfacial rheology is that the surfactant concentration at the interface remains essentially constant during the latter.

6.2.2 Interfacial Viscoelasticity

The interfacial viscoelasticity of protein films depends on numerous factors, including the type of protein, the cosolutes present, and the thermal history of the solution. Due to the variety of amino acids contained in a typical protein, a range of intermolecular interactions are possible at the interface, including hydrogen bonding, hydrophobic contacts, electrostatics, disulfide bond formation, and van der Waals interactions (Prins et al. 1998). Numerous studies have found that globular proteins, such as β-lactoglobulin (β-lg), ovalbumin, and lysozyme, tend to form more viscoelastic films, compared with less ordered proteins such as β-casein (Bos and van Vliet 2001). This is explained by the flexible β-casein not transmitting force across the interface as efficiently as the more rigid, globular proteins (Pereira et al. 2003). Recently, dilute mixtures of β-casein and β-lg adsorbed at the air–water interface were characterized via dilatational and shear interfacial rheology (Ridout, Mackie, and Wilde 2004). Simple models adapted from three-dimensional rheology were found to be capable of discerning the nature of the adsorbed layer using both techniques. Shear measurements were more sensitive to the interactions between the adsorbed proteins, while the dilatational data reflected both the interactions between the proteins and the composition.

Electrostatic interactions play a significant role in both protein adsorption and interfacial rheology. Foaming properties have been reported as optimal for a range of proteins near their isoelectric points (pI) (Davis, Foegeding, and Hansen 2004; Hammershoj, Prins, and Qvist 1999; Phillips, Schulman, and Kinsella 1990; Zhu and Damodaran 1994). Accordingly, protein adsorption at the interface is generally most rapid at this pH because electrostatic repulsion is minimized for the net neutrally charged proteins (Dickinson 1999). Furthermore, the viscoelasticity of the interfacial films generally peaks for a range of proteins near their pIs (Davis, Foegeding, and Hansen 2004; Dickinson 1999; Hammershoj, Prins, and Qvist 1999; Pereira et al. 2003; Pezennec et al. 2000). The addition of NaCl to the WPI solutions at pH levels above or below their pIs increased the protein adsorption, as evidenced by the dynamic surface tension measurements (Davis, Foegeding, and Hansen 2004). This was explained by the salt counterions screening the charged protein molecules. Concomitant increases in dilatational elasticity were observed for these adsorbed layers with the addition of NaCl, although this increase was minimal for the WPI at a pH of 3.0 compared with the same solution at a pH of 7.0. The weakened dilatational elasticity of the WPI layers adsorbed at a pH of 3.0 potentially explains the notable decreases in the foam yield stress (Davis, Foegeding, and Hansen 2004) and the foam stability (Zhu and Damodaran 1994) for the WPI foams formed at an acidic pH. From the above discussion of electrostatic interactions, it is clear that the age-tested benefit of adding acid to egg white improves the foaming properties by decreasing the egg white pH (7–8.5 for fresh eggs and up to 9.7 for stored eggs) such that it is closer to the pIs of the egg proteins (Li-Chan and Nakai 1989).

Small amounts of positive, multivalent cations can significantly increase the dilatational viscoelasticity of the adsorbed whey proteins (net negatively charged) via specific electrostatic bridging interactions with corresponding improvements in various foaming properties (Davis, Foegeding, and Hansen 2004; Sarker, Wilde, and Clark 1996). Copper has a unique association with egg white foams. For centuries, whipping egg whites in a copper bowl has been recommended as a way to improve the foam stability (McGee, Long, and Briggs 1984). This effect was speculated to be due to the increased stability of the ovotransferrin to denaturation when the copper is bound (McGee, Long, and Briggs 1984). A more recent investigation found that the surface tension of the egg whites was essentially unchanged with the addition of copper, although the dilatational elasticity was significantly increased, primarily explaining the increase in the foam stability (Sagis et al. 2001). It is possible, although unproven, that there is a connection between the increased stability to denaturation and the interfacial dilatational elasticity.

Heating is an important processing step for many products that consist of protein foams. The effect of heat (20°C–80°C) was examined for adsorbed WPI using pendant drop tensiometry (the maximum diameter and the ratio between the maximum diameter and the diameter at a distance from the drop apex to the maximum diameter are evaluated and formulas are used to derive the surface tension, based on the Young–Laplace equation) (Patino, Nino, and Sanchez 1999). The competing effects between the increased fluidity of the adsorbed layer and the increased elasticity from the interfacial gelation were noted. The rheological changes were heat-dependent.

Foamed food products often contain high contents of sugars and proteins. An understanding of the behaviors of these compounds both within the solution and at the air–water interface is desired to better predict or control the bulk foaming properties. The short-time adsorption rates of bovine serum albumin (BSA) were found to increase in the presence of increasing concentrations (up to 1 M) of sucrose (Nino and Patino 2002). A potential explanation was that the protein molecule would be more compact in sugar solutions due to preferential hydration, and hence, would adsorb more rapidly. It was also noted that the increased viscosity of the concentrated sugar solutions should limit the diffusion at the interface, meaning that the protein adsorption in the sugar solutions would be a balance of these two phenomena. By contrast, the adsorption of ovalbumin was found to decrease in the presence of sucrose, as observed by the dynamic surface tension measurements (Antipova, Semenova, and Belyakova 1999). Light scattering and mixing calorimetry data suggested that the ovalbumin participated in hydrogen bonding with the sucrose molecule, increasing its hydrophilicity and hence decreasing its surface activity.

6.3 Peptides: Hydrolysates as Foaming Agents

Enzymatic hydrolysis is a common means of improving the foaming potential of the protein ingredients (Kilara and Panyam 2003). A common approach for evaluating the foaming performance of the protein hydrolysates is the whipping of dilute hydrolysate solutions (≤0.05% w/v) in graduated cylinders, after which the initial foam height and its decrease with time are taken as measurements of the foam capacity and the foam stability, respectively. Accordingly, the foaming capacity and the foam stability were markedly improved for a variety of hydrolysates, as compared with their unhydrolyzed counterparts (Caessens et al. 1999; van der Ven et al. 2002). Chromatographic characterizations suggest that this improvement is partially attributable to the reduced size of peptides, as compared with proteins, which promotes a more rapid adsorption at the air–water interface (Caessens et al. 1999; Turgeon et al. 1992). However, too much hydrolysis may be detrimental for functionality purposes. For example, the hydrolysates containing high percentages of high molecular weight fragments (>7 kDa) were most correlated to an improved foam stability in a comparative study of 44 different hydrolysates (van der Ven et al. 2002).

There is also substantial evidence from the literature focusing on the emulsion properties of hydrolysates, suggesting that extensive hydrolysis can be detrimental to functionality (Kilara and Panyam 2003). The relative hydrophobicity of hydrolysates has also been correlated with improved foamability and foam stability (Caessens et al. 1999; van der Ven et al. 2002).

The direct characterization of the hydrolyzed proteins at model air–water or oil–water interfaces has received far less attention than the foaming or emulsifying tests of these materials. The tryptic peptides derived from β-lg that were most effective at decreasing interfacial tension were those containing distinct zones of hydrophobic and hydrophilic regions within a minimum molecular weight allowing this distribution (Turgeon et al. 1992). Increasing levels of hydrolysis (up to 86%) for β-lg variant A with a protease specific for glutamic and aspartic acid residues decreased the interfacial shear elasticity and viscosity of these materials, but resulted in an improved foam overrun and stability, as determined via a small-scale foaming test (Ipsen et al. 2001). This was surprising because an improved foaming performance is typically associated with increases in the interfacial rheological moduli. Dilatational rheological tests of an amphipathic peptide isolated from a tryptic hydrolysis of β-casein showed surface behavior similar to that of the intact protein (Girardet et al. 2002). Very little has been reported on the interfacial dilatational rheological behavior of unfractionated mixtures of protein hydrolysates, which is the typical form of these ingredients.

β-Lactoglobulin was hydrolyzed with three different proteases and subsequently evaluated for its foaming potential. Hydrolysis of β-lg with alcalase and pepsin produced fractions that formed foams with significantly increased τ_0, compared with unhydrolyzed β-lg. Trypsin hydrolysis only slightly improved the foam τ_0. Heating the hydrolysates at 75°C/30 min and 90°C/15 min successfully terminated the enzymatic activity for the three hydrolysates. However, heating at 75°C/30 min better preserved the foam τ_0 and this was generally reflected in a higher *E-yield stress* and/or a low interfacial phase angle (Davis, Doucet, and Foegeding 2005).

Limited hydrolysis of the ovomucin and other food proteins is a tool to increase the solubility and surface hydrophobicity, which may result in a higher foaming capacity and possibly—after further studies—also in an improved foam stability. The fractionation of egg albumen by isoelectric precipitation followed by sieving resulted in an ovomucin isolate. Hydrolysis with flavorzyme and neutrase also increased the surface hydrophobicity, primarily during the first 6 h, that is, the exposure of the hydrophobic residues was higher than the increase in the polar groups in the peptides formed. By contrast, hydrolysis with pronase E and alcalase reduced the surface hydrophobicity compared with that of the unhydrolyzed ovomucin. The resulting foam capacity of the ovomucin increased with the degree of hydrolysis and reached optimum foam overrun at 15%–40% degree of hydrolysis, but decreased again after more extensive hydrolysis. The foam capacity showed a high correlation with the initial drop in surface tension. There was no significant effect of hydrolysis on the foam stability against liquid drainage (Hammershoj, Nebel, and Carstens 2008).

Martínez et al. (2009) studied the interfacial and foaming properties of the soy protein (SP) hydrolysates. The hydrolysis of the SPs increased the surface activity at bulk concentrations where the SP adopted a condensed conformation at the monolayer. The hydrolysis also improved the dilatational elasticity and viscoelasticity of the films at bulk concentrations below that corresponding to the collapse of the SP monolayer (2% bulk protein).

6.4 Rheological Properties of Foams

As previously mentioned, foams are often characterized in culinary arts by their textural (rheological) properties. Whipping to a "soft peak" and forming "soft or hard" meringues are common examples. Therefore, in addition to overrun and stability, empirically assessed rheological properties are used to determine the stages of formation and the quality of the foams. In understanding the foam rheological properties, some assumptions are made concerning the foam structure, generally starting with the air phase consisting of hexagonally close-packed, monodispersed, cylindrical drops (Princen 1983) that assume a two-dimensional foam of hexagonal cell structure at an air volume fraction of 1 (Khan, Schnepper, and Armstron 1988; Princen 1983). When deformed, the hexagonal cells "hop" to a location adjacent to the previous one (Khan, Schnepper, and Armstron 1988; Princen 1983). That model predicts that the yield stress (τ_0) of foams and concentrated emulsions depend on the interfacial surface tension (γ), the radius of the undeformed drops (R), the air phase fraction, and the dimensionless contribution of each drop to the yield stress (Princen 1985).

The radius of the undeformed drops is replaced with the surface-volume or Sauther mean drop radius (R_{32}), and Y is used as an experimentally derived function to estimate the yield stress contribution from each unit cell.

Several authors have adapted the vane method of Dzuy and Boger (1983, 1985) for measuring the yield stress of foams. The advantage of this method is that the shear plane occurs along a circumference outlined by the diameter of the vane blades, thereby circumventing the problems related to the slip in the traditional rheometer testing cells (e.g., parallel plates). The egg white foams were shown to have yield stress values ranging from 100 to 150 Pa, while the WPI foams, of equal or higher protein concentrations, had values ranging from 55 to 80 Pa. The yield stress of the egg white foams was relatively stable, while the yield stress of the WPI foams decayed over the first 3 min after formation (Pernell et al. 2002a). The air phase volume, surface tension, and bubble size of the egg white foams and WPI foams were very similar (Pernell et al. 2002a). Based on the model of Princen (1989), the only variable remaining is the fitted parameter of Y. When compared with the Y values from a concentrated (ranging from 0.833 to 0.975) paraffin oil emulsion, the Ys of the egg white foams and WPI foams were higher at lower air phase fractions. That is to say, the adjustable factor Y that was determined for the egg white foams at =0.86–0.89 was similar to the paraffin oil emulsion values at =0.94–0.98.

A number of factors affect the yield stress of the foams made with whey proteins. The addition of salts (NaCl or $CaCl_2$), glycine, and lactose to the WPI, α-lactalbumin, or β-lg tends to alter the yield stress and overrun (Luck, Bray, and Foegeding 2002). However, deviations were seen, with $CaCl_2$ generally showing higher than predicted yield stress values. Since $CaCl_2$ is known to affect the interfacial properties of the proteins, the relationship between the interfacial rheological properties and the yield stress was studied by Davis, Foegeding, and Hansen (2004). The WPI solutions at varying pH (3, 5, and 7) and concentrations of salts (NaCl or $CaCl_2$) were whipped into foams, and the foam yield stress was compared with the surface tension and the interfacial dilatational elasticity. It was observed that no relationship was found between the yield stress and the surface tension; however, one was established between the foam yield stress and the interfacial dilatational elasticity. This correlates with recent observations for highly concentrated protein emulsions in which the dimensionless bulk elasticity, $G'/(\gamma/r)$, was correlated with the dilatational elasticity of the various protein interfaces (Dimitrova and Leal-Calderon 2004).

Pernell et al. (2002a) suggested that the relationship between the air phase fraction and the yield stress for egg white foams may result in the bubbles being at a higher phase fraction than calculated. This could be due to an inaccurate determination of the air phase fraction or the possibility of some connectivity between the protein in the lamellae and that on the bubble surface. Determining the connectivity between the proteins in the lamellae and on the bubble surface is not simple. It possibly could be done based on the changes in the rheological properties. The yield stress of concentrated water-in-oil emulsions shows a marked increase when the dispersed phase fraction goes from 65% to 70% (Jager-Lezer et al. 1998), suggesting that determining the change in the yield stress in foams as it transitions out of the close-packed region may be a way to determine the close-packed air phase fraction. However, the instability of the protein-based foams makes this approach difficult. Additional research is needed in this area.

Mild heating of the protein solutions has been associated with improved foaming properties. Heating the protein solutions under proper ionic and pH conditions causes denaturation and aggregation such that soluble aggregates or whey protein polymers (based on covalent intermolecular linking) are formed (Vardhanabhuti and Foegeding 1999). The whey protein polymers have a much higher intrinsic viscosity than the native globular proteins (Vardhanabhuti and Foegeding 1999). As a way to understand the effect of mild heating, the foaming properties of mixtures of whey protein polymers and that of the native whey protein mixtures were investigated (Davis, Foegeding, and Hansen 2004). Increasing the relative percentage of the whey protein polymers in the solution decreased the air phase volume, but made the foams more stable by decreasing the drainage rate. The increase in stability was directly associated with an increase in protein solution viscosity, thereby slowing the drainage rate. Increasing the percentage of the whey protein polymers up to 50%, increased the yield stress, but the stability of the yield stress measurement decreased (Davis, Foegeding, and Hansen 2004). Moreover, it was found that the changes in the foam yield stress coincided with the changes in the interfacial dilatational elasticity. It appears that the interconnecting of proteins at the air–water interface forming an elastic network is important to the yield stress of foams.

6.5 Conclusion

Protein foams are an integral component of many foods, such as meringue, ice cream, nougat, and cake. In a protein, the desirable levels of foam must be investigated and achieved in order to attribute and use this protein in food foams. Investigation of protein foams is more complex because of the number of variables and the relative instability of these systems. Hence, protein foams must also maintain stability during their formation and during the several processing steps involved in the making of the food. All the factors and interactions of potential proteins with foaming tendencies with good stability must be investigated. Continued exploration through research of new proteins with enhanced foam characteristics and stability will provide many applications for such proteins in the functional food industry.

References

Antipova, A. S., Semenova, M. G., and Belyakova, L. E. 1999. Effect of sucrose on the thermodynamic properties of ovalbumin and sodium caseinate in bulk solution and at air–water interface. *Colloids and Surfaces B: Biointerfaces* 12: 261–70.

Bos, M. A. and van Vliet, T. 2001. Interfacial rheological properties of adsorbed protein layers and surfactants: A review. *Advances in Colloid and Interface Science* 91: 437–71.

Caessens, P., Visser, S., Gruppen, H., and Voragen, A. G. J. 1999. Beta-lactoglobulin hydrolysis. 1. Peptide composition and functional properties of hydrolysates obtained by the action of plasmin, trypsin, and *Staphylococcus aureus* V8 protease. *Journal of Agricultural and Food Chemistry* 47: 2973–79.

Campbell, G. M. and Mougeot, E. 1999. Creation and characterisation of aerated food products. *Trends in Food Science and Technology* 10: 283–96.

Davis, J. P. and Foegeding, E. A. 2004. Foaming and interfacial properties of polymerized whey protein isolate. *Journal of Food Science* 69: C404–10.

Davis, J. A., Foegeding, E. A., and Hansen, F. K. 2004. Electrostatic effects on the yield stress of whey protein isolate foams. *Colloids and Surfaces B: Biointerfaces* 34: 13–23.

Davis, J. P., Doucet, D., and Foegeding, E. A. 2005. Foaming and interfacial properties of hydrolyzed β-lactoglobulin. *Journal of Colloid and Interface Science* 288: 412–22.

Dickinson, E. 1986. Competitive protein adsorption. *Food Hydrocolloids* 1: 3–23.

Dickinson, E. 1992. *An Introduction to Food Colloids*. New York: Oxford University Press.

Dickinson, E. 1999. Adsorbed protein layers at fluid interfaces: Interactions, structure and surface rheology. *Colloids and Surfaces B: Biointerfaces* 15: 161–76.

Dimitrova, T. D. and Leal-Calderon, F. 2004. Rheological properties of highly concentrated protein-stabilized emulsions. *Advances in Colloid and Interface Science* 108–109: 49–61.

Dutta, A., Chengara, A., Nikolov, A., Wasan, D. T., Chen, K., and Campbell, B. 2004. Texture and stability of aerated food emulsions: Effects of buoyancy and Ostwald ripening. *Journal of Food Engineering* 62: 169–75.

Dzuy, N. Q. and Boger, D. V. 1983. Yield stress measurement for concentrated suspensions. *Journal of Rheology* 27: 321–49.

Dzuy, N. Q. and Boger, D. V. 1985. Direct yield stress measurement with the vane method. *Journal of Rheology* 29: 335–47.

Girardet, J. M., Humbert, G., Creusot, N., Chardot, V., Campagna, S., Courthaudon, J. L., et al. 2002. Dilational rheology of mixed β-casein/Tween 20 and β-casein (f114–169)/Tween 20 films at oil–water interface. *Journal of Colloid and Interface Science* 245: 219.

Graham, D. E. and Phillips, M. C. 1979. Proteins at liquid interfaces. I. Kinetics of adsorption and surface denaturation. *Journal of Colloid and Interface Science* 70: 403–14.

Halling, P. J. 1981. Protein-stabilized foams and emulsions. *CRC Critical Reviews in Food Science and Nutrition* 15: 155–203.

Hammershoj, M., Nebel, C., and Carstens, J. H. 2008. Enzymatic hydrolysis of ovomucin and effect on foaming properties. *Food Research International* 41: 522–31.

Hammershoj, M., Prins, A., and Qvist, K. B. 1999. Influence of pH on surface properties of aqueous egg albumen solutions in relation to foaming behaviour. *Journal of the Science of Food and Agriculture* 79: 859–68.

Ipsen, R., Otte, J., Sharma, R., Nielsen, A., Hansen, L. G., and Qvist, K. B. 2001. Effect of limited hydrolysis on the interfacial rheology and foaming properties of β-lactoglobulin A. *Colloid Surfaces B: Biointerfaces* 21: 173.

Jager-Lezer, N., Tranchant, J.-F., Alard, V., Vu, C., Tchoreloff, P. C., and Grossiord, J.-L. 1998. Rheological analysis of highly concentrated w/o emulsions. *Rheologica Acta* 37: 129–38.

Kato, A. and Nakai, S. 1980. Hydrophobicity determined by a fluorescent probe method and its correlation with surface properties of proteins. *Biochimica et Biophysica Acta* 624: 13.

Khan, S. A., Schnepper, C. A., and Armstron, R. C. 1988. Foam rheology. III. Measurement of shear flow properties. *Journal of Rheology* 32: 69–92.

Kilara, A. and Panyam, D. 2003. Peptides from milk proteins and their properties. *CRC Critical Reviews in Food Science and Nutrition* 43: 607–33.

Lau, K. C. and Dickson, E. 2005. Instability and structural change in an aerated system containing egg albumen and invert sugar. *Food Hydrocolloids* 19(1): 111–21.

Li-Chan, E. and Nakai, S. 1989. Biochemical basis for the properties of egg white. *CRC Critical Reviews in Poultry Biology* 2: 21–58.

Lucassen, J. 1981. Dynamic properties of free liquid films and foams. In *Anionic Surfactants – Physical Chemistry of Surfactant Action*, ed. E. H. Lucassen-Reijnders, pp. 217–65. New York: Marcel Dekker.

Lucassen-Reynders, E.-H. 1993. Interfacial viscoelasticity in emulsions and foams. *Food Structure* 12: 1–12.

Luck, P. J., Bray, N., and Foegeding, E. A. 2002. Factors determining yield stress and overrun of whey protein foams. *Journal of Food Science* 67: 1677–81.

Martínez, K. D., Sánchez, C. C., Patino, J. M. R., and Pilosof, A. M. R. 2009. Interfacial and foaming properties of soy protein and their hydrolysates. *Food Hydrocolloids* 23: 2149–57.

McGee, H., Long, S. R., and Briggs, W. R. 1984. Why whip egg whites in copper bowls? *Nature* 308: 667–68.

Moro, A., Gatti, C., and Delorenzi, N. 2001. Hydrophobicity of whey protein concentrates measured by fluorescence quenching and its relation with surface functional properties. *Journal of Agricultural and Food Chemistry* 49: 4784–89.

Murray, B. S. 2002. Interfacial rheology of food emulsifiers and proteins. *Current Opinion in Colloid and Interface Science* 7: 426–31.

Nino, M. R. R. and Patino, J. M. R. 2002. Effect of the aqueous phase composition on the adsorption of bovine serum albumin to the air–water interface. *Industrial and Engineering Chemistry Research* 41: 1489–95.

Patino, J. M. R., Nino, M. R. R., and Sanchez, C. C. 1999. Dynamic interfacial rheology as a tool for the characterization of whey protein isolates gelation at the oil–water interface. *Journal of Agricultural and Food Chemistry* 47: 3640–48.

Patino, J. M. R., Nino, M. R. R., and Sanchez, C. C. 2003. Protein–emulsifier interactions at the air–water interface. *Current Opinion in Colloid and Interface Science* 8: 387–95.

Pereira, L. G. C., Theodoly, O., Blanch, H. W., and Radke, C. J. 2003. Dilatational rheology of BSA conformers at the air/water interface. *Langmuir* 19: 2349–56.

Pernell, C. W., Foegeding, E. A., and Daubert, C. R. 2000. Measurement of the yield stress of protein foams by vane rheometry. *Journal of Food Science* 65: 110–14.

Pernell, C. W., Foegeding, E. A., Luck, P. J., and Davis, J. P. 2002a. Properties of whey and egg white protein foams. *Colloids and Surfaces A: Physicochemical and Engineering Aspects* 204: 9–21.

Pezennec, S., Gauthier, F., Alonso, C., Graner, F., Croguennec, T., and Brule, G. 2000. The protein net electric charge determines the surface rheological properties of ovalbumin adsorbed at the air–water interface. *Food Hydrocolloids* 14: 463–72.

Phillips, L. G., Yang, S. T., Schulman, W., and Kinsella, J. E. 1989. Effects of lysozyme, clupeine, and sucrose on the foaming properties of whey protein isolate and β-lactoglobulin. *Journal of Food Science* 54: 743–47.

Phillips, L. G., Schulman, W., and Kinsella, J. E. 1990. pH and heat-treatment effects on foaming of whey-protein isolate. *Journal of Food Science* 55: 1116–19.

Princen, H. M. 1983. Rheology of foams and highly concentrated emulsions. I. Elastic properties and yield stress of a cylindrical model system. *Journal of Colloid and Interface Science* 91: 160–75.

Princen, H. M. 1985. Rheology of foams and highly concentrated emulsions. II. Experimental study of the yield stress and wall effects for concentrated oil-in-water emulsions. *Journal of Colloid and Interface Science* 105: 150–71.

Princen, H. M. and Kiss, A. D. 1989. Rheology of foams and highly concentrated emulsions. IV. An experimental study of the shear viscosity and yield stress of concentrated emulsions. *Journal of Colloid and Interface Science* 128: 176–87.

Prins A. 1986. Some physical aspects of aerated milk products. *Netherlands Milk and Dairy Journal* 40: 203–15.

Prins A. 1988. Principles of foam stability. In *Advances in Food Emulsions and Foams*, eds. E. Dickinson and G. Stainsby, 91–122. New York: Elsevier.

Prins, A., Bos, M., Boerboom, F. J. G., and van Kalsbeek, H. K. A. I. 1998. Relation between surface rheology and foaming behaviour of aqueous protein solutions. In *Proteins at Liquid Interfaces*, eds. D. Möbius and R. Miller, pp. 221–66. Amsterdam: Elsevier Sciences.

Ridout, M. J., Mackie, A. R., and Wilde, P. J. 2004. Rheology of mixed beta-casein/beta-lactoglobulin films at the air–water interface. *Journal of Agricultural and Food Chemistry* 52: 3930–37.

Sagis, L. M. C., de Groot-Mostert, A. E. A., Prins, A., and van der Linden, E. 2001. Effect of copper ions on the drainage stability of foams prepared from egg white. *Colloids and Surfaces A: Physicochemical and Engineering Aspects* 180: 163–72.

Sarker, D. K., Wilde, P. J., and Clark, D. C. 1996. Enhancement of the stability of protein-based food foams using trivalent cations. *Colloids and Surfaces A: Physicochemical and Engineering Aspects* 114: 227–36.

Townsend, A. and Nakai, S. 1983. Relationships between hydrophobicity and foaming characteristics of food proteins. *Journal of Food Science* 48: 588–94.

Turgeon, S. L., Gauthier, S. F., and Paquin, P. 1991. Interfacial and emulsifying properties of whey peptide fractions obtained with a 2-step ultrafiltration process. *Journal of Agricultural and Food Chemistry* 39: 673–76.

Turgeon, S. L., Gauthier, S. F., Molle D., and Leonil J. 1992. Interfacial properties of tryptic peptides of beta-lactoglobulin. *Journal of Agricultural and Food Chemistry* 40(4): 669–675.

van der Ven, C., Gruppen, H., de Bont, D. B. A., and Voragen, A. G. J. 2002. Correlations between biochemical characteristics and foam-forming and stabilizing ability of whey and casein hydrolyzates. *Journal of Agricultural and Food Chemistry* 50: 2938–46.

Vardhanabhuti, B. and Foegeding, E. A. 1999. Rheological properties and characterization of polymerized whey protein isolates. *Journal of Agricultural and Food Chemistry* 47: 3649–55.

Wilde, P. J. and Clark, D. C. 1996. Foam formation and stability. In *Methods of Testing Protein Functionality*, ed. G. M. Hall, pp. 110–52. London: Blackie Academic.

Wilde, P., Mackie, A., Husband, F., Gunning, P., and Morris, V. 2004. Proteins and emulsifiers at liquid interfaces. *Advances in Colloid and Interface Science* 108–109: 63–71.

Zhu, H. M. and Damodaran, S. 1994. Proteose peptones and physical factors affect foaming properties of whey protein isolate. *Journal of Food Science* 59: 554–60.

Chapter 7

Chemical and Enzymatic Protein Modifications and Functionality Enhancement

Hitomi Kumagai

Contents

7.1 Introduction

Food proteins in their native form do not necessarily possess the desirable functions as to digestibility, nutritional value, and processing properties, such as emulsifying, foaming, and gelling properties. Structural modification is often effective to improve these functions. Modification can be performed chemically, physically, and/or enzymatically. This chapter focuses on the methods for chemical and enzymatic modification of proteins and the functions improved by modification. Protein modification often makes use of the functional groups in the side chains of amino acid residues, such as the ε-amino group of lysine residue, the thiol group of cysteine residue, the carboxyl group of glutamic acid and aspartic acid residues, the phenolic hydroxyl group of tyrosine residue, the alcoholic hydroxyl group of serine and threonine residues, the imidazole group of histidine residue, the thioether group of methionine residue, and the indole group of tryptophan residue. The reactivity of these functional groups differs according to the reaction conditions, such as pH and reagent type, their location in the protein structure, and molecular weight of the protein. In this chapter, details of the reaction of each chemical and enzymatic modification are reviewed and their possible applications are discussed.

7.2 Acylation

Acylation (Figure 7.1) is the reaction that covalently attaches the acyl groups (R–CO–), such as the acetyl and succinyl groups, to the amino groups of proteins and sometimes the hydroxyl, imidazole, and thiol groups, using dicarboxylic acid anhydride.

Various food proteins have been modified by acylation so far (Table 7.1). Acetic anhydride (Franzen and Kinsella 1976; Barman, Hansen, and Mossey 1977; Groninger and Miller 1979; Eisele and Brekke 1981; Eisele, Brekke, and McCurdy 1981; King, Ball, and Garlich 1981; Ma 1984; Goulet et al. 1987; Ponnampalam et al. 1988; Krause and Schwenke 1996; Schwenke, Dahme, and Wolter 1998; Shahidi and Wanasundara 1998; Pomianowski, Borowski, and Danowska-Oziewicz 1999; Lawal and Adebowale 2006; Szymkiewicz and Jedrychowski 2008), succinic anhydride (Gounaris and Perlmann 1967; Grant 1973; Thompson and Reyes 1980; Ma and Holme 1982; Gueguen et al. 1990; Sitohy, Sharobeem, and Abdel-Ghany 1992; Bae and Jang 1999), maleic anhydride (Butler et al. 1969), *cis, cis, cis, cis*-tetrahydrofuran-2, 3, 4, 5-tetracarboxylic deanhydride (Eisele, Brekke, and McCurdy 1981), 1, 2, 4-benzenetricarboxylic anhydride (Eisele, Brekke, and McCurdy 1981), and citraconic anhydride (Brinegar and Kinsella 1980) are used for acetylation, succinylation, maleylation, tetrahydrofuranylation, benzeneylation, and citraconylation, respectively. *N*-Carboxy-amino acid anhydrides (Bjarnason-Baumann, Pfaender, and Siebert 1977) or an *N*-hydroxysuccinimide ester of *tert*-butyloxycarbonyl

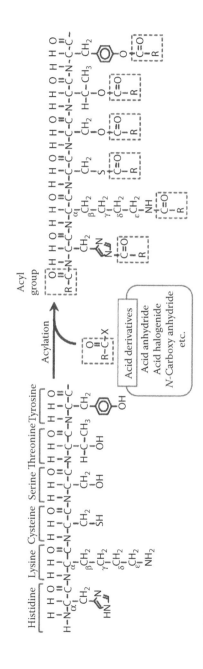

Figure 7.1 Acylation of proteins.

Table 7.1   Various Food Proteins Modified by Acylation

Soybean protein	Franzen and Kinsella (1976); Barman, Hansen, and Mossey (1977); Brinegar and Kinsella (1980)
Fish protein	Groninger and Miller (1979)
Myofibrillar protein	Eisele and Brekke (1981); Eisele, Brekke, and McCurdy (1981); Pomianowski, Borowski, and Danowska-Oziewicz (1999)
Oat protein	Ma (1984); Goulet et al. (1987); Ponnampalam et al. (1988)
Faba bean protein	Krause and Schwenke (1996)
Jack bean protein	Lawal and Adebowale (2006)
Pea protein	Szymkiewicz and Jedrychowski (2008)
Rapeseed protein	Schwenke, Dahme, and Wolter (1998); Gueguen et al. (1990)
Peanut protein	Beuchat (1977)
Flax protein	Shahidi and Wanasundara (1998)
Wheat protein	Grant (1973); Sitohy, Sharobeem, and Abdel-Ghany (1992)
Corn protein	Sitohy, Sharobeem, and Abdel-Ghany (1992)
Rice bran protein	Bae and Jang (1999)
Whey protein	Thompson and Reyes (1980)
Egg protein	Ma and Holme (1982); Sitohy, Sharobeem, and Abdel-Ghany (1992)
Milk protein	Puigserver et al. (1979a, 1979b)
Yeast protein	Kinsella and Shetty (1979)

amino acid (Puigserver et al. 1979a, 1979b; Ferjancic-Biagini, Giardina, and Puigserver 1998) can be used to incorporate an amino acid into the amino group of a protein.

The ε-amino group of lysine residue is most readily acylated (Kinsella and Shetty 1979; Shukla 1982). The phenolic hydroxyl group of tyrosine residue is less reactive because of its high pK. The imidazole group of histidine residue and the thiol group of cysteine residue are rarely acylated because the reaction products are hydrolyzed in an aqueous solution. The alcoholic hydroxyl group of serine and threonine residues is not easily acylated in an aqueous solution because it is a weak nucleophile.

7.2.1 Acetylation and Succinylation

Acetylation and succinylation (Figure 7.2) are the most popular acylation forms. The reaction is performed with acetic anhydride and succinic anhydride at pH around 8. Acetic anhydride for acetylation is more reactive than succinic anhydride for succinylation (Eisele, Brekke, and McCurdy 1981). ε-*N*-Succinylated lysine is stable, while *O*-succinylated tyrosine is easily decomposed. *O*-Succinylated serine and threonine can be deacylated with hydroxylamine (Gounaris and Perlmann 1967).

As acetylation mostly modifies the cationic amino group with the neutral acetyl group, the shifting of the isoelectric point to a lower pH occurs, which slightly enhances the solubility in acidic pH (Kinsella and Shetty 1979). However, it may

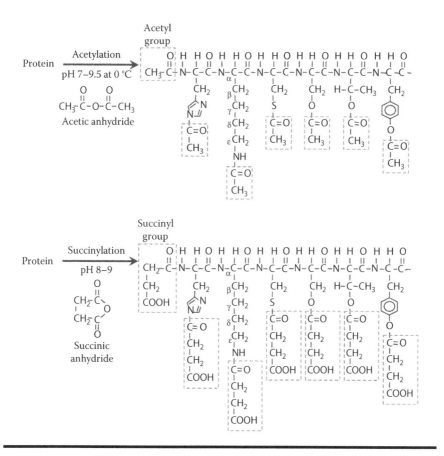

Figure 7.2 Acetylation and succinylation of proteins.

reduce the gelling property because of the reduction of electrostatic attraction between proteins. On the other hand, as succinylation enhances the net negative charge, electrostatic repulsion occurs between the anionic carboxyl group in the succinyl group and the side chain of aspartic and glutamic acid residues, which leads to an increase in the dissociation of proteins into subunits and penetration of water molecules. Succinylation also enhances hydration and aqueous solubility (Kinsella and Shetty 1979).

Acetylation improves solubility in water (Barman, Hansen, and Mossey 1977; Eisele and Brekke 1981; Ma 1984; Ponnampalam et al. 1988; Lawal and Adebowale 2006), emulsifying property (Eisele and Brekke 1981; Ma 1984; Ponnampalam et al. 1988; Krause and Schwenke 1996; Lawal and Adebowale 2006), and foaming capacity (Lawal and Adebowale 2006), but decreases gelling property (Barman, Hansen, and Mossey 1977). Succinylation is often more effective on the change in functionality than acetylation (Kester and Richardson 1984; Ma 1984; Ponnampalam et al. 1988). Succinylation improves solubility (Eisele and Brekke 1981; Ma 1984; Ponnampalam et al. 1988; Sitohy, Sharobeem, and Abdel-Ghany 1992; Bae and Jang 1999), emulsifying property (Franzen and Kinsella 1976; Thompson and Reyes 1980; Eisele and Brekke 1981; Ma 1984; Ponnampalam et al. 1988; Gueguen et al. 1990; Sitohy, Sharobeem, and Abdel-Ghany 1992; Bae and Jang 1999; Lawal and Adebowale 2006), foaming capacity (Franzen and Kinsella 1976; Gueguen et al. 1990; Sitohy, Sharobeem, and Abdel-Ghany 1992; Krause 2002; Lawal and Adebowale 2006), and gelling property (Ma and Holme 1982; Bae and Jang 1999). Both succinylation and acetylation decrease digestibility (Eisele, Brekke, and McCurdy 1981; Pomianowski, Borowski, and Danowska-Oziewicz 1999), protein efficiency ratio (Groninger and Miller 1979; Goulet et al. 1987), and bioavailability (King, Ball, and Garlich 1981), and have no specific biological activity in vivo.

7.2.2 Incorporation of Amino Acid

When an N-carboxy-amino acid anhydride or an N-hydroxysuccinimide ester of a *tert*-butyloxycarbonyl amino acid is used, an amino acid can be incorporated into the ε-amino group of the N-terminal amino acid by a peptide bond or into the ε-amino group of lysine by an isopeptide bond (Figure 7.3) (Ferjancic-Biagini, Giardina, and Puigserver 1998; Puigserver et al. 1979a). The protein efficiency ratio of L-methionylcasein, in which L-methionine is covalently attached to the ε-amino group of lysine in casein, is almost equivalent to that of casein supplemented with free L-methionine, indicating the cleavage of the isopeptide bond between the ε-amino group of lysine and methionine in vivo (Puigserver et al. 1979b). As the isopeptide bond between the ε-amino group of lysine and the carboxyl group of the amino acid can be cleaved in vivo, the incorporation of limiting essential amino acids into a protein is useful in improving nutritional value. Although some reaction by-products need to be removed after esterification, the incorporation of

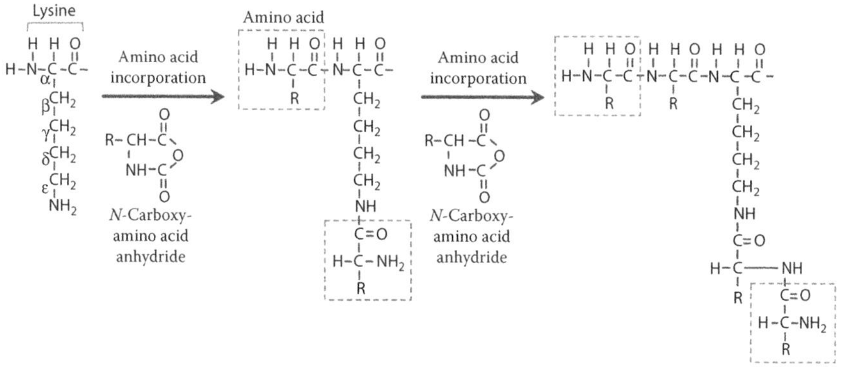

Figure 7.3   Amino acid incorporation into proteins by acylation.

amino acids into a protein is more beneficial than fortification with amino acids because most of the free amino acids have an undesirable taste; additionally, they may react with sugars via the Maillard reaction during processing.

7.3 Alkylation

Reductive alkylation (Figure 7.4) is the reaction that attaches alkyl groups, such as methyl and butyl, to the α- or ε-amino groups of proteins. Aldehydes or ketones are used for the alkyl group source together with sodium borohydride or sodium cyanoborohydride at pH between 7 and 10, aldehydes being more reactive than ketones (Means 1977). Cyanoborohydride is more efficient with fewer side reactions than borohydride (Kester and Richardson 1984). A condensation

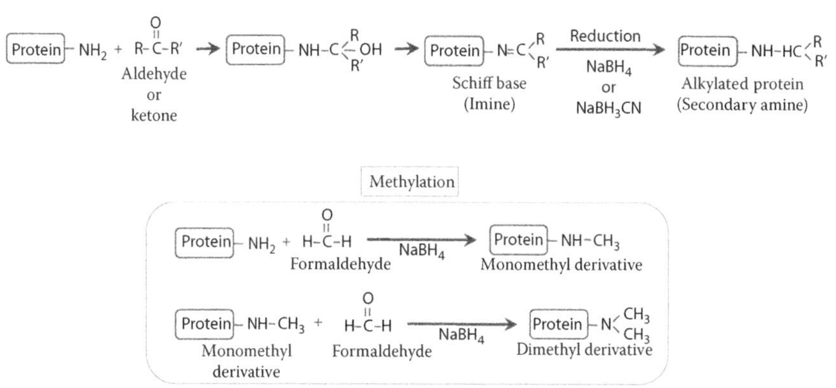

Figure 7.4   Reductive alkylation of proteins.

reaction proceeds between the amino group of a protein and a carbonyl compound (aldehyde or ketone) to produce an imine (Schiff base). Then, it is reduced by a reducing agent, such as sodium borohydride, to form an alkylated protein (secondary amine). As for sodium borohydride, pH 9 is considered to be most suitable because it is unstable below pH 9, and the protein structure may be affected above pH 9.

Dimethyl derivatives are mainly produced by the reaction of proteins with formaldehyde as the monomethyl derivative formed is further methylated by the same reaction. Dimethyl derivatives are also produced by using acetone, but the percentage of amino groups in a modified protein is about 60% of that reacted using formaldehyde (Means 1977). Acetaldehyde is much less reactive and only small amounts of diethyl derivatives are formed, even after extensive reaction.

A methylated protein mostly retains its chemical and biological properties, preserving conformation without changing the distribution of charged groups (Kester and Richardson 1984). The longer alkyl substitution increases the hydrophobicity of a protein (Fretheim, Iwai, and Feeney 1979) and reduces the formation of hydrogen bonds (Means 1977). Alkylation of a protein often retards Maillard reaction, improves the emulsifying property, enhances water absorption, and makes the protein resistant to proteolytic hydrolysis (Kester and Richardson 1984).

Alkylation of the carboxyl group of a protein occurs with alcohol and trichloroacetic acid during Coomassie blue staining (Haebel et al. 1998). Methyl alcohol causes methylation, while ethyl alcohol causes ethylation. The carboxyl group of the glutamic acid side chain is more preferentially alkylated than that of the aspartic acid side chain.

7.4 Glycosylation

Different from enzymatic glycosylation, which attaches mono-saccharides or oligosaccharides to the alcoholic hydroxyl group of serine and threonine residues or to the amide group of asparagine residue, chemical glycosylation of a food protein

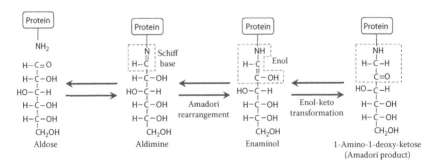

Figure 7.5 Glycosylation of proteins by Maillard reaction.

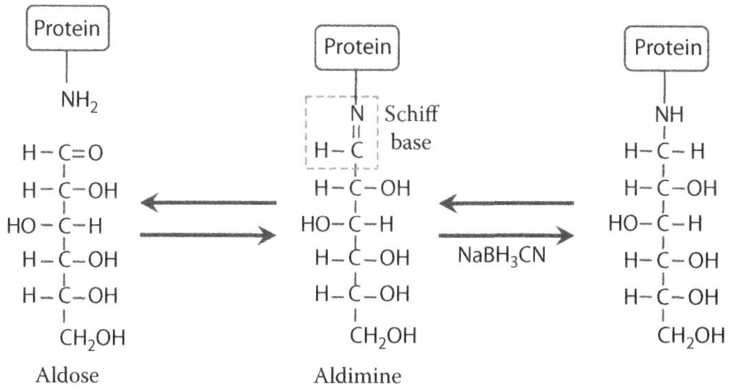

Figure 7.6 Glycosylation of proteins by reductive alkylation.

mostly attaches saccharides to the ε-amino group of lysine residue or to the α-amino group of the *N*-terminal amino acid via the Maillard reaction (Figure 7.5) (Kato, Watanabe, and Sato 1981; Kato et al. 1995; Aoki et al. 1997; Wahyuni, Ishizaki, and Tanaka 1999; Saeki 1999; Katayama, Shima, and Saeki 2002; Kato 2002; Paraman, Hettiarachchy, and Schaefer 2007; Jiménez-Castaño, Villamiel, and López-Fandiño 2007; Nakamura et al. 2008) or via reductive alkylation (Figure 7.6) (Schwartz and Gray 1977; Lee et al. 1979; Courthaudon, Colas, and Lorient 1989; Baniel et al. 1992).

Not only monosaccharides (Lee et al. 1979; Kato, Watanabe, and Sato 1981; Saeki 1999; Katayama, Shima, and Saeki 2002; Paraman, Hettiarachchy, and Schaefer 2007) and disaccharides (Courthaudon, Colas, and Lorient 1989; Baniel et al. 1992), but also polysaccharides, such as chitosan, galactomannan, pectin, arabinogalactan, xyloglucan, dextran, dextrin, cyclodextrin, xanthan gum (Saeki 1999; Kato 2002; Jiménez-Castaño, Villamiel, and López-Fandiño 2007; Paraman, Hettiarachchy, and Schaefer 2007; Nakamura et al. 2008), and even glucose 6-phosphate (Kato et al. 1995; Aoki et al. 1997; Wahyuni, Ishizaki, and Tanaka 1999), can be attached to a protein via the Maillard reaction. During the reaction, the aldehyde group of an aldose or a polysaccharide first attaches to the amino group of a protein to form aldimine (Schiff base). Then, it is transformed into enaminol (ene-diol) through Amadori rearrangement and this Amadori intermediate is converted into 1-amino-1-deoxy-ketose.

Glycosylation can also be performed by reductive alkylation above pH 5 using aldose (hydroxyaldehyde) like glucose, or ketose (hydroxyketone) like fructose, together with sodium cyanoborohydoride (Schwartz and Gray 1977; Marsh, Denis, and Wriston 1977). Reactivity for the glycosylation is higher with higher pH values (Schwartz and Gray 1977). Pentose usually reacts faster than hexose, and monosaccharide reacts faster than disaccharide and oligosaccharide.

Various food proteins, such as ovalbumin (Kato, Watanabe, and Sato 1981; Kato et al. 1983, 1995), β-lactoglobulin (Waniska and Kinsella 1988; Aoki et al.

1997; Jiménez-Castaño, Villamiel, and López-Fandiño 2007), casein (Lee et al. 1979; Courthaudon, Colas, and Lorient 1989), bovine serum albumin (Schwartz and Gray 1977; Jiménez-Castaño, Villamiel, and López-Fandiño 2007), pea legumin (Baniel et al. 1992), buckwheat protein (Nakamura et al. 2008), rice endosperm protein (Paraman, Hettiarachchy, and Schaefer 2007), fish albumin (Wahyuni, Ishizaki, and Tanaka 1999), fish myofibrillar protein (Saeki 1999), and shellfish muscle protein (Katayama, Shima, and Saeki 2002) have been glycosylated.

Some functional properties change by glycosylation. The solubility near the isoelectric point increases by the glycosylation of β-lactoglobulin with dextran (Jiménez-Castaño, Villamiel, and López-Fandiño 2007), myofibrillar proteins with glucose (Katayama, Shima, and Saeki 2002), and casein with monosaccharides and disaccharides, such as glucose, galactose, fructose, lactose, and maltose (Courthaudon, Colas, and Lorient 1989), but decreases by the glycosylation of β-lactoglobulin with maltose (Waniska and Kinsella 1988). Thermal stability is enhanced by glycosylation with dextran (Saeki 1999; Jiménez-Castaño, Villamiel, and López-Fandiño 2007). Emulsifying and foaming properties are improved by the glycosylation of β-lactoglobulin with glucose or maltose (Waniska and Kinsella 1988), pea legumin with galactose (Baniel et al. 1992), and buckwheat protein with polysaccharides (Nakamura et al. 2008). The emulsifying property of rice endosperm protein is enhanced by glycosylation with glucose or xanthan (Paraman, Hettiarachchy, and Schaefer 2007). Modification of casein with monosaccharides or disaccharides reduces the digestibility by digestive enzymes and nutritional value (Lee et al. 1979). The effect of glycosylation is usually much higher for glucose 6-phosphate conjugate than for glucose conjugate (Kato et al. 1995; Wahyuni, Ishizaki, and Tanaka 1999).

7.5 Phosphorylation

Phosphorylation (Figure 7.7) occurs from chemically and enzymatically attaching phosphate to proteins by *O*- or *N*-esterification reactions. Chemical phosphate linkage occurs with the hydroxyl group of serine, threonine, and tyrosine residues, the ε-amino group of lysine residue, the imidazole group of histidine residue, and the guanidine group of arginine (Matheis and Whitaker 1984; Shih 1996). The amino or hydroxyl groups of proteins can also react with the phosphoryl group of a phosphorylated protein forming a cross-linkage (Huang and Kinsella 1986b; Matheis 1991). *O*-Phosphorylation is stable in acidic pH, while *N*-phosphorylation is labile in acidic pH but stable in alkaline pH (Vojdan and Whitaker 1996).

For chemical phosphorylation, phosphorous oxychloride ($POCl_3$) (Woo, Creamer, and Richardson 1982; Woo and Richardson 1983; Hirotsuka et al. 1984; Chobert, Sitohy, and Whitaker 1989; Matheis 1991; Vojdan and Whitaker 1996; Schwenke et al. 2000), phosphoric acid (H_3PO_4) with trichloroacetonitrile

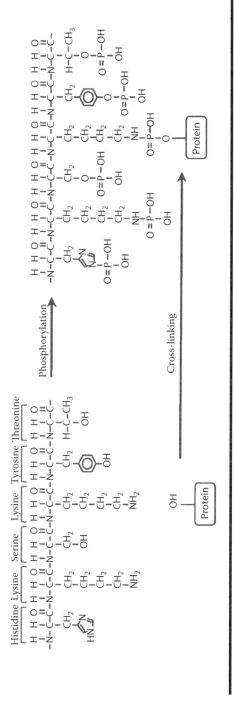

Figure 7.7 Phosphorylation and cross-linking of proteins.

(Cl₃CCN) (Yoshikawa, Sasaki, and Chiba 1981), sodium trimetaphosphate ((NaPO₃)₃; STMP) (Sung et al. 1983), phosphorus pentoxide (P₂O₅) in phosphoric acid or phosphate (Matheis and Whitaker 1984), and drying by heating in phosphoric acid or phosphate (Li et al. 2003) can be used.

When phosphorous oxychloride (POCl₃) is dissolved in an aqueous solution, it reacts with water, producing phosphate, which decreases the pH of the solution. To prevent protein denaturation due to the decrease in pH and heat generation, phosphorous oxychloride is usually dissolved in an organic solvent, such as *n*-hexane and carbon tetrachloride, and a small portion of the protein solution is added to it at pH 5–9 in an ice bath (Shih 1996). Phosphorylation by phosphorous oxychloride induces cross-linking between the hydroxyl groups of proteins (Matheis 1991; Schwenke et al. 2000). Polyphosphates, such as diphosphates or triphosphates, and intermolecular cross-linkage are also observed (Woo, Creamer, and Richardson 1982; Matheis et al. 1983; Matheis and Whitaker 1984; Huang and Kinsella 1986a; Schwenke et al. 2000).

Phosphoric acid (H₃PO₄) reacts with trichloroacetonitrile (Cl₃CCN) to produce trichloroacetimidoyl phosphate that can react with the hydroxyl group of serine and threonine (Figure 7.8). The reaction is performed in dimethyl sulfoxide as a solvent (Yoshikawa, Sasaki, and Chiba 1981).

Phosphorylation by STMP (NaPO₃)₃ is carried out in an alkaline condition, and the reaction is dependent on pH (Sung et al. 1983). Therefore, some of the undesirable products, such as lysinoalanine, may be produced during the reaction. STMP does not cause cross-linking and its hydrolysis produces phosphoric acid.

When phosphorus pentoxide (P₂O₅) in phosphate is used, a protein needs to be kept in the mixture for several days (Matheis and Whitaker 1984).

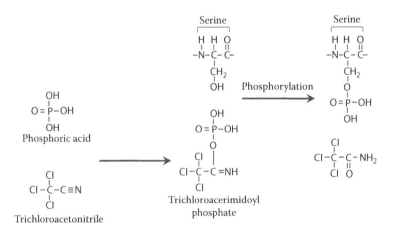

Figure 7.8 Phosphorylation by phosphoric acid with trichloroacetonitrile.

Phosphorylation occurs when a protein dissolved in phosphate buffer is lyophilized and heated for several days (Li et al. 2003; Enomoto et al. 2008; Hayashi et al. 2009). This phosphorylation by dry heating is more likely to occur with lower pH and higher temperature.

Phosphorylation increases the negative charges of proteins, which causes some changes in the functional properties (Matheis and Whitaker 1984; Kester and Richardson 1984). Various proteins have been modified by phosphorylation, such as a soybean protein (Sung et al. 1983; Hirotsuka et al. 1984), α-lactoalbumin (Vojdan and Whitaker 1996), β-lactoglobulin (Woo and Richardson 1983), egg white protein (Li et al. 2004; Hayashi et al. 2009), ovalbumin (Li et al. 2005), zein (Chobert, Sitohy, and Whitaker 1989), rapeseed (Schwenke et al. 2000; Krause 2002), and casein (Yoshikawa, Sasaki, and Chiba 1981; Matheis et al. 1983).

Solubility in water is often enhanced by phosphorylation (Chobert, Sitohy, and Whitaker 1987), but cross-linking reduces it (Matheis et al. 1983). Phosphorylation slows down digestibility, but the degree of hydrolysis of phosphorylated casein is almost the same as that of a nonphosphorylated one (Matheis et al. 1983). On the other hand, little effect was observed on the digestibility of soybean protein by phosphorylation (Sung et al. 1983). The emulsifying and foaming properties often change by phosphorylation (Matheis et al. 1983; Hirotsuka et al. 1984; Sung et al. 1983; Chobert, Sitohy, and Whitaker 1989; Vojdan and Whitaker 1996; Krause 2002; Li et al. 2004; Hayashi et al. 2009). The gelling property is improved by phosphorylation in the presence of calcium ion (Matheis et al. 1983; Woo and Richardson 1983). Transparent gels with high water-holding capacity are prepared with the protein phosphorylated by dry heating in the presence of phosphate (Enomoto et al. 2008).

7.6 Cross-Linking

Cross-linking is the reaction that covalently attaches one functional group of a side chain to another, intramolecularly or intermolecularly. The linkage is mostly a disulfide bond between cysteine residues, a monosulfide bond between cysteine and some other residues, and an isopeptide bond between lysine and some other residues.

7.6.1 Disulfide Bond Linkage

The linkage between the thiol groups of cysteine residues to form cystine is one of the popular cross-linking reactions in a protein (Figure 7.9). The properties of a protein are often changed by the disulfide bond linkage. The formed disulfide bonds stabilize the protein structure and make it resistant to enzymatic digestion. The disulfide bond linkage enhances the heat stability, viscosity, and

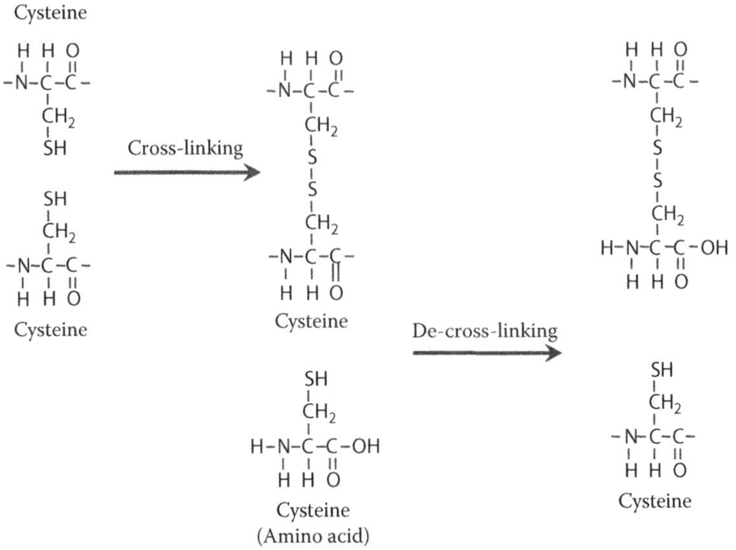

Figure 7.9 **Cross-linking of proteins by disulfide bonds and their de-cross-linking by the addition of cysteine.**

foaming and gelling properties of β-lactoglobulin (Kester and Richardson 1984). β-Lactoglobulin and κ-casein form disulfide bonds by heating, which prevents gelation (Purkayastha, Tessier, and Rose 1967), while the gelation of egg albumin depends on the sulfhydryl–disulfide interchange (Ma and Holme 1982). Disulfide bond formation affects the rheological properties of wheat flour dough (Jones, Phillips, and Hird 1974). Disulfide bonds restrict the elastic deformation of wheat flour dough, while the viscous deformation increases with increasing thiol groups (Bloksma 1975).

Disulfide bonds can be cleaved by thiols, such as mercaptoethanol or dithiothreitol, which serve as reductants. The amino acid cysteine can also be used for de-cross-linking by forming disulfide bonds between the cysteine residue of a protein and the amino acid, cysteine. Protein unfolding often occurs by this de-cross-linking, and becomes more susceptible to the reaction (Kumagai et al. 2007).

7.6.2 Monosulfide Bond Linkage (Thioether Linkage)

Cysteine residue is involved in the reaction with dehydroalanine or β-methyl-dehydroalanine to form a monosulfide bond (thioether bond) (Figure 7.10). When a protein is heated at alkaline pH, cystine, *O*-phosphorylserine, or *O*-glycosylserine become dehydroalanine, and *O*-phosphorylthreonine and *O*-glycosylthreonine become β-methyldehydroalanine by the β-elimination reaction (Lee et al. 1977; Friedman

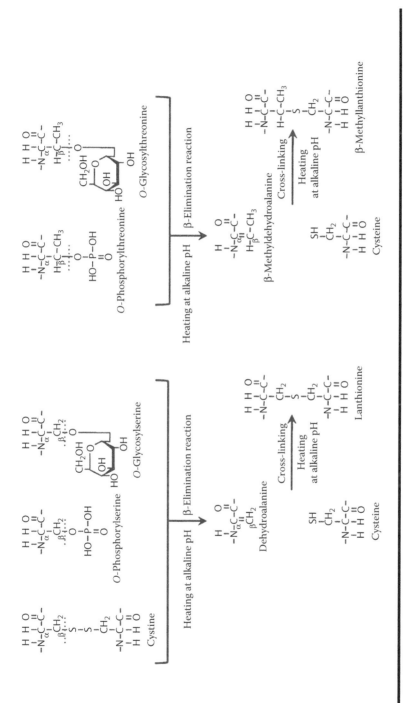

Figure 7.10 Formation of lanthionine and β-methyllanthionine by β-elimination reaction and cross-linking of dehydroalanine and β-methyldehydroalanine residues with cysteine residue.

1999). Then, these β-unsaturated carbonyl compounds proceed to cross-link with the thiol group of a cysteine residue, forming lanthionine and β-methyllanthionine from dehydroalanine and β-methyldehydroalanine, respectively.

Lanthionine and/or β-methyllanthionine are contained in lantibiotics, peptide antibiotics, such as nisin, cinnamycin, duramycin, epidermin, subtilin, mersacidin, and lacticin (Ingram 1969; Friedman 1999; Kleinnijenhuis et al. 2003; Paul and van der Donk 2005), as well as in body organs and tissues, such as the aorta, bone, collagen, dentin, and eye cataracts. As they are toxic to Gram-positive bacteria, such as *Streptococcus* and *Streptomyces*, but nontoxic to humans, lantibiotics have been used to preserve foods such as dairy products.

7.6.3 Amine Linkage

Dehydroalanine and β-methyldehydroalanine also react with the ε-amino group of lysine to form lysinoalanine and β-methyllysinoalanine, respectively (Figure 7.11) (Watanabe and Klostermeyer 1977; Friedman 1999). Although dehydroalanine reacts less favorably with the ε-amino group than the thiol group (Friedman 1999), lysinoalanine is produced more than lanthionine in some cases (Watanabe and Klostermeyer 1977). The amount of lysinoalanine in processed foods varies between 10 and 50,000 µg/g (Friedman 1999). The formation of lysinoalanine is accelerated by high pH, high temperature, and a long processing time, although lysinoalanine is both formed and degraded at very high pH, while it is suppressed by sodium sulfite, ammonia, biogenic amines, ascorbic acid, citric acid, malic acid, glucose, dephosphorylation of *O*-phosphoryl ester, and acylation of the ε-amino group of lysine. Lysinoalanine formation enhances the strength of collagen fibers (Friedman 1999). However, it is often considered to be an unfavorable reaction because it not only decreases one of the most limiting essential amino acids, lysine, but it also lowers the digestibility and protein efficiency ratio of protein (Sarwar et al. 1999; Sarwar and Sepehr 2003).

7.6.4 Isopeptide Bond Linkage

Another cross-linking reaction involving lysine is the formation of ε-(β-aspartyl) lysine and ε-(γ-glutamyl) lysine (Figures 7.12 and 7.13). Each is produced by the reaction of lysine with aspartic acid and glutamic acid, respectively, when heated (Hurrell and Carpenter 1976). ε-(γ-Glutamyl) lysine is also formed by transglutaminase (Ikura et al. 1980a, 1980b; Nonaka et al. 1989; Aboumahmoud and Savello 1990; Larré et al. 1994), which catalyzes the acyl transfer reaction between the γ-carboxyamide of glutamine and the ε-amino group of lysine. Transglutaminase also catalyzes the amine-incorporating reaction and the deamidation reaction. Cross-linking reaction occurs more favorably at neutral and weak alkaline pH (Aboumahmoud and Savello 1990; Larré et al. 1994). Isopeptide bonds in ε-(β-aspartyl) lysine and ε-(γ-glutamyl) lysine are digestible (Hurrell and Carpenter

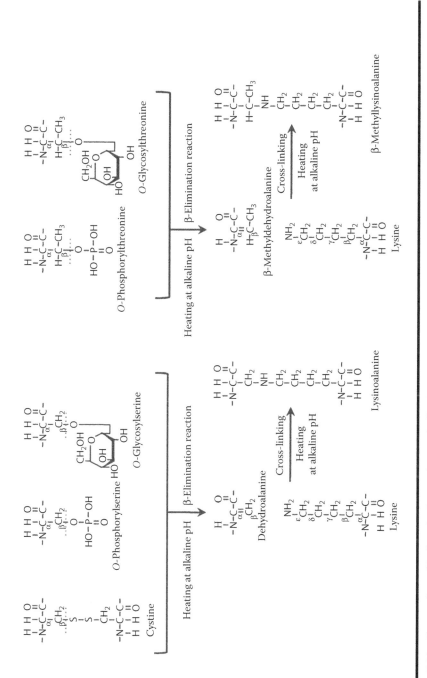

Figure 7.11 **Formation of lysinoalanine and β-methyllysinoalanine by β-elimination reaction and cross-linking of dehydroalanine and β-methyldehydroalanine residues with lysine residue.**

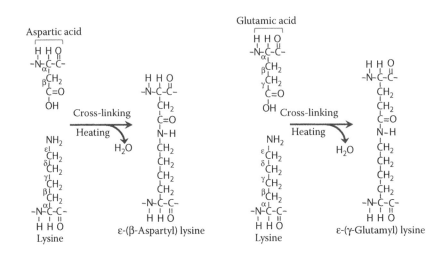

Figure 7.12 **Cross-linking of aspartic acid and glutamic acid residue with lysine residue.**

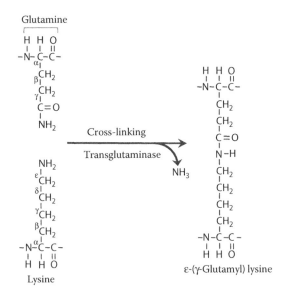

Figure 7.13 **Cross-linking of glutamine residue with lysine residue by transglutaminase.**

1976) by enzymes such as γ-glutamylamine cyclotransferase and 5-oxoprolinase in the kidney (Yasumoto and Suzuki 1990).

The solubility and emulsifying activity of αs1-casein (Motoki, Nio, and Takinami 1984) and wheat gliadin (Larré et al. 1994) increase by polymerization and deamidation with transglutaminase. A heterologous polymer produced by αs1-casein and soybean 11S globulin shows higher emulsifying activity than its equivalent mixture (Motoki, Nio, and Takinami 1987). With higher protein concentration, polymerized αs1-casein and soybean globulin form self-supporting gels (Nonaka et al. 1989).

7.7 Hydrolysis

Hydrolysis (Figure 7.14) is a reaction by which a molecule of water breaks a bond in a protein. Protein hydrolysis mostly indicates the cleavage of the peptide bond between amino acid residues. As hydrolysis increases the ionizable groups of the α-amino and carboxyl groups, and reduces the size, the produced peptides become more soluble and more amenable to chemical modification. However, hydrolysis may produce bitter-tasting peptides and reduce the processing properties.

Chemical hydrolysis is commonly performed by acid or base. Peptide bonds on either side of aspartic acid are most easily cleaved by acid hydrolysis (Han, Richard, and Biserte 1983). Serine and threonine residues at the *N*-terminal side are cleaved faster than other amino acid residues. Acid treatment often hydrolyzes acid amide in glutamine and asparagine residues and converts them into glutamic and aspartic acids. Alkali hydrolyzes peptide bonds faster than acid (Shih 1992), but alkali treatment often causes unfavorable reactions, such as lysinoalanine formation.

Acid treatment of soybean protein and wheat gluten enhances their solubility and emulsifying property, but reduces their foaming property (Wu, Nakai, and Powrie 1976). However, this effect may be due to the deamidation that

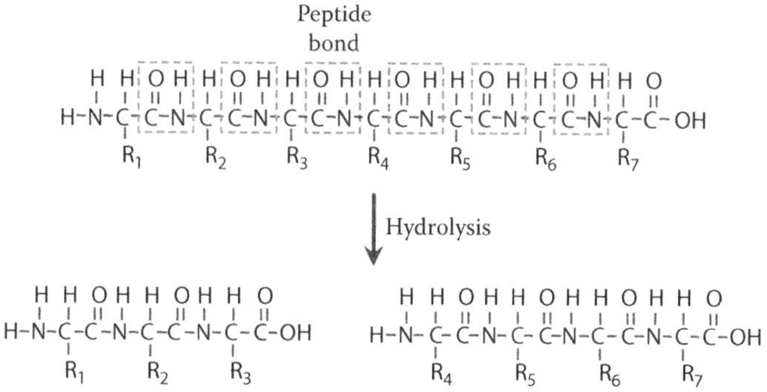

Figure 7.14 Peptide bond hydrolysis of proteins.

simultaneously occurs with peptide bond hydrolysis. Acid-treated ovalbumin is polymerized and has higher emulsifying and foaming properties than untreated ovalbumin (Matsudomi et al. 1985a).

Proteins can be hydrolyzed enzymatically by using proteases. The proteases, papain, pronase, pepsin, bromelain, alcalase, trypsin, and chymotrypsin are popularly used to hydrolyze food proteins, such as ovalbumin, soybean protein, casein, and whey protein (Kuehler and Stine 1974; Mohri and Matsushita 1984; Kim, Park, and Rhee 1990; Kumagai et al. 2002). Hydrolysis of protein often changes its solubility, water-binding property, and foaming and emulsifying properties (Kuehler and Stine 1974; Mohri and Matsushita 1984; Chobert, Sitohy, and Whitaker 1988; Kim, Park, and Rhee 1990; Kumagai et al. 2002).

7.8 Deamidation

Deamidation (Figure 7.15) is the reaction that converts glutamine and asparagine into glutamic acid and aspartic acid, respectively. Deamidation can be performed both chemically and enzymatically.

The most popular chemical modification is acid treatment (Matsudomi et al. 1985a; Ma and Khanzada 1987; Bollecker et al. 1990; Wagner and Guéguen 1995), although polymerization may occur with some proteins (Matsudomi et al. 1985b). During acid deamidation, the asparagine residue is much more favorable than the glutamine residue (Robinson, Scotchler, and McKerrow 1973). The addition of a sulfate group that attaches to a long alkyl chain, such as sodium dodecyl sulfate (SDS), or to resin in the acid solution enhances the deamidation reaction (Shih and Kalmar 1987; Shih 1987). However, certain levels of peptide bond hydrolysis are expected to obtain high levels of deamidation by acid treatment, which may lead to the production of a bitter taste and off-flavors, and the reduction of processing properties, such as gelling, foaming, and emulsification. Anions from salts, such as phosphate and bicarbonate, deamidate proteins at pH 8 and 100°C, though acetate, sulfate, and chloride anions are ineffective (Shih 1991). A carboxylated

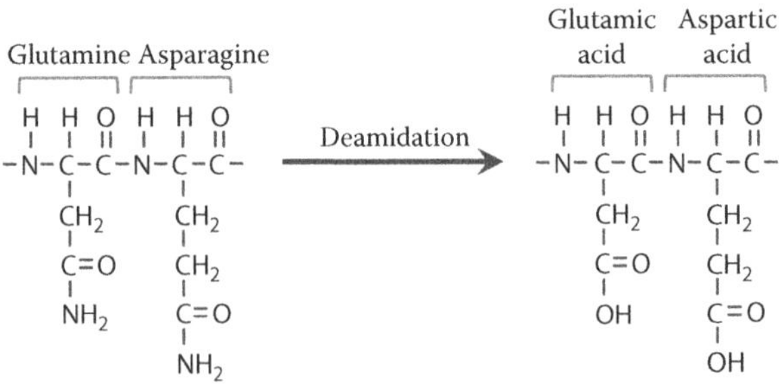

Figure 7.15 Deamidation of proteins.

cation-exchange resin is effective in deamidating proteins at neutral pH without causing peptide bond hydrolysis (Kumagai et al. 2002). Deamidation also occurs by heating (Zhang, Lee, and Ho 1993) or extrusion (Izzo, Lincoln, and Ho 1993), causing denaturation and degradation of the protein.

As for enzymatic deamidation, glutaminases, such as proteinglutaminase (Gu et al. 2001; Yong et al. 2004; Yong, Yamaguchi, and Matsumura 2006), peptidoglutaminase (Hamada et al. 1988; Hamada and Marshall 1988, 1989; Hamada 1991a, 1991b, 1992), and transglutaminase (Motoki et al. 1986; Ohtsuka et al. 2001) catalyze the deamidation reaction. However, glutaminases are only specific to glutamine residues and are not able to deamidate asparagine residue. Moreover, transglutaminase causes a cross-linking reaction through the replacement of the amide group by the amino group of proteins unless the amino group is blocked prior to the reaction by transglutaminase (Motoki et al. 1986). Some proteases, such as pepsin, trypsin, and chymotrypsin, have a function to deamidate proteins in the alkaline pH (Matsudomi et al. 1986; Kato et al. 1987a, 1987b; Shih 1990). The susceptibility to enzymatic deamidation may be enhanced by structure alteration caused by heating, proteolysis (Hamada 1992), or a chelating agent such as EDTA (Gu et al. 2001).

As cereal, pulse, and seed proteins are abundant in acid amides, wheat protein (Matsudomi, Kato, and Kobayashi 1982; Matsudomi et al. 1986; Bollecker et al. 1990; Hamada 1991b; Izzo, Lincoln, and Ho 1993; Zhang, Lee, and Ho 1993; Yong, Yamaguchi, and Matsumura 2006; Kumagai et al. 2007), soybean protein (Shih 1987, 1990, 1991; Kato et al. 1987a, 1987b; Hamada et al. 1988; Hamada and Marshall 1988, 1989; Hamada 1991a, 1991b; Zhang, Lee, and Ho 1993; Wagner and Guéguen 1995, 1999; Kumagai et al. 2002, 2004), corn protein (Hamada 1991b; Yong et al. 2004), egg protein (Kato et al. 1987a, 1987b; Hamada 1991b), milk protein (Motoki et al. 1986; Hamada 1991b; Gu et al. 2001), oilseed protein (Shih and Kalmar 1987), canola protein (Igor, Diosady, and Rubin 1993), and oat protein (Ma and Khanzada 1987) are mainly deamidated to improve their functionality.

Solubility, foaming, and emulsifying properties are often improved by deamidation (Bollecker et al. 1990; Matsudomi et al. 1985a, 1986; Motoki et al. 1986; Kato et al. 1987a; Ma and Khanzada 1987; Hamada and Marshall 1989; Wagner and Guéguen 1995, 1999; Yong et al. 2004, Yong, Yamaguchi, and Matsumura 2006; Kumagai et al. 2007). Deamidated protein prepared by cation-exchange resins is odorless and colorless because odorants and coloring substances are absorbed onto the resins. As protein deamidated by carboxylated cation-exchange resins retains its structure and is hydrophilic, it forms water-retainable gel without the addition of calcium. In addition, deamidation enhances the digestibility and reduces the allergenicity of wheat gliadin (Kumagai et al. 2007) because its epitope structure contains consecutive sequences of glutamine residues. When phytate, a calcium absorption inhibitor, is removed, the deamidated soybean protein enhances calcium absorption from the small intestine (Kumagai et al. 2004) and promotes bone formation (Kumagai, Koizumi, and Kumagai 2008).

7.9 Plastein Reaction

Plastein reaction is an enzymatic synthesis that forms a high-molecular-weight protein–like substance called plastein by transpeptidation and condensation (Yamashita et al. 1970b, 1970c; Watanabe and Arai 1992). Plastein formation usually proceeds by the following three-step process (Doucet et al. 2003). First, protein hydrolysate is produced at low substrate concentrations (3%–5%) with protease, such as papain, pepsin, α-chymotrypsin, or trypsin. Then, the hydrolysate is concentrated by evaporation, lyophilization, or spray drying. Finally, plastein is formed by incubating a high concentration of hydrolysate with protease. Papain (Yamashita, Arai, and Fujimaki 1976), α-chymotrypsin (Fujimaki et al. 1970a) or pepsin (Williams, Brownsell, and Andrews 2001) are effective for the biosynthesis. Hydrophobic amino acids, except the β-branched amino acids, are effectively incorporated into the protein (Aso et al. 1974a; Eriksen and Fagerson 1976). However, even β-branched amino acids can be incorporated if previously esterified with a hydrophobic alcohol (Aso et al. 1977).

The factors mainly affecting plastein formation are substrate size, substrate concentration, and reaction pH (Fujimaki et al. 1971; Fujimaki, Arai, and Yamashita 1977; Tsai et al. 1972; Watanabe and Arai 1992). The substrate must be a peptide–protein hydrolysate with a low molecular weight of less than 1000 and a substrate concentration in the range of 20%–50%. Plastein reaction occurs in the narrow range of pH 4–7, which is different from the optimum pH for protein hydrolysis (Yamashita et al. 1971b; Watanabe and Arai 1992).

Plastein reaction has been applied to various food proteins, such as soybean protein (Fujimaki et al. 1970a, 1970b, 1971; Yamashita et al. 1970a–d, 1971a, 1975; Yamashita, Arai, and Fujimaki 1976; Tsai et al. 1972; Fujimaki 1978), wheat protein (Fujimaki et al. 1970a; Fujimaki 1978), fish protein (Fujimaki et al. 1970a; Fujimaki 1978), egg protein (Yamashita et al. 1971a; Aso et al. 1977; Edwards and Shipe, 1978), milk protein (Yamashita et al. 1970d; Doucet et al. 2003), corn protein (Aso et al. 1974b), and mycoprotein (Williams, Brownsell, and Andrews 2001).

The plastein produced by hydrolysis and resynthesis shows similar nutritional value to the original protein (Yamashita et al. 1970a). The nutritional value of protein can be improved by the incorporation of limiting amino acid esters, such as methionine ester into soybean protein, lysine ester into wheat gluten, lysine, threonine, and tryptophan esters into zein, and tyrosine ester into fish protein after the removal of phenylalanine (Yamashita et al. 1970d, 1971a; Aso et al. 1974b; Fujimaki, Arai, and Yamashita 1977; Fujimaki 1978). The plastein reaction mostly occurs between peptides and not between a peptide and a free amino acid (Yamashita et al. 1970d; Edwards and Shipe 1978). Therefore, in order to incorporate limiting amino acids like methionine into protein, the use of protein hydrolysate rich in methionine is effective rather than that of free methionine (Yamashita et al. 1971a).

The plastein reaction is used for debittering (Fujimaki et al. 1970b; Fujimaki, Arai, and Yamashita 1977) because bitter-tasting peptides are incorporated into plastein (Yamashita et al. 1970c; Williams, Brownsell, and Andrews 2001). Odor can be removed by ether extraction after pepsin treatment (Fujimaki et al. 1970a). Water-holding capacity increases and gel is often formed, while the emulsifying property and viscosity decrease with the plastein reaction (Tsai et al. 1972). When glutamic acid is incorporated into soybean protein, solubility in water is enhanced over the wide range of pH 1–9, and does not change even after heating for 60 min (Yamashita et al. 1975).

Amino acid ester can be directly incorporated into protein by one-step aminolysis reaction (Arai, Yamashita, and Fujimaki 1978; Watanabe and Arai 1992). High protein concentration and the reaction at pH 9–10 with a cysteine proteinase, such as papain, are required for this reaction (Watanabe and Arai 1992). When a lipophilic long-chain amino acid alkyl ester is incorporated into hydrophilic protein, such as gelatin and succinylated protein, the plastein produced has an amphiphilic function and works as a surfactant. High whipping and foam stability are obtained with gelatin attached L-leucine C_6 and C_8 alkyl esters, while emulsifying activity becomes higher with longer chain alkyl esters of leucine (Watanabe et al. 1981b). Gelatin-Leu-OC_{12} can be used as an emulsifier to make ice cream, mayonnaise-like food, and bakery products (Watanabe et al. 1981a). Moreover, it can be used as an antifreeze agent to retard freezing (Arai, Watanabe, and Tsuji 1984).

7.10 Conclusion

Modification is often useful to improve the functional properties of proteins. Chemical modification is generally more effective than enzymatic or physical modification with regard to structural change. However, the reaction with the side chain of essential amino acids, such as lysine, threonine, tryptophan, histidine, and methionine, may reduce the nutritional value of proteins. Moreover, the production of undesirable by-products and the residue of harmful reagents will restrict chemical modification. On the other hand, enzymatic modification does not require any harmful reagent, and the reaction proceeds in water. Factors that limit the use of enzymes for protein modification are price and residual enzymatic activity after the reaction that may affect the color and flavor of the product.

Some of the modifications presented here improve nutritional value by incorporating limiting essential amino acids into protein. Some improve the taste and flavor by debittering and deodorizing, and also improve the texture by forming a gel or changing the surface properties. As taste and flavor degradation is mainly caused by the production of small peptides, and texture is also important for food acceptability, the modifications that do not cause peptide bond hydrolysis may be promising for food application. Although bioavailability, safety, and acceptability of modified proteins need to be carefully evaluated, chemical and enzymatic modifications are promising tools to provide proteins with desirable functionality and widen their usage.

References

Aboumahmoud, R. and Savello, P. 1990. Crosslinking of whey protein by transglutaminase. *Journal of Dairy Science* 73(2): 256–63.

Aoki, T., Kitahata, K., Fukumoto, T., Sugimoto, Y., Ibrahim, H. R., Kimura, T., Kato, Y., and Matsuda, T. 1997. Improvement of functional properties of β-lactoglobulin by conjugation with glucose-6-phosphate through the Maillard reaction. *Food Research International* 30(6): 401–6.

Arai, S., Yamashita, M., and Fujimaki, M. 1978. Nutritional improvement of food proteins by means of the plastein reaction and its novel modification. *Advances in Experimental Medicine and Biology* 105: 663–80.

Arai, S., Watanabe, M., and Tsuji, R. F. 1984. Enzymatically modified gelatin as an antifreeze protein. *Agricultural and Biological Chemistry* 48(8): 2173–5.

Aso, K., Yamashita, M., Arai, S., and Fujimaki, M. 1974a. Hydrophobic force as a main factor contributing to plastein chain assembly. *Journal of Biochemistry* 76(2): 341–7.

Aso, K., Yamashita, M., Arai, S., and Fujimaki, M. 1974b. Tryptophan-, threonine-, and lysine-enriched plasteins from zein. *Agricultural and Biological Chemistry* 38(3): 679–80.

Aso, K., Yamashita, M., Arai, S., Suzuki, J., and Fujimaki, M. 1977. Specificity for incorporation of alpha-amino acid esters during the plastein reaction by papain. *Journal of Agricultural and Food Chemistry* 25(5): 1138–41.

Bae, D. and Jang, I. S. 1999. Development of new food protein through chemical modification of rice bran proteins. *Agricultural Chemistry and Biotechnology* 42(4): 180–5.

Baniel, A., Caer, D., Colas, B., and Gueguen, J. 1992. Functional properties of glycosylated derivatives of the 11S storage protein from pea (*Pisum sativum* L.). *Journal of Agricultural and Food Chemistry* 40(2): 200–5.

Barman, B. G., Hansen, J. R., and Mossey, A. R. 1977. Modification of the physical properties of soy protein isolate by acetylation. *Journal of Agricultural and Food Chemistry* 25(3): 638–41.

Beuchat, L. R. 1977. Functional and electrophoretic characteristics of succinylated peanut flour protein. *Journal of Agricultural and Food Chemistry* 25(2): 258–61.

Bjarnason-Baumann, B., Pfaender, P., and Siebert, G. 1977. Enhancement of the biological value of whey protein by covalent addition into peptide linkage of limiting essential amino acids. *Nutrition and Metabolism* 21(Suppl. 1): 170–1.

Bloksma, A. H. 1975. Thiol and disulfide groups in dough rheology. *Cereal Chemistry* 52(3, Pt II): 170–83.

Bollecker, S., Viroben, G., Popineau, Y., and Gueguen, J. 1990. Acid deamidation and enzymic modification at pH 10 of wheat gliadins: Influence on their functional properties. *Sciences des Aliments* 10(2): 343–56.

Brinegar, A. C. and Kinsella, J. E. 1980. Reversible modification of lysine in soybean proteins, using citraconic anhydride: Characterization of physical and chemical changes in soy protein isolate, the 7S globulin, and lipoxygenase. *Journal of Agricultural and Food Chemistry* 28(4): 818–24.

Butler, P. J. G., Harris, J. I., Hartley, B. S., and Leberman, R. 1969. Use of maleic anhydride for the reversible blocking of amino groups in polypeptide chains. *Biochemical Journal* 112(5): 679–89.

Chobert, J. M., Sitohy, M., and Whitaker, J. R. 1987. Specific limited hydrolysis and phosphorylation of food proteins for improvement of functional and nutritional properties. *Journal of the American Oil Chemists' Society* 64(12): 1704–11.

Chobert, J. M., Sitohy, M., and Whitaker, J. R. 1988. Solubility and emulsifying properties of caseins modified enzymatically by *Staphylococcus aureus* V8 protease. *Journal of Agricultural and Food Chemistry* 36(1): 220–4.

Chobert, J. M., Sitohy, M., and Whitaker, J. R. 1989. Covalent attachment of phosphate and amino acids to zein. Functional and nutritional properties. *Sciences des Aliments* 9(4): 749–61.

Courthaudon, J. L., Colas, B., and Lorient, D. 1989. Covalent binding of glycosyl residues to bovine casein: Effects on solubility and viscosity. *Journal of Agricultural and Food Chemistry* 37(1): 32–6.

Doucet, D., Gauthier, S. F., Otter, D. E., and Foegeding, E. A. 2003. Enzyme-induced gelation of extensively hydrolyzed whey proteins by alcalase: Comparison with the plastein reaction and characterization of interactions. *Journal of Agricultural and Food Chemistry* 51(20): 6036–42.

Edwards, J. H. and Shipe, W. F. 1978. Characterization of plastein reaction products formed by pepsin, α-chymotrypsin, and papain treatment of egg albumin hydrolyzates. *Journal of Food Science* 43(4): 1215–8.

Eisele, T. A. and Brekke, C. J. 1981. Chemical modification and functional properties of acylated beef heart myofibrillar proteins. *Journal of Food Science* 46(4): 1095–102.

Eisele, T. A., Brekke, C. J., and McCurdy, S. M. 1981. Nutritional properties and metabolic studies of acylated beef heart myofibrillar proteins. *Journal of Food Science* 47(1): 43–8, 51.

Enomoto, H., Li, C.-P., Morizane, K., Ibrahim, H. R., Sugimoto, Y., Ohki, S., Ohtomo, H., and Aoki, T. 2008. Improvement of functional properties of bovine serum albumin through phosphorylation by dry-heating in the presence of pyrophosphate. *Journal of Food Science* 73(2): C84–91.

Eriksen, S. and Fagerson, I. S. 1976. The plastein reaction and its applications: A review. *Journal of Food Science* 41(3): 490–3.

Ferjancic-Biagini, A., Giardina, T., and Puigserver, A. 1998. Acylation of food proteins and hydrolysis by digestive enzymes: A review. *Journal of Food Biochemistry* 22(4): 331–45.

Franzen, K. L. and Kinsella, J. E. 1976. Functional properties of succinylated and acetylated soy protein. *Journal of Agricultural and Food Chemistry* 24(4): 788–95.

Fretheim, K., Iwai, S., and Feeney, R. E. 1979. Extensive modification of protein amino groups by reductive addition of different sized substituents. *International Journal of Peptide and Protein Research* 14(5): 451–6.

Friedman, M. 1999. Chemistry, biochemistry, nutrition, and microbiology of lysinoalanine, lanthionine, and histidinoalanine in food and other proteins. *Journal of Agricultural and Food Chemistry* 47(4): 1295–319.

Fujimaki, M. 1978. Nutritional improvement of food proteins by enzymatic modification, especially by plastein synthesis reaction. *Annales de la Nutrition et de l'Alimentation* 32 (2–3): 233–41.

Fujimaki, M., Arai, S., and Yamashita, M. 1977. Enzymatic protein degradation and resynthesis for protein improvement. *Advances in Chemistry Series* 160 (Food Proteins, Symp.): 156–84.

Fujimaki, M., Kato, H., Arai, S., and Yamashita, M. 1971. Application of microbial proteases to soybean and other materials to improve acceptability, especially through the formation of plastein. *Journal of Applied Bacteriology* 34(1): 119–31.

Fujimaki, M., Yamashita, M., Arai, S., and Kato, H. 1970a. Enzymic modification of proteins in foodstuffs. I. Enzymic proteolysis and plastein synthesis application for preparing bland protein-like substances. *Agricultural and Biological Chemistry* 34(9): 1325–32.

Fujimaki, M., Yamashita, M., Arai, S., and Kato, H. 1970b. Plastein reaction. Its application to debittering of proteolyzates. *Agricultural and Biological Chemistry* 34(3): 483–4.

Goulet, G., Ponnampalam, R., Amiot, J., Roy, A., and Brisson, G. J. 1987. Nutritional value of acylated oat protein concentrates. *Journal of Agricultural and Food Chemistry* 35(4): 589–92.

Gounaris, A. D. and Perlmann, G. E. 1967. Succinylation of pepsinogen. *Journal of Biological Chemistry* 242(11): 2739–45.

Grant, D. R. 1973. Modification of wheat flour proteins with succinic anhydride. *Cereal Chemistry* 50(4): 417–28.

Groninger, H. S. and Miller, R. 1979. Some chemical and nutritional properties of acylated fish protein. *Journal of Agricultural and Food Chemistry* 27(5): 949–55.

Gu, Y. S., Matsumura, Y., Yamaguchi, S., and Mori, T. 2001. Action of protein-glutaminase on α-lactalbumin in the native and molten globule states. *Journal of Agricultural and Food Chemistry* 49(12): 5999–6005.

Gueguen, J., Bollecker, S., Schwenke, K. D., and Raab, B. 1990. Effect of succinylation on some physicochemical and functional properties of the 12S storage protein from rapeseed (*Brassica napus* L.). *Journal of Agricultural and Food Chemistry* 38(1): 61–9.

Haebel, S., Albrecht, T., Sparbier, K., Walden, P., Körner, R., and Steup, M. 1998. Electrophoresis-related protein modification. Alkylation of carboxy residues revealed by mass spectrometry. *Electrophoresis* 19(5): 679–86.

Hamada, J. S. 1991a. Ultrafiltration for recovery and reuse of peptidoglutaminase in protein deamidation. *Journal of Food Science* 56(6): 1731–4.

Hamada, J. S. 1991b. Peptidoglutaminase deamidation of proteins and proteins hydrolysates for improved food use. *Journal of the American Oil Chemists' Society* 68(7): 459–62.

Hamada, J. S. 1992. Effects of heat and proteolysis on deamidation of food proteins using peptidoglutaminase. *Journal of Agricultural and Food Chemistry* 40(5): 719–23.

Hamada, J. S. and Marshall, W. E. 1988. Enhancement of peptidoglutaminase deamidation of soy protein by heat treatment and/or proteolysis. *Journal of Food Science* 53(4): 1132–4, 1149.

Hamada, J. S. and Marshall, W. E. 1989. Preparation and functional properties of enzymically deamidated soy proteins. *Journal of Food Science* 54(3): 598–601, 635.

Hamada, J. S., Shih, F. F., Frank, A. W., and Marshall, W. E. 1988. Deamidation of soy peptides and proteins by *Bacillus circulans* peptidoglutaminase. *Journal of Food Science* 53(2): 671–2.

Han, K. K., Richard, C., and Biserte, G. 1983. Current developments in chemical cleavage of proteins. *International Journal of Biochemistry* 15(7): 875–84.

Hayashi, Y., Nagano, S., Enomoto, H., Li, C.-P., Sugimoto, Y., Ibrahim, H. R., Hatta, H., Takeda, C., and Aoki, T. 2009. Improvement of foaming property of egg white protein by phosphorylation through dry-heating in the presence of pyrophosphate. *Journal of Food Science* 74(1): C68–72.

Hirotsuka, M., Taniguchi, H., Narita, H., and Kito, M. 1984. Functionality and digestibility of a highly phosphorylated soybean protein. *Agricultural and Biological Chemistry* 48(1): 93–100.

Huang, Y. T. and Kinsella, J. E. 1986a. Phosphorylation of yeast protein: Reduction of ribonucleic acid and isolation of yeast protein concentrate. *Biotechnology and Bioengineering* 28(11): 1690–8.

Huang, Y. T. and Kinsella, J. E. 1986b. Functional properties of phosphorylated yeast protein: Solubility, water-holding capacity, and viscosity. *Journal of Agricultural and Food Chemistry* 34(4): 670–4.

Hurrell, R. F. and Carpenter, K. J. 1976. Mechanisms of heat damage in proteins. 7. The significance of lysine-containing isopeptides and of lanthionine in heated proteins. *British Journal of Nutrition* 35(3): 383–95.

Igor, S. O., Diosady, L. L., and Rubin, L. J. 1993. Catalytic deamidation of canola proteins. *Acta Alimentaria* 22(4): 325–36.

Ikura, K., Kometani, T., Sasaki, R., and Chiba, H. 1980a. Crosslinking of soybean 7S and 11S proteins by transglutaminase. *Agricultural and Biological Chemistry* 44(12): 2979–84.

Ikura, K., Kometani, T., Yoshikawa, M., Sasaki, R., and Chiba, H. 1980b. Crosslinking of casein components by transglutaminase. *Agricultural and Biological Chemistry* 44(7): 1567–73.

Ingram, L. C. 1969. Synthesis of the antibiotic, nisin: Formation of lanthionine and β-methyllanthionine. *Biochimica et Biophysica Acta* 184(1): 216–9.

Izzo, H. V., Lincoln, M. D., and Ho, C. T. 1993. Effect of temperature, feed moisture, and pH on protein deamidation in an extruded wheat flour. *Journal of Agricultural and Food Chemistry* 41(2): 199–202.

Jiménez-Castaño, L., Villamiel, M., and López-Fandiño, R. 2007. Glycosylation of individual whey proteins by Maillard reaction using dextran of different molecular mass. *Food Hydrocolloids* 21(3): 433–43.

Jones, I. K., Phillips, J. W., and Hird, F. J. R. 1974. Estimation of rheologically important thiol and disulfide groups in dough. *Journal of the Science of Food and Agriculture* 25(1): 1–10.

Katayama, S., Shima, J., and Saeki, H. 2002. Solubility improvement of shellfish muscle proteins by reaction with glucose and its soluble state in low-ionic-strength medium. *Journal of Agricultural and Food Chemistry* 50(15): 4327–32.

Kato, A. 2002. Industrial applications of Maillard-type protein-polysaccharide conjugates. *Food Science and Technology Research* 8(3): 193–9.

Kato, Y., Aoki, T., Kato, N., Nakamura, R., and Matsuda, T. 1995. Modification of ovalbumin with glucose 6-phosphate by amino-carbonyl reaction. Improvement of protein heat stability and emulsifying activity. *Journal of Agricultural and Food Chemistry* 43(2): 301–5.

Kato, Y., Matsuda, T., Watanabe, K., and Nakamura, R. 1983. Immunochemical studies on the denaturation of ovalbumin stored with glucose. *Journal of Food Science* 48(3): 769–72.

Kato, A., Tanaka, A., Lee, Y., Matsudomi, N., and Kobayashi, K. 1987a. Effects of deamidation with chymotrypsin at pH 10 on the functional properties of proteins. *Journal of Agricultural and Food Chemistry* 35(2): 285–8.

Kato, A., Tanaka, A., Matsudomi, N., and Kobayashi, K. 1987b. Deamidation of food proteins by protease in alkaline pH. *Journal of Agricultural and Food Chemistry* 35(2): 224–7.

Kato, Y., Watanabe, K., and Sato, Y. 1981. Effect of Maillard reaction on some physical properties of ovalbumin. *Journal of Food Science* 46(6): 1835–9.

Kester, J. J. and Richardson, T. 1984. Modification of whey proteins to improve functionality. *Journal of Dairy Science* 67(11): 2757–74.

Kim, S. Y., Park, P. S. W., and Rhee, K. C. 1990. Functional properties of proteolytic enzyme modified soy protein isolate. *Journal of Agricultural and Food Chemistry* 38(3): 651–6.

King, A. J., Ball, H. R., and Garlich, J. D. 1981. A chemical and biological study of acylated egg white. *Journal of Food Science* 46(4): 1107–10.

Kinsella, J. E. and Shetty, K. J. 1979. Chemical modification for improving functional properties of plant and yeast proteins. In *Functionality and Protein Structure*, ed. A. Pour-El, ACS Symposium Series 92, 37–63. Washington, DC: American Chemical Society.

Kleinnijenhuis, A. J., Duursma, M. C., Breukink, E., Heeren, R. M. A., and Heck, A. J. R. 2003. Localization of intramolecular monosulfide bridges in lantibiotics determined with electron capture induced dissociation. *Analytical Chemistry* 75(13): 3219–25.

Krause, J.-P. 2002. Comparison of the effect of acylation and phosphorylation on surface pressure, surface potential and foaming properties of protein isolates from rapeseed (*Brassica napus*). *Industrial Crops and Products* 15(3): 221–8.

Krause, J.-P. and Schwenke, K. D. 1996. Relationships between adsorption and emulsifying of acetylated protein isolates from faba beans (*Vicia faba* L.). *Nahrung* 40(1): 12–7.

Kuehler, C. A. and Stine, C. M. 1974. Effect of enzymatic hydrolysis on some functional properties of whey protein. *Journal of Food Science* 39(2): 379–82.

Kumagai, H., Ishida, S., Koizumi, A., Sakurai, H., and Kumagai, H. 2002. Preparation of phytate-removed deamidated soybean globulins by ion exchangers and characterization of their calcium-binding ability. *Journal of Agricultural and Food Chemistry* 50(1): 172–6.

Kumagai, H., Koizumi, A., and Kumagai, H. 2008. Promotion of bone formation by phytate-removed deamidated soybean glicinin. In *Functional Foods and Health*, eds. T. Shibamoto, K. Kanazawa, F. Shahidi, and C.-T. Ho, ACS Symposium Series 993, 419–28. Washington, DC: American Chemical Society.

Kumagai, H., Koizumi, A., Suda, A., Sato, N., Sakurai, H., and Kumagai, H. 2004. Enhanced calcium absorption in the small intestine by a phytate-removed deamidated soybean globulin preparation. *Bioscience, Biotechnology, and Biochemistry* 68(7): 1598–600.

Kumagai, H., Seto, H., Norimatsu, Y., Ishii, K., and Kumagai, H. 2002. Changes in activity coefficient γ_w of water and the foaming capacity of protein during hydrolysis. *Bioscience, Biotechnology, and Biochemistry* 66(7): 1455–61.

Kumagai, H., Suda, A., Sakurai, H., Kumagai, H., Arai, S., Inomata, N., and Ikezawa, Z. 2007. Improvement of digestibility, reduction in allergenicity, and induction of oral tolerance of wheat gliadin by deamidation. *Bioscience, Biotechnology, and Biochemistry* 71(4): 977–85.

Larré, C., Chiarello, M., Blanloeil, Y., Chenu, M., and Gueguen, J. 1994. Gliadin modifications catalyzed by guinea pig liver transglutaminase. *Journal of Food Biochemistry* 17(4): 267–82.

Lawal, O. S. and Adebowale, K. O. 2006. The acylated protein derivatives of *Canavalia ensiformis* (jack bean): A study of functional characteristics. *LWT – Food Science and Technology* 39(8): 918–29.

Lee, H. S., Osuga, D. T., Nashef, A. S., Ahmed, A. I., Whitaker, J. R., and Feeney, R. E. 1977. Effects of alkali on glycoproteins. β-Elimination and nucleophilic addition reactions of substituted threonyl residues of antifreeze glycoprotein. *Journal of Agricultural and Food Chemistry* 25(5): 1153–8.

Lee, H. S., Sen, L. C., Clifford, A. J., Whitaker, J. R., and Feeney, R. E. 1979. Preparation and nutritional properties of caseins covalently modified with sugars. Reductive alkylation of lysines with glucose, fructose, or lactose. *Journal of Agricultural and Food Chemistry* 27(5): 1094–8.

Li, C.-P., Hayashi, Y., Shinohara, H., Ibrahim, H. R., Sugimoto, Y., Kurawaki, J., Matsudomi, N., and Aoki, T. 2005. Phosphorylation of ovalbumin by dry-heating in the presence of pyrophosphate: Effect on protein structure and some properties. *Journal of Agricultural and Food Chemistry* 53(12): 4962–7.

Li, C.-P., Ibrahim, H. R., Sugimoto, Y., Hatta, H., and Aoki, T. 2004. Improvement of functional properties of egg white protein through phosphorylation by dry-heating in the presence of pyrophosphate. *Journal of Agricultural and Food Chemistry* 52(18): 5752–8.

Li, C.-P., Salvador, A. S., Ibrahim, H. R., Sugimoto, Y., and Aoki, T. 2003. Phosphorylation of egg white proteins by dry-heating in the presence of phosphate. *Journal of Agricultural and Food Chemistry* 51(23): 6808–15.

Ma, C. Y. 1984. Functional properties of acylated oat protein. *Journal of Food Science* 49(4): 1128–31.

Ma, C. Y. and Holme, J. 1982. Effect of chemical modifications on some physicochemical properties and heat coagulation of egg albumen. *Journal of Food Science* 47(5): 1454–9.

Ma, C. Y. and Khanzada, G. 1987. Functional properties of deamidated oat protein isolates. *Journal of Food Science* 52(6): 1583–7.

Marsh, J. W., Denis, J., and Wriston, J. C., Jr. 1977. Glycosylation of *Escherichia coli* L-asparaginase. *Journal of Biological Chemistry* 252(21): 7678–84.

Matheis, G. 1991. Phosphorylation of food proteins with phosphorus oxychloride – improvement of functional and nutritional properties: A review. *Food Chemistry* 39(1): 13–26.

Matheis, G., Penner, M. H., Feeney, R. E., and Whitaker, J. R. 1983. Phosphorylation of casein and lysozyme by phosphorus oxychloride. *Journal of Agricultural and Food Chemistry* 31(2): 379–87.

Matheis, G. and Whitaker, J. R. 1984. Chemical phosphorylation of food proteins: An overview and a prospectus. *Journal of Agricultural and Food Chemistry* 32(4): 699–705.

Matsudomi, N., Kato, A., and Kobayashi, K. 1982. Conformation and surface properties of deamidated gluten. *Agricultural and Biological Chemistry* 46(6): 1583–6.

Matsudomi, N., Sasaki, T., Kato, A., and Kobayashi, K. 1985a. Conformational changes and functional properties of acid-modified soy protein. *Agricultural and Biological Chemistry* 49(5): 1251–6.

Matsudomi, N., Sasaki, T., Tanaka, A., Kobayashi, K., and Kato, A. 1985b. Polymerization of deamidated peptide fragments obtained with the mild acid hydrolysis of ovalbumin. *Journal of Agricultural and Food Chemistry* 33(4): 738–42.

Matsudomi, N., Tanaka, A., Kato, A., and Kobayashi, K. 1986. Functional properties of deamidated gluten obtained by treating with chymotrypsin at alkali pH. *Agricultural and Biological Chemistry* 50(8): 1989–94.

Means, G. E. 1977. Reductive alkylation of amino groups. *Methods in Enzymology* 47: 469–78.

Mohri, M. and Matsushita, S. 1984. Improvement of water absorption of soybean protein by treatment with bromelain. *Journal of Agricultural and Food Chemistry* 32(3): 486–90.

Motoki, M., Nio, N., and Takinami, K. 1984. Functional properties of food proteins polymerized by transglutaminase. *Agricultural and Biological Chemistry* 48(5): 1257–61.

Motoki, M., Nio, N., and Takinami, K. 1987. Functional properties of heterologous polymer prepared by transglutaminase between milk casein and soybean globulin. *Agricultural and Biological Chemistry* 51(1): 237–9.

Motoki, M., Seguro, K., Nio, N., and Takinami, K. 1986. Glutamine-specific deamidation of αs1-casein by transglutaminase. *Agricultural and Biological Chemistry* 50(12): 3025–30.

Nakamura, S., Suzuki, Y., Ishikawa, E., Yakushi, T., Jing, H., Miyamoto, T., and Hashizume, K. 2008. Reduction of *in vitro* allergenicity of buckwheat Fag e 1 through the Maillard-type glycosylation with polysaccharides. *Food Chemistry* 109(3): 538–45.

Nonaka, M., Tanaka, H., Okiyama, A., Motoki, M., Ando, H., Umeda, K., and Matsuura, A. 1989. Polymerization of several proteins by calcium-independent transglutaminase derived from microorganisms. *Agricultural and Biological Chemistry* 53(10): 619–23.

Ohtsuka, T., Umezawa, Y., Nio, N., and Kubota, K. 2001. Comparison of deamidation activity of transglutaminases. *Journal of Food Science* 66(1): 25–9.

Paraman, I., Hettiarachchy, N. S., and Schaefer, C. 2007. Glycosylation and deamidation of rice endosperm protein for improved solubility and emulsifying properties. *Cereal Chemistry* 84(6): 593–9.

Paul, M. and van der Donk, W. A. 2005. Chemical and enzymatic synthesis of lanthionines. *Mini-Reviews in Organic Chemistry* 2(1): 23–37.

Pomianowski, J. F., Borowski, J., and Danowska-Oziewicz, M. 1999. The characteristics of selected physico-chemical properties of chemically modified myofibrillar protein. *Nahrung* 43(2): 90–4.

Ponnampalam, R., Goulet, G., Amiot, J., Chamberland, B., and Brisson, G. J. 1988. Some functional properties of acetylated and succinylated oat protein concentrates and a blend of succinylated oat protein and whey protein concentrates. *Food Chemistry* 29(2): 109–18.

Puigserver, A. J., Sen, L. C., Gonzales-Flores, E., Feeney, R. E., and Whitaker, J. R. 1979a. Covalent attachment of amino acids to casein. 1. Chemical modification and rates of *in vitro* enzymic hydrolysis of derivatives. *Journal of Agricultural and Food Chemistry* 27(5): 1098–104.

Puigserver, A. J., Sen, L. C., Clifford, A. J., Feeney, R. E., and Whitaker, J. R. 1979b. Covalent attachment of amino acids to casein. 2. Bioavailability of methionine and *N*-acetylmethionine covalently linked to casein. *Journal of Agricultural and Food Chemistry* 27(6): 1286–93.

Purkayastha, R., Tessier, H., and Rose, D. 1967. Thio-disulfide interchange in formation of β-lactoglobulin-κ-casein complex. *Journal of Dairy Science* 50(5): 764–6.

Robinson, A. B., Scotchler, J. W., and McKerrow, J. H. 1973. Rates of nonenzymic deamidation of glutaminyl and asparaginyl residues in pentapeptides. *Journal of the American Chemical Society* 95(24): 8156–9.

Saeki, H. 1999. Functional improvement of fish myofibrillar protein by glycosylation. *Recent Research Developments in Agricultural & Food Chemistry* 3(1): 149–57.

Sarwar, G., L'Abbé, M. R., Trick, K., Botting, H. G., and Ma, C. Y. 1999. Influence of feeding alkaline/heat processed proteins on growth and protein and mineral status of rats. *Advances in Experimental Medicine and Biology* 459: 161–77.

Sarwar, G. G. and Sepehr, E. 2003. Protein digestibility and quality in products containing antinutritional factors are adversely affected by old age in rats. *Journal of Nutrition* 133(1): 220–5.

Schwartz, B. A. and Gray, G. R. 1977. Proteins containing reductively aminated disaccharides. Synthesis and chemical characterization. *Archives of Biochemistry and Biophysics* 181(2): 542–9.

Schwenke, K. D., Dahme, A., and Wolter, T. 1998. Heat-induced gelation of rapeseed proteins: Effect of protein interaction and acetylation. *Journal of the American Oil Chemists' Society* 75(1): 83–7.

Schwenke, K. D., Mothes, R., Dudek, S., and Göernitz, E. 2000. Phosphorylation of the 12S globulin from rapeseed (*Brassica napus* L.) by phosphorous oxychloride: Chemical and conformational aspects. *Journal of Agricultural and Food Chemistry* 48(3): 708–15.

Shahidi, F. and Wanasundara, P. K. J. P. D. 1998. Effect of acylation on flax protein functionality. In *Functional Properties of Proteins and Lipids*, eds. J. R. Whitaker, F. Shahidi, A. L. Munguia, R. Y. Yada, and G. Fuller, ACS Symposium Series 708, 96–120. Washington, DC: American Chemical Society.

Shih, F. F. 1987. Deamidation of protein in a soy extract by ion exchange resin catalysis. *Journal of Food Science* 52(6): 1529–31.

Shih, F. F. 1990. Deamidation during treatment of soy protein with protease. *Journal of Food Science* 55(1): 127–9, 132.

Shih, F. F. 1991. Effect of anions on the deamidation of soy protein. *Journal of Food Science* 56(2): 452–4.

Shih, F. F. 1992. Modification of food proteins by non-enzymic methods. In *Biochemistry of Food Proteins*, ed. B. J. F. Hudson, 235–48. London: Elsevier.

Shih, F. F. 1996. Deamidation and phosphorylation for food protein modification. In *Surface Activity of Proteins*, ed. S. Magdassi, 91–113. New York: Dekker.

Shih, F. F. and Kalmar, A. D. 1987. SDS-catalyzed deamidation of oilseed proteins. *Journal of Agricultural and Food Chemistry* 35(5): 672–5.

Shukla, T. P. 1982. Chemical modification of food proteins. In *Mechanisms and Functionality*, ed. J. P. Cherry, ACS Symposium Series 206, 275–300. Washington, DC: American Chemical Society.

Sitohy, M. Z., Sharobeem, S. F., and Abdel-Ghany, A. A. 1992. Functional properties of some succinylated protein preparations. *Acta Alimentaria* 21(1): 31–8.

Sung, H. Y., Chen, H. J., Liu, T. Y., and Su, J. C. 1983. Improvement of the functionalities of soy protein isolate through chemical phosphorylation. *Journal of Food Science* 48(3): 716–21.

Szymkiewicz, A. and Jedrychowski, L. 2008. Effect of acylation and enzymatic modification on pea proteins allergenicity. *Polish Journal of Food and Nutrition Sciences* 58(3): 345–50.

Thompson, L. U. and Reyes, E. S. 1980. Modification of heat-coagulated whey protein concentrates by succinylation. *Journal of Dairy Science* 63(5): 715–21.

Tsai, S.-J., Yamashita, M., Arai, S., and Fujimaki, M. 1972. Effect of substrate concentration on plastein productivity and some rheological properties of the products. *Agricultural and Biological Chemistry* 36(6): 1045–9.

Vojdani, F. and Whitaker, J. R. 1996. Phosphorylation of proteins and their functional and structural properties. In *Macromolecular Interactions in Food Technology*, eds. N. Parris, A. Kato, L. K. Creamer, and J. Pearce, *ACS Symposium Series* 650, 210–29. Washington, DC: American Chemical Society.

Wagner, J. R. and Guéguen, J. 1995. Effects of dissociation, deamidation, and reducing treatment on structural and surface active properties of soy glycinin. *Journal of Agricultural and Food Chemistry* 43(8): 1993–2000.

Wagner, J. R. and Guéguen, J. 1999. Surface functional properties of native, acid-treated, and reduced soy glycinin. 2. Emulsifying properties. *Journal of Agricultural and Food Chemistry* 47(6): 2181–7.

Wahyuni, M., Ishizaki, S., and Tanaka, M. 1999. Modification of fish water soluble proteins with glucose-6-phosphate through the Maillard reaction-II. Improvement of functional properties of fish water soluble proteins with glucose-6-phosphate through the Maillard reaction. *Fisheries Science* 65(4): 618–22.

Waniska, R. D. and Kinsella, J. E. 1988. Foaming and emulsifying properties of glycosylated beta-lactoglobulin. *Food Hydrocolloids* 2(6): 439–9.

Watanabe, K. and Klostermeyer, H. 1977. Formation of dehydroalanine, lanthionine, and lysinoalanine during heat treatment of β-lactoglobulin A. *Zeitschrift für Lebensmittel-Untersuchung und -Forschung* 164(2): 77–9.

Watanabe, M. and Arai, S. 1992. The plastein reaction: Fundamentals and applications. In *Biochemistry of Food Proteins*, ed. B. J. F. Hudson, 271–305. London: Elsevier.

Watanabe, M., Shimada, A., Yazawa, E., Kato, T., and Arai, S. 1981a. Proteinaceous surfactants produced from gelatin by enzymatic modification: Application to preparation of food items. *Journal of Food Science* 46(6): 1738–40.

Watanabe, M., Toyokawa, H., Shimada, A., and Arai, S. 1981b. Proteinaceous surfactants produced from gelatin by enzymatic modification: Evaluation for their functionality. *Journal of Food Science* 46(5): 1467–9.

Williams, R. J. H., Brownsell, V. L., and Andrews, A. T. 2001. Application of the plastein reaction to mycoprotein: I. Plastein synthesis. *Food Chemistry* 72(3): 329–35.

Woo, S. L., Creamer, L. K., and Richardson, T. 1982. Chemical phosphorylation of bovine β-lactoglobulin. *Journal of Agricultural and Food Chemistry* 30(1): 65–70.

Woo, S. L. and Richardson, T. 1983. Functional properties of phosphorylated β-lactoglobulin. *Journal of Dairy Science* 66(5): 984–7.

Wu, C. H., Nakai, S., and Powrie, W. D. 1976. Preparation and properties of acid-solubilized gluten. *Journal of Agricultural and Food Chemistry* 24(3): 504–10.

Yamashita, M., Arai, S., and Fujimaki, M. 1976. Plastein reaction for food protein improvement. *Journal of Agricultural and Food Chemistry* 24(6): 1100–4.

Yamashita, M., Arai, S., Gonda, M., Kato, H., and Fujimaki, M. 1970a. Enzymatic modification of proteins in foodstuffs. II. Nutritive properties of soy plastein and its bio-utility evaluation in rats. *Agricultural and Biological Chemistry* 34(9): 1333–7.

Yamashita, M., Arai, S., Kokubo, S., Aso, K., and Fujimaki, M. 1975. Synthesis and characterization of a glutamic acid enriched plastein with greater solubility. *Journal of Agricultural and Food Chemistry* 23(1): 27–30.

Yamashita, M., Arai, S., Matsuyama, J., Gonda, M., Kato, H., and Fujimaki, M. 1970b. Enzymatic modification of proteins in foods. III. Phenomenal survey on α-chymotryptic plastein synthesis from peptic hydrolyzate of soy protein. *Agricultural and Biological Chemistry* 34(10): 1484–91.

Yamashita, M., Arai, S., Matsuyama, J., Kato, H., and Fujimaki, M. 1970c. Enzymatic modification of proteins in foods. IV. Bitter dipeptides as plastein-building blocks debittering of peptic proteolyzate with α-chymotrypsin. *Agricultural and Biological Chemistry* 34(10): 1492–1500.

Yamashita, M., Arai, S., Tsai, S.-J., and Fujimaki, M. 1970d. Supplementing sulfur-containing amino acids by plastein reaction. *Agricultural and Biological Chemistry* 34(10): 1593–6.

Yamashita, M., Arai, S., Tsai, S.-J., and Fujimaki, M. 1971a. Plastein reaction as a method for enhancing the sulfur-containing amino acid level of soybean protein. *Journal of Agricultural and Food Chemistry* 19(6): 1151–4.

Yamashita, M., Tsai, S.-J., Arai, S., Kato, H., and Fujimaki, M. 1971b. Enzymatic modification of proteins in foods. V. Plastein yields and their pH dependence. *Agricultural and Biological Chemistry* 35(1): 86–91.

Yasumoto, K. and Suzuki, F. 1990. Aspartyl- and glutamyllysine crosslinks formation and their nutritional availability. *Journal of Nutritional Science and Vitaminology* 36(4), Suppl. 1: S71–77.

Yong, Y. H., Yamaguchi, S., Gu, Y. S., Mori, T., and Matsumura, Y. 2004. Effects of enzymatic deamidation by protein-glutaminase on structure and functional properties of α-zein. *Journal of Agricultural and Food Chemistry* 52(23): 7094–100.

Yong, Y. H., Yamaguchi, S., and Matsumura, Y. 2006. Effects of enzymatic deamidation by protein-glutaminase on structure and functional properties of wheat gluten. *Journal of Agricultural and Food Chemistry* 54(16): 6034–40.

Yoshikawa, M., Sasaki, R., and Chiba, H. 1981. Effects of chemical phosphorylation of bovine casein components on the properties related to casein micelle formation. *Agricultural and Biological Chemistry* 45(4): 909–14.

Zhang, J., Lee, T. C., and Ho, C.-T. 1993. Thermal deamidation of proteins in a restricted water environment. *Journal of Agricultural and Food Chemistry* 41(11): 1840–3.

Chapter 8

Heat-Induced Casein–Whey Protein Interactions

Sundaram Gunasekaran and Oscar Solar

Contents

8.1 Introduction

The heat treatment of milk is a common practice in the food processing industry, primarily to reduce the content of both the pathogenic and the spoilage microorganisms that are typically present in raw milk to a level that ensures safe consumption and/or further processing of the milk. Thus, the thermal treatment of milk is a necessary step to ensure the safety and shelf stability of milk and milk-based products. When this heat treatment reaches a level that is sufficient to produce thermal denaturation in heat-sensitive proteins, such as whey proteins (WPs), the physical properties are affected and structural conformational changes take place, leading to intermolecular and intramolecular protein aggregation and interactions between casein (CN), the primary milk protein, and WPs. Although these interactions could be seen as undesirable because they greatly affect the texture of the final products, such as cheese and yogurt, they have been considered beneficial, for example, to reduce syneresis and to retain more of the WPs and their well-known nutritional and therapeutic health benefits (Guyomarc'h 2006; Hinrichs 2001; Yalcin 2006).

A thorough knowledge of the main factors and levels affecting the heat-induced CN–WP interactions in bovine milk systems is crucial in understanding how they can be used advantageously, not only to control and/or improve the quality of processed milk-based products but also to enhance the myriad of nutritional and health benefits of WPs. In this chapter, the relevant research performed within the last decade regarding the mechanisms and factors affecting the heat-induced CN–WP interactions in bovine milk systems and the techniques used to quantify these interactions are presented. Among the factors studied are the time–temperature and pH during milk heating, protein concentration and ratio, and how these affect the properties of milk and milk gels.

8.2 Milk Proteins

Although the qualitative composition of fresh bovine milk can be considered roughly constant, its quantitative composition is affected by several factors, such as the genetics, feed, climate, illness history, and stage of lactation of the cow (Walstra et al. 1999). The protein content of fresh milk is about 3.3% w/w, of which CN micelles and WPs constitute about 2.6% and 0.7% w/w, respectively. Therefore, it is possible to say that the CN:WP ratio in natural bovine milk is roughly 80:20.

8.2.1 Caseins

CN micelles are units comprising essentially four different CNs: $\alpha s1$, $\alpha s2$, $(\beta + \gamma)$, and κ, which are found in a molar ratio of 4:1:3:1.3, respectively (de Kruif and May 1991). Both $\alpha s2$- and κ-CNs have one disulfide linkage, the former in the Cys36–Cys41 and the latter in the Cys11–Cys88 position; therefore, under specific conditions, both proteins have the potential to interact with other molecules containing disulfide bonds. The average protein composition in bovine milk is presented in Table 8.1.

The structure of the CN micelle is still unknown, but some models have been proposed. The classical submicelle model proposed by Morr, Slattery, Schmidt, and

Table 8.1   Protein Composition in Bovine Milk

	αs1-CN	αs2-CN	β-CN	γ-CN	κ-CN	β-LG	α-LA
Concentration (g/kg milk)	10.0	2.6	9.3	0.8	3.3	3.2	1.2
Molecular weight (kDa)	23.6	25.3	24.0	~11.7	19.5	18.3	14.2
No. of amino acids	199	207	209	–	169	162	123
Cysteine residues	–	2	–	–	2	5	8
Free thiol group	–	–	–	–	–	1 (Cys121)	–
Isoelectric point	4.6 for CN micelles					5.1	4.2–4.5

Walstra (Walstra et al. 1999) has been widely accepted and supported. This model defines the CN micelle as roughly a spherical particle of about 100 nm in diameter with an external hairy surface comprising the κ-CN C-terminals and inner subunits called submicelles, formed by either αs-CN and β-CN or αs-CN and κ-CN (Figure 8.1). This model also assumes that the submicelles could be linked together by colloidal calcium phosphate (CCP) clusters and hydrophobic interactions. The external hairy layer comprising the C-terminals of κ-CN protrudes from the micelle surface, preventing any further aggregation with other submicelles and stabilizing the micelle against flocculation via steric and electrostatic repulsion. However, the existence of well-defined submicelle units has been questioned by Holt and coworkers (Holt 1992; Holt and Horne 1996), who proposed a different model that suggests a fairly open and fluid structure without submicelles (Figure 8.1). The Holt model assumes that different CNs within the micelle are held together via calcium phosphate cross-links, which are in the form of nanoclusters. Horne (2006) argued that both the submicelle and the Holt models are particular cases of the dual-binding model that he had proposed (Figure 8.1). The dual-binding model states that both the attractive hydrophobic interactions and the electrostatic repulsion forces play a balanced role in the association of the CNs and the micelle structure, where calcium phosphate nanoclusters are also crucial for cross-linking and for the aggregative stability of the micelle.

8.2.2 Whey Proteins

WPs, also called milk serum proteins, include the globular proteins β-lactoglobulin (β-LG), α-lactalbumin (α-LA), bovine serum albumin (BSA), immunoglobulin (Ig),

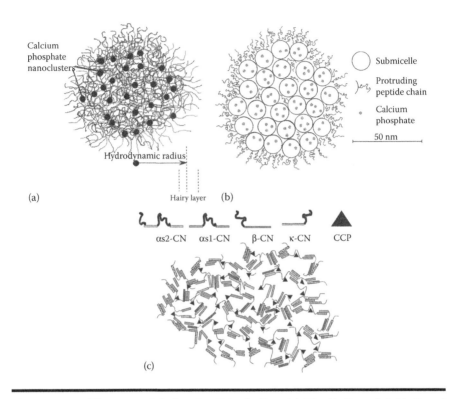

Figure 8.1   Different models for the CN micelles. (a) The Holt model (Holt and Horne 1996). (b) The submicelle model (Walstra 1999). (c) The dual-binding model (Horne 1998) in a redrawn version. (From Lucey, J. A., *Journal of Dairy Science* 85, 281–94, 2002.)

and other minor proteins, such as serum CN, not associated with micelles and enzymes. The native structure of β-LG (Figure 8.2) is well known from both x-ray crystallographic studies and high-resolution nuclear magnetic resonance studies (Creamer et al. 2004). It is conformed by 162 amino acid residues and has two disulfide linkages (Cys106–Cys119 and Cys66–Cys160) plus one very reactive free sulfhydryl group (Cys121), which plays an important role in the self-aggregation and association with proteins containing other disulfide linkages. A more detailed description of the properties, structure, and function of β-LG can be found elsewhere (Kontopidis, Holt, and Sawyer 2004). After β-LG, the second major WP is α-LA, which consists of 123 amino acid residues. α-LA has four disulfide linkages, which can interact with the β-LG free thiol group, when induced. Additional information on milk components and their properties is available from other sources (Walstra et al. 1999).

Recently, Considine et al. (2007) reviewed the milk protein interactions as a function of heat and high hydrostatic pressure and detailed the denaturation mechanisms of the WPs and the protein–protein interactions in milk and model

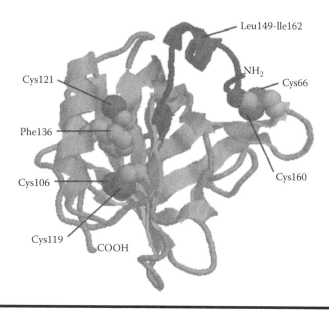

Figure 8.2   Diagram of the three-dimensional structure of β-LG. (From Creamer, L. K., et al., *Journal of Agricultural and Food Chemistry*, 52, 7660–68, 2004.)

systems. The objective of this review is to present in-depth information on the factors affecting specifically the heat-induced CN–WP interactions in bovine milk systems.

8.3   Quantification of Protein Interactions

In the study of heat-induced protein interactions in liquid systems, the analytical techniques are mostly based on the separation and subsequent analysis of the soluble and nonsoluble fractions; whereas in the analysis of protein interactions in milk gels, the proteins conform to a network, which requires different analysis methods.

The use of one-dimensional (1D) and two-dimensional (2D) sodium dodecyl sulfate polyacrylamide gel electrophoresis (SDS-PAGE) to study the protein interactions in WP concentrate solutions was first reported by Havea et al. (1998). Later, Havea (2006) studied the protein interactions in 5% milk protein concentrate (MPC) powder solutions at a pH of 6.9 using 1D and 2D SDS-PAGE and transmission electron microscopy (TEM) techniques to determine the types of bonds involved in the formation of WP aggregates and the insoluble fraction in heat-induced systems. Since SDS is well known for its ability to dissociate both inter- and intra-noncovalent protein–protein linkages, it is used in such studies. Comparing the 1D and 2D PAGE analyses of both the sediment and the supernatant fractions of samples treated with SDS against standardized MPC samples, it is possible to determine the interacting protein molecules and whether this interaction is based on covalent (disulfide-linked) or noncovalent bonds. The 1D and 2D PAGE are also useful in conjunction with complementary methods, such as TEM, rheological tests, or size-exclusion high-performance liquid chromatography (SE-HPLC), to determine the protein–protein bond types (Havea, Carr, and Creamer 2004; Havea, Singh, and Creamer 2000, 2001, 2002; Havea et al. 1998).

Using diffusing wave spectroscopy (DWS), Alexander and Dalgleish (2005) studied the acid-induced aggregation of CN micelles in the presence of different denatured WP preparations from milk. This technique was used to measure the particle size change under different treatments, to study the aggregation phenomena and the interactions between the CN micelles and the soluble WPs in nondiluted milk samples. The authors concluded that both the CN micelles and the WPs are former constituents of milk gels; therefore, both proteins contribute to the properties of the network as a whole system. Dalgleish, Alexander, and Corredig (2004) studied the acid gelation of milk using ultrasonic spectroscopy (US) and DWS; the first was used for determining the hydration and particle compressibility, and the second was used for particle sizing.

Corredig and Dalgleish (1999) studied the heat-induced CN–WP interactions in skim milk previously added with purified α-LA and β-LG using SDS-PAGE. In this study, the quantification of the amount of α-LA and β-LG associated with the CN micelle was performed by relating these proteins to the amount of κ-CN present in micellar pellets over a heating time at 80°C. The ratios of α-LA/κ-CN and β-LG/κ-CN were assumed as indices of the interaction between both proteins. This methodology has been described in detail in Dalgleish, van Mourik, and Corredig (1997). The technique to correlate the association of α-LA and β-LG with κ-CN using the SDS-PAGE method has also been published (Oldfield, Singh, and Taylor 2005a; Oldfield et al. 2000; Oldfield, Taylor, and Singh 2005b).

Vasbinder and de Kruif (2003) quantified the heat-induced CN–WP interactions in milk using the rennet fractionation of WPs combined with capillary electrophoresis (EC) to quantify the distribution of WPs in aggregates and as a coating. The SDS-agarose gel EC was used to characterize the aggregates present in a supernatant by size and also to study the interactions between the CN and the WP. This technique was previously described by Alting et al. (2000) and was used to study the disulfide bond formation in acid-induced gels of a preheated WP isolate. The interaction of denatured WPs with CN micelles can also be measured as the increase in the size of the CN micelle using sensitive sizing techniques (Anema and Li 2003a).

Vasbinder et al. (2004) investigated the gelation properties of CN–WP mixtures at a pH of 4.6 and 32°C, using DWS for a continuous evaluation of the mobility of the particles in the turbid systems during gel formation, while confocal scanning laser microscopy (CSLM) was used to evaluate the microstructure of the gels by staining CN micelles and WPs with different stains. The WP-free milk was labeled covalently with Texas Red-X carboxylic acid and the WP solutions were labeled covalently with Oregon Green 488 carboxylic acid succinimidyl ester, as described by Haugland (2002). Large deformation rheological tests were performed in the absence and presence of 5 mM *N*-ethylmaleimide (NEM), a chemical that prevents the formation of disulfide bridges on gels.

In the analysis of milk gels, microstructural and rheological studies have been preferred, since the structure and strength of the gel have a direct relationship with the entanglements formed by the interacting proteins within the solid matrix. Rheological tests in gels can also be a useful tool for the quantification of protein cross-linking, in conjunction with the addition of some reagents to destroy the covalent or noncovalent bonding, such as NEM and SDS, respectively. Some typical settings used for the rheological analysis of protein–protein interactions are listed in Table 8.2.

8.4 Heat-Induced CN–WP Interactions in Liquid Milk Systems

The protein–protein interactions in bovine milk systems have been extensively studied because of their importance to the dairy processing industry. The influence of factors such as the time–temperature, pressure, pH, and protein concentration on milk protein interactions is well documented; however, the exact reaction mechanisms and the stoichiometry taking place in protein associations are complex in nature and are still unknown. In the case of heat-induced protein interactions, the mechanisms and kinetics are different in model solutions (purified selected proteins) compared with those observed in milk systems, in which several proteins interact together.

The heat-induced protein interactions in milk proteins are well understood qualitatively, and several models have been proposed to explain the possible aggregation mechanisms (Havea, Singh, and Creamer 2000, 2001; Patel et al. 2006; Schokker et al. 1999; Vasbinder and de Kruif 2003a; Vasbinder et al. 2001). Nevertheless, further research is needed to fully understand them quantitatively as a function of specific process conditions and how these interactions may be intentionally triggered to improve and/or control the desired functional properties of foods.

The heat-induced denaturation of WPs is a critical step in the association of WPs with CN. Most researchers have investigated the heat-induced denaturation of the major WPs at temperatures above 60°C; however, certain denaturation reactions of WPs could start at temperatures as low as 40°C (Parris and Baginski 1991). Above approximately 70°C, conformational changes take place in the secondary and tertiary structures of β-LG, including the unfolding and exposure of its free thiol group at Cys121, and the subsequent protein aggregation proceeds via

Table 8.2 Rheological Tests and Settings Used in the Study of Protein Interactions in Milk Gels

Equipment Used	Test Conditions[a]	Milk System Used	References
Bohlin VOR rheometer. Two coaxial cylinders (25 and 27.5 mm diameter)	MHT: 15–30 min at 75°C–90°C; pH ~6.7	SKP was reconstituted to 107 g solids/L, heat-treated, acidified with 13 g/L GDL, and 13 mL was transferred to the rheometer. Samples were oscillated at 0.1 Hz every 10 min for 16 h and the maximum strain applied was <0.01. After ~16 h, a frequency sweep test was performed on the gels	Lucey et al. (1997)
	GC: GDL at 30°C; pH ~4.5–4.6		
Bohlin VOR rheometer. Two coaxial cylinders (25 and 27.5 mm diameter)	MHT: 30 min at 80°C; pH ~7.0	SKP was reconstituted to 107 g solids/L, heat-treated, acidified with 13 g/L GDL, and 17 mL was transferred to the rheometer. Set point 2 mm (gap). Samples were oscillated at 0.1 Hz every 10 min for 15 h and the strain applied was 0.01. After ~15 h, a frequency sweep (up and down) test was performed on gels from 0.001 to 1.0 Hz. NEM was added to some samples	Lucey et al. (1998a)
	GC: GDL at 30°C; pH ~4.6		
Bohlin VOR rheometer. Two coaxial cylinders (25 and 27.5 mm diameter)	Idem as (Lucey et al. 1997)	As described by Lucey et al. (1997a). Gels were subjected to a low (0.00185 sec⁻¹) constant shear rate up to the yielding of the gel	Lucey, Munro, and Singh (1999)

Bohlin CVO rheometer. C25 cup-and-bob	30 min at 80°C; pH 6.65	As described by Lucey et al. (1997a). A strain of 0.01 and a frequency of 0.1 Hz. Measurements were made every 5 min for 6 h. After 6 h, the frequency was varied from 0.01 to 10 Hz	Bikker et al. (2000)
	GDL at 30°C; pH ~4.2		
Bohlin CVO rheometer. C25 cup-and-bob	30 min at 80°C; pH 6.65	As described by Bikker et al. (2000)	Graveland-Bikker and Anema (2003)
	GDL at 30°C; pH ~4.6		
Carrimed CSL500 rheometer, conical concentric cylinder (8.60 and 9.33 mm)	10 min at 90°C; pH 6.67	Milk was prepared by adding 10.45 g in 100 mL water (final protein concentration of 3.5% w/w). Measurements were made every 12 min at 1 rad/sec at a constant strain of 1%	Vasbinder et al. (2003b)
	GDL at 20°C; pH 4.6		
Bohlin VOR rheometer. C25 cup-and-bob	10 min at 95°C; pH 6.7	Milk was reconstituted to 10% w/v. Measurements (13 mL of sample) were made every 30 sec for 9 h at a constant strain of 1% at a frequency of 0.1 Hz, and the stress generated was sensed with a calibrated torsion bar	Guyomarc'h et al. (2003b)

(continued)

Table 8.2 (Continued) Rheological Tests and Settings Used in the Study of Protein Interactions in Milk Gels

Equipment Used	Test Conditions[a]	Milk System Used	References
Carrimed CSL100 rheometer. Cone (4 cm, 4°) and plate geometry	1.5% GDL at 40°C; pH 4.4 30 min at 90°C; pH 6.5–7.1	Milk was reconstituted to 10% w/w total solids. 1.2 mL of sample was oscillated at 0.1 Hz at a constant strain of 0.01, and measurements were taken every 5 min for 5.5 h. After 5.5 h, the frequency was varied from 0.01 to 10 Hz	Anema et al. (2004)
4 sec at 80°C–140°C UHT	2% w/w GDL at 30°C; pH ~4.2 Rennet pH ~6.5		Waungana, Singh, and Bennett (1996)
Bohlin CVO rheometer. Cup C25 concentric cylinder	Cooking at 80°C at a constant shear rate of 15 sec⁻¹; pH 5.7	Sample 1: Processed cheese sample (48% moisture, 26% fat, 12% protein, 3% sodium phosphate, and 11% lactose). Sample 2: Processed cheese sample (80% moisture, 16% protein, 2% sodium phosphate, and 2% lactose)	Lee et al. (2003)

[a] MHT: milk heat treatment; GC: gelation conditions; SKP: skim milk powder.

disulfide–thiol exchange reactions and hydrophobic interactions (Anema and Li 2003a, 2003b; Hoffmann and van Mil 1997; Vasbinder, Alting, and de Kruif 2003a; Vasbinder and de Kruif 2003). The heat denaturation process does not follow the same pattern and rate for all WPs, since significant structural differences exist among the WPs, for example, the number of disulfide bonds and free thiol groups. Conversely, due to their small secondary and tertiary structures, CN molecules are hardly denatured at the temperatures that denature WPs.

It is important to bear in mind that the milk proteins containing cysteine residues are αs2-CN, κ-CN, α-LA, and β-LG; therefore, all of these molecules have the potential to interact if induced. The principal heat-induced interactions between CNs and WPs have been observed with the denatured β-LG and κ-CN via the free thiol group and disulfide bonding.

The interactions of both α-LA and β-LG with the CN micelles in whole milk increase with time and temperature when the milk is heated between 75°C and 90°C, where α-LA is not able to interact with the CN on its own, but the formation of intermediate complexes consisting of β-LG/α-LA allows its association with the micelles via κ-CN binding, the latter only having a limited number of available sites for interactions with β-LG (Corredig and Dalgleish 1999; Dalgleish, van Mourik, and Corredig 1997). The denaturation kinetics of β-LG and α-LA and its association with the CN micelles can be comparable to skim milk heated at 75°C; however, at temperatures between 80°C and 130°C, the denaturation rate of β-LG is much greater than α-LA and almost no association between α-LA and the CN micelle is observed (Oldfield et al. 1998).

8.4.1 Role of Thiol Binding

The interactions among milk proteins are strongly mediated by the presence of the free thiol groups on β-LG and the disulfide linkages, the latter being present in both WPs and CNs. Extensive research has been performed regarding the influence of these groups in protein–protein aggregation in either the heated milk or the model systems (Alting et al. 2000, 2003, 2004; Anema 2000; Creamer et al. 2004; Havea, Carr, and Creamer 2004; Hoffmann and van Mil 1997; Lowe et al. 2004; Patel et al. 2006; Vasbinder et al. 2003b).

The known effects of heat on a β-LG structure are its initial reversible dissociation from a dimer to monomers, followed by the partial unfolding and exposure of the free sulfhydryl group on Cys121 and its subsequent interaction with the intramolecular or intermolecular disulfide bonds with either β-LG molecules or κ-CN. The formation of these intermolecular disulfide bonds under heating conditions makes the modification in the tertiary structure of β-LG irreversible (Hoffmann and van Mil 1997). The disulfide self-aggregation of purified β-LG increases with the increase in pH in a range from 6.4 to 8.0 for model solutions, although this reaction still occurs at a pH of 6.0, but to a lesser extent (Hoffmann and van Mil 1999).

Heat treatment of β-LG (80°C for 15 min) causes redistribution of its intermolecular disulfide bonds, thereby shifting the free Cys121 of the native β-LG to Cys160, this site being the most suitable for covalent cross-linking in the heat-induced aggregation of WPs (Creamer et al. 2004).

Examination of the specific thiol groups involved in the β-LG and κ-CN interactions in heated milk has been difficult due to the large number of protein species present, which could give different results compared with those obtained from model systems. Cho, Singh, and Creamer (2003) studied the β-LG:κ-CN association in a heated model system (80°C for 60 min, pH 6.7) at different protein ratios. They determined that the β-LG free thiol group (Cys121), which is exposed when heated, plays a central role in β-LG:κ-CN heat-induced interactions, and when β-LG is heated in the presence of κ-CN, it denatures quickly. They also found that a 1:1 ratio of κ-CN to β-LG generates the maximum quantity of disulfide-bonded complexes. Lowe et al. (2004) reported that in heated skim milk (90°C for 20 min, no pH control) and in a model system containing only β-LG and κ-CN, the association sites between these two proteins were different for each system studied. In milk systems, no evidence of a β-LG:κ-CN association involving the free Cys121 or the thiol groups Cys106 and Cys119 of β-LG was found, in contrast to the model system where peptide associations involving either Cys119 or Cys121 of β-LG and Cys residues of κ-CN were observed. There is evidence that during heat treatment, redistribution of the disulfide bonds within the milk proteins takes place; however, the specific thiol groups involved in the interactions between β-LG and κ-CN remain unclear. Actually, during the heat-induced thiol–disulfide interchange reactions, the free thiol of the native β-LG (Cys121) could be rearranged into any one of its five cysteine residues (Lowe et al. 2004).

Under the heating conditions of up to 90°C, αs2-CN is virtually absent in the protein aggregates formed in milk, possibly due to this protein not being at the surface of the micelle and its disulfide bond being inaccessible to react with the denatured β-LG (Patel et al. 2006).

8.4.2 Effect of pH

The pH value greatly affects the CN–WP interactions in heated milk, some of which are summarized in Table 8.3. In general, the CN–WP association in heated milk seems to be faster and higher as the pH decreases from 6.8 to 5.8 (Corredig and Dalgleish 1996). At pH values less than 6.7, a greater quantity of denatured WPs are associated with CN micelles, whereas at pH values higher than 6.7, the β-LG/κ-CN complexes dissociate from the micelle surface due to the dissociation of κ-CN (Singh and Waungana 2001). The pH value greatly influences a CN micelle dissociation in heated, reconstituted skim milk (10% w/w), measured as the increase in the level of different types of soluble CN (Anema and Klostermeyer 1997). These authors studied the behavior of CN

Table 8.3 Effect of pH on CN–WP Interactions in Heated Milk

Shift in pH	Temperature Range (°C)	Interaction Observed	References
6.8 ± 5.8	75–90	CN–WP association: increase and rate is faster. Denaturation of β-LG is more T and pH sensitive than for α-LA.	Corredig and Dalgleish (1996)
6.7 ± 7.1	20–70	CN micelle dissociation: increase	Anema (1998); Anema and Klostermeyer (1997)
		CN–WP association: decrease	
	100–120	CN micelle dissociation: increase	Anema (1998)
		CN–WP association: decrease	
6.7 ± 6.1	20–70	CN micelle dissociation: decrease	Anema (1998); Anema and Klostermeyer (1997)
		CN–WP association: increase	
	100–120	CN micelle dissociation: decrease	Anema (1998)
		CN–WP association: increase	
6.7 ± 6.5	80–100	CN–WP association: increase	Anema and Li (2003b)
6.48 ± 6.83	75–90	κ-CN dissociation: increase	Oldfield et al. (2000)
		CN–WP association: decrease	
6.90 ± 6.35	80	CN–WP association: increase	Vasbinder and de Kruif (2003)
		WP aggregation: decrease	

and WPs in the temperature and pH ranges from 20°C to 90°C and 6.3 to 7.1, respectively. At pH values higher than 6.7 and over the whole range of temperatures studied, the dissociation of κ-CN increased almost linearly with temperature, but the opposite behavior was seen at pH values less than 6.7, where a slightly negative slope in the κ-CN dissociation vs. the temperature curve was observed. Furthermore, at pH values higher than 6.7, both the soluble αs-CN and β-CN levels increased in the dissociation level as the temperature increased (a similar pattern compared with κ-CN), but for those proteins where a maximum dissociation peak was observed at near 70°C at a pH higher than 6.7, the soluble levels decreased as the temperature rose to 90°C. These results suggest that αs-CN and β-CN have the same dissociation pattern as that of κ-CN at the pH and temperature ranges of 6.1–6.7 and 20°C–90°C, respectively, but substantial differences occur at the pH and temperature ranges of 6.7–7.1 and 70°C–90°C, respectively. In the same study, both α-LA and β-LG remained almost completely soluble up to 60°C, regardless of the pH; however, at near 70°C, the level of these soluble proteins decreased with increasing temperature up to 90°C, indicating that thermal denaturation had proceeded. These results suggest that the association between the CN micelles and the WPs increased when decreasing the pH within the range studied, where two critical values were found to be important when combined: a pH of 6.7 for the micellar CN dissociation and a 60°C–70°C temperature range for the denaturation of WPs. The pH and temperature dependence on CN dissociation was later studied in heated reconstituted skim milk as a function of the total solids concentration (10%, 17.5%, and 25% w/w) and at the same pH range (6.3–7.1) but at a wider temperature range from 20°C to 120°C (Anema 1998). Comparable patterns were found for κ-CN in the whole range of temperature, pH, and concentration studied, that is, a generally increasing level of dissociation from the CN micelle for increasing temperature, pH, and total solids concentration throughout the range studied. The change in dissociation behavior was once again observed for WPs at near 60°C–80°C, which was assumed to be due to the interaction between the κ-CN and the denatured WPs. It was suggested that the pH dependence observed on the CN micelle dissociation indicates that electrostatic interactions may be playing an important role in maintaining the CN micelle structure, as do the CCP and hydrophobic interactions. Further studies indicated that the association of the CN micelles with the denatured WPs when heated was very sensitive to the temperature and to small changes in the pH. Anema and Li (2003b) determined that in reconstituted skim milk heated to around 80°C–100°C, up to 80%, 50%, and 30% of the total WPs associated with the CN micelles at a pH of 6.5, 6.6, and 6.7, respectively. This pattern concurred with that reported by Oldfield et al. (2000), who determined that after heating skim milk at 90°C for 10 min, the extent of the maximum association of β-LG with the CN micelles was 90%, 80%, and 60% at a pH of 6.48, 6.60, and 6.83, respectively. The authors assumed that the low association

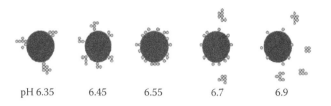

pH 6.35   6.45   6.55   6.7   6.9

Figure 8.3 A schematic representation of the interactions between the CN micelles (large circles) and the denatured WPs (small circles) as a function of the pH during the heat treatment of milk (for 10 min at 80°C). The WPs are either present in aggregates or covalently associated with the CN micelle. The native whey proteins are not included in the figure. (From Vasbinder, A. J. and de Kruif, C. G. *International Dairy Journal* 13, 669–77, 2003.)

value for β-LG with the CN micelles at a pH of 6.83 compared with those obtained at a pH of 6.48 and 6.60 was due to the greater dissociation values of κ-CN at a pH higher than 6.7, that is, the dissociation point stated by Anema and Klostermeyer (1997).

Vasbinder and de Kruif (2003) studied the CN–WP interactions in heated milk (10 min at 80°C) as a function of the pH in the range of 6.9–6.35, which were in agreement with previous research. They observed an inverse relationship between the CN–WP association and the pH shift, that is, an increase in the CN–WP association as the pH decreased while the WP aggregation decreased within the same pH range. Heating to a pH higher than 6.6 leads to the WPs partially coating the CN micelles and forming separate WP aggregates, whereas at a pH less than 6.6, all the WPs are attached to the CN micelle (Figure 8.3). It was demonstrated that during the heating of milk, small changes in the pH can cause important changes in the CN–WP association and whey aggregation, thereby affecting the gelation behavior.

8.4.3 Effect of Protein Concentration

The heat-induced interactions of the CN micelles and the WPs are affected by the protein concentration. In skim milk that is heated between 70°C and 90°C, β-LG is apparently able to interact to a limited extent with the micellar κ-CN, since a further addition of 2 g/L of β-LG did not affect its association level with the CN micelles. Also, a roughly constant ratio of about 0.6 was observed for β-LG/κ-CN, which suggests the existence of a limited number of available binding sites in the κ-CN for β-LG association and a final molar ratio of <1 (Dalgleish, van Mourik, and Corredig 1997). In a similar experiment (75°C–90°C, pH 5.8–6.8), a molar ratio of about 1.5 was found for WP/κ-CN association (Corredig and Dalgleish 1996). The existence of only a discrete number of binding sites available for β-LG/κ-CN interaction was subsequently confirmed (Corredig and Dalgleish

1999). Conversely, higher concentrations of α-LA in skim milk (up to 2 g/L) caused an increase in the level of α-LA associated with the CN micelles. In this work, it was concluded that when β-LG is present at less than its normal concentration in milk, it is able to bind efficiently to the CN micelles. When denatured, α-LA and β-LG present in similar amounts in skim milk, associate on a 1:1 molar basis, and subsequently bind κ-CN, resulting in a complex containing one of each molecule.

Beaulieu, Pouliot, and Pouliot (1999) studied the effect of different CN:WP ratios on model solutions at a constant protein content (32.5 g/L) over the thermal aggregation of the WPs (95°C for 5 min, pH 6.7). They found that complexes consisting of CN and WPs were formed, and as the CN:WP ratio was reduced from 80:20 to 20:80, the particle size increased; however, from a CN:WP ratio of 60:40 to 0:100, the distribution of the aggregates became heterogeneous and difficult to measure. The new particles formed were assumed to be mainly composed of WPs, as a proof of the saturation of the binding sites available in the CN micelle, which conformed to the results previously reported (Corredig and Dalgleish 1996, 1999; Dalgleish, van Mourik, and Corredig 1997). The 80:20 CN:WP ratio yields a β-LG/κ-CN mass ratio of ~1, and according to Cho, Singh, and Creamer (2003), at this level most of the disulfide bonds between these molecules are formed, while a further addition of WPs will not result in further CN–WP association.

Guyomarc'h, Law, and Dalgleish (2003a) found that in heated milk (95°C for 10 min, pH 6.67), the molar ratio of the associated β-LG:α-LA was approximately 3:1, regardless of the CN:WP ratio (tested within a range from 6.6 to 3.0). Considering three β-LG plus one α-LA as WPs, the molar ratio of the associated WP:κ-CN increased from 2.2 to 2.9 as the CN:WP ratio decreased from 6.6 (~87:13) to 3.0 (75:25), indicating that within the range tested, the lower the CN:WP ratio, the higher their interactions.

8.4.4 Heat-Induced Mechanisms

The contributions of the team of New Zealand researchers, including Creamer, Singh, Anema, Havea, and Patel, have been crucial in understanding how the heat-induced protein interactions proceed in the milk and model systems. This team (Patel et al. 2006) has recently proposed a mechanism for heat-induced protein–protein interactions in skim milk (Figure 8.4). According to this, the initial step consists of the β-LG dimer dissociation and a redistribution of the disulfide bonds in the β-LG monomers where the initial free thiol group in Cys121 is shifted to give a free Cys160, which due to its proximity to the C terminal of the protein is able to react with the intermolecular disulfide bonds in either β-LG or κ-CN. The disulfide bond interchange can also occur between β-LG and α-LA. The wide range of κ-CN polymeric species (6–12 monomers per aggregate) allows the existence of multiple intermolecular disulfide bonds,

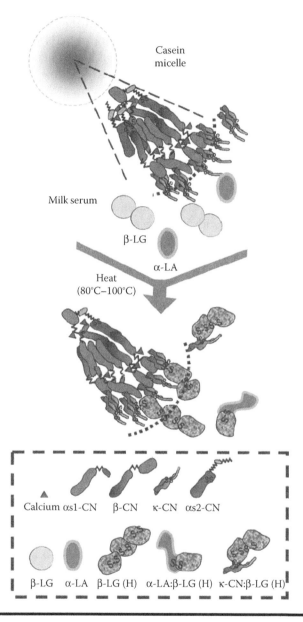

Figure 8.4   A pictorial representation of the likely effect of heating milk at 90°C. The native β-LG dimer dissociates and the monomer undergoes internal disulfide bond interchange to give reactive monomer polymers that react with κ-CN at the surface (outer region) of the CN micelle. The native β-LG monomers can also form an adduct with α-LA, which then gives rise to α-LA dimers and β-LG:α-LA dimers. In the severely heat-treated samples, αs2-CN also forms disulfide bonds with other proteins. (From Patel, H. A., et al., *Journal of Agricultural and Food Chemistry*, 54, 3409–20, 2006.)

plus the fact that it is a surface component within the micelle that facilitates interactions with the WPs. This is not the case for αs2-CN when heated at less than 100°C, which is present as a monomer and as a dimer; although it has a disulfide group inside the micelle, it cannot reach the free thiol group of the denatured β-LG.

8.5 Heat-Induced CN–WP Interactions in Milk Gel Systems

The heat treatment of milk prior to gel induction at temperatures sufficiently high to induce some level of denaturation in the WPs results in firm gels of a high storage modulus (G′) with an increase in the pH of gelation and a reduction of the gelation time, compared with those prepared from unheated milk (Anema et al. 2004; Guyomarc'h et al. 2003b; Lucey et al. 1997, 1998a). Lucey et al. (1997) studied the strength of the glucono-delta-lactone (GDL)-induced gels at 30°C as a function of heating milk (75°C–90°C for up to 30 min). As a result, the gels formed from the milk heated at 90°C exhibited G′ values of about 360–450 Pa compared with 15 Pa for those from unheated milk. Also, it was observed that the higher the denaturation level reached during the heat treatment of milk (time–temperature combination), the shorter the gelation time and the higher the gelation pH. This behavior was attributed to the association of the denatured WPs with κ-CN via disulfide bonding and the susceptibility of protein aggregation due to the reduction in the net repulsive charge of proteins as the pH gets closer to the isoelectric point (pI). The increase in gelation pH could be explained because the pI of different WPs (for β-LG: pI = 5.1) is higher than that of CN micelles (pI = 4.6), and as the CN and WPs are forming complexes, the gelation pH tends to increase since the WPs tend to aggregate at a higher pH.

As observed in liquid milk systems, the interprotein or intraprotein disulfide linkages formed also play an important role in the physical and rheological properties and strength of milk gels. Lucey et al. (1998a) demonstrated that blocking –SH groups with 5, 10, and 20 mM of NEM resulted in the reduction of G′ for gels prepared from heated acid skim milk (pH ~ 4.6), suggesting that the disulfide bonds between the κ-CN and the WPs correlated with an increase in firmness, measured as G′. In this study, different combinations of denatured and native WPs were prepared, and the heat-induced association of the WPs with the CN micelles was determined by analyzing the ultracentrifuged supernatants with SDS-PAGE and comparing them with the control samples. The assumption was that the denatured WPs associated with the CN micelles would be sedimented along with the CN micelles, although it is probable that large aggregates of the WP would also sediment on centrifugation, as reported by other authors (Waungana, Singh, and Bennett 1996). As shown in their previous work, it was also observed that the G′

value was a direct function of the heat treatment of milk, and the more the WP is associated with the CN micelles, the higher the G′ values in the set gels. As the concentrations of NEM decreased in the milk samples (i.e., higher CN–WP interactions), the G′ and gelation pH increased, and the gelation time decreased in the set gels tested.

In a subsequent study (Lucey, Munro, and Singh 1999), it was observed that the addition of the native WPs to heated milk did not contribute to the firmness of acid milk gels (thus with the cross-linking increasing), suggesting that the native WPs acted as an inert filler in acid milk gels. The opposite behavior was observed when the addition of the native WPs was followed by heat treatment. The microstructure of the gels prepared from heated milk (~80°C) greatly differs from that of those prepared from unheated milk (Lucey et al. 1998b). The former appeared to have thinner but more numerous branches and a higher "apparent interconnectivity," possibly due to the aggregation of the WPs and a further association with the CN micelles; the latter appeared to have a more tortuous microstructure with less apparent interconnectivity.

Schorsch et al. (2001) studied the effect of heating the WPs alone or with the CN micelles (45 g CN and 10 g WP/kg), analyzing the rheological and microstructural properties. The respective treatments were: CN/native WP co-heated at 20°C for 30 min (treatment A), CN/native WP co-heated at 80°C for 30 min (treatment B), WP heated alone (80°C for 30 min) and then further co-heated at 20°C for 30 min with CN (treatment C), and WP heated alone (80°C for 30 min) and then further co-heated at 80°C for 30 min with CN (treatment D). Treatment A represents the unheated condition where no interactions between the CN and the WP take place. For treatment B, they observed results similar to those previously reported by Lucey et al. (1998a), that is, stronger gels with an increase in G′ as a result of the CN–WP interactions and cross-linking. This was explained as the denaturation of the WPs and their complexation with κ-CN via a sulfydryl interchange together with hydrophobic interactions and calcium bridging. When the WPs are heated alone at 80°C without the presence of the CN micelles, they can form self-aggregates mainly linked by disulfide bonds, and after further addition of unheated CN (treatment C), the denatured WPs are unable to covalently bind κ-CN, resulting in more particulated gels that are less firm and exhibit more syneresis compared with those gels produced by co-heated CN–WP systems. Nevertheless, when preheated WP is further co-heated at 80°C in a second step with the CN micelles (treatment D), a network comprising WP self-aggregates and CN–WP associates is formed, leading in terms of strength to a gel with intermediate properties between those observed in the gels formed from treatments D and B. A scheme of this behavior is depicted in Figure 8.5. The direct relationship between G′ and the association of β-LG with κ-CN in heated milk gels, as well as the inverse relationship of G′ with syneresis, was confirmed later (Cayot et al. 2003), where the highest value of G′ was observed in milk heated at 90°C for 1–4 min.

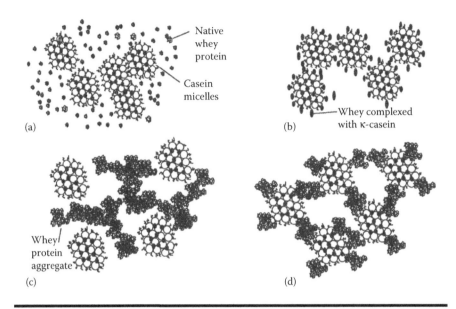

Figure 8.5 Schematic representations of the CN–WP acid gels. (a) Unheated CN-native WP; (b) CN-native WP after heating at 80°C, heat treatment of a mixture of both WP and CN micelles; (c) CN-predenatured WP, heat treatment of only WPs; (d) CN-predenatured WP after heating, heat treatment of WPs and a mixture with CN micelles. (From Schorsch, C., et al., *Journal of Dairy Research*, 68, 471–81, 2001.)

Vasbinder et al. (2003b) studied the acid gelation (GDL at 20°C up to pH 4.6 in 24 h) of three types of milk (fresh, reconstituted, and WP-free reconstituted skim milks) heated for 10 min at 90°C. They used Ellman's reagent (DTNB) to quantify the number of accessible thiol groups in milk after heating, and the addition of 5 mM NEM was sufficient to block all detectable thiol groups in the different milks studied. The reactive thiol groups in reconstituted and fresh skim milk after heat treatment were found to be 0.07 and 0.10 mM, respectively. Heated reconstituted skim milk without NEM formed stronger gels (G′ ~ 500 Pa) compared with the same milk with NEM (G′ ~ 400 Pa) and unheated milk (G′ < 100 Pa), revealing the clear contribution of both the disulfide bonds and the heat treatment of milk in G′, since unheated milk set gels with or without the addition of NEM presented similar mechanical properties. The authors attributed the increase in gel hardness and G′ to the additional disulfide cross-links formed and the amount of reactive thiol groups present after heating the milk.

Guyomarc'h et al. (2003b) monitored the acid gelation of five heated dairy mixes, each containing 4.7% of the total protein at different CN:WP ratios: 3.0, 3.9, 4.8, 5.7, and 6.6. The G′ values of all the samples increased as a function of time, but the lower the CN:WP ratio, the higher the G′ plateau (G$_p$′). After more than 9 h, G$_p$′ ranged from 300 to 450 Pa as the CN:WP ratio decreased from 6.6 to 3.0. The gelation point increased to a pH of ~5.3 compared with the typical value for unheated micelles (pI = 4.7). A possible cause for this is the overall pI of the heated mixture of β-LG (pI = 5.3), κ-CN (pI = 5.4), and α-LA (pI = 4.8).

Anema et al. (2004) studied the acid gelation of milk heated at 90°C for 30 min, considering the pH (from 6.5 to 7.1) of milk during heating as a factor. The acidification was triggered by adding GDL at 30°C, and the pH was monitored over time until it reached a value of 4.2 after 6 h. An increased pH at heating decreased the gelation time, increased the gelation pH, and increased the final G′ of the acid set gels prepared from the heated samples. This shows that not only does the heat treatment of milk prior to gelation affect the gel properties, but also the pH of the milk during heating. It is important to consider that the association of CN–WP in the heated milk decreased and the G′ of the final gels increased as the pH increased from 6.5 to 7.1, indicating that there is no relationship between the level of WP associated with the CN micelles and the final G′ of the acid gels when all pH values are considered. The higher G′ values were mostly influenced by the soluble denatured WPs rather than by the denatured WPs associated with the CN micelles. These results do not concur with those of Lucey et al. (1998a).

The behavior observed in acid milk gels, that is, an increase in the gel strength for heated milk, is the opposite of what had been observed in rennet milk gels at a high pH (Singh and Waungana 2001). During rennet coagulation of UHT heated milk (4 sec at 80°C–140°C), increasing amounts of β-LG associated with the CN micelles are correlated with decreasing values of G′, and the milk containing more than 50% β-LG associated with the CN micelles has very low values of G′ and a reduced strength of rennet gels. The G′ response over time after the addition of rennet (pH 6.5 at 32°C) as a function of temperature from 80°C to 140°C decreased drastically as the heating temperature increased and even at 140°C, no gel was formed in 120 min. This behavior was attributed to the β-LG/κ-CN complex formation, which may protrude from the micelle surface and would interfere with the close approach of the reactive sites in the micelle formed by the rennet action. The presence of β-LG aggregates in the serum after heating can also interfere with gel formation, possibly reducing the number and strength of the contacts between the adjacent aggregate chains within the gel network. The rennet coagulation properties and the gel firmness could be improved by reducing the pH of the heated milk during coagulation to values below 6.2, for example, pH ~ 5.5, followed by reneutralization to a higher pH; this is called "pH cycling." This behavior can be explained by the disruption of the micelles at a low pH and the subsequent exposure of the hidden κ-CN and allowing its participation in the gelation process. The research findings on CN–WP interactions as a function of the pH are summarized in Table 8.4.

Table 8.4 Gelation Properties of Milk Heated at Different Temperatures at Selected pH

Milk Heat Treatment	Gelation (Final pH)	Interactions Observed	References
15–30 min at 75°C–90°C; natural pH ~ 6.7	GDL at 30°C; pH ~ 4.5–4.6	G′ is consistently higher for milk heated between 80°C and 90°C (300–450 Pa) compared with milk heated at 75°C and unheated milk. An increase in the gelation pH up to 5.29 was observed in heated milk	Lucey et al. (1997)
30 min at 80°C; pH ~ 7.0	GDL at 30°C; pH ~ 4.6	G′ is consistently higher for heated milk (390–430 Pa) compared with unheated milk (<20 Pa). An increase in the gelation pH up to 5.27 was observed in heated milk.	Lucey et al. (1998a)
15–30 min at 75°C–90°C; natural pH ~ 6.7	GDL at 30°C; pH ~ 4.5–4.6	G′ is consistently higher for heated milk (~400 Pa) and for milk with +1% WPC added and then heated (~620 Pa), compared with either unheated milk or milk with +1% WPC added without further heating (<20 Pa). An increase in the gelation pH up to 5.24 was observed in heated milk	Lucey, Munro, and Singh (1999)
30 min at 80°C; pH 6.65	GDL at 30°C; pH ~ 4.2	Acid gels prepared from heated milk had markedly higher G′ values, a reduced gelation time, and an increase in the gelation pH than gels from unheated milk. The addition of different genetic variants of β-LG was tested	Bikker et al. (2000)

30 min at 80°C; pH 6.65	GDL at 30°C; pH ~ 4.6	The addition of increasing levels of β-LG to WP depleted the milk prior to the heat treatment and acidification caused marked increases in the G' value and the gelation pH and a reduction in the gelation time	Graveland-Bikker and Anema (2003)
10 min at 90°C; pH 6.67	GDL at 20°C; pH 4.6	Heated reconstituted skim milk without NEM formed stronger gels (G' ~ 500 Pa) compared with the same milk with NEM (G' ~ 400 Pa) and unheated milk (G' < 100 Pa)	Vasbinder et al. (2003b)
10 min at 95°C; pH 6.7	GDL at 40°C; pH 4.4	G' value increases faster and its value is consistently higher for heated milk (150–300 Pa) than that observed in unheated milk (<30 Pa). An increase in the gelation point to a pH of 5.3 was observed in heated milk.	Guyomarc'h et al. (2003b)
30 min at 90°C; pH 6.5–7.1	GDL at 30°C; pH ~ 4.2	An increased pH on heating the milk decreased the gelation time, increased the gelation pH, and increased the final G' of the acid set gels prepared from the heated samples	Anema et al. (2004)
4 sec at 80°C–140°C; UHT	Rennet pH ~ 6.5	G' decreased gradually with an increase in the extent of the denaturation of β-LG and its association with CN micelles	Waungana, Singh, and Bennett (1996)

8.6 Effect of Shear Rate on Heat-Induced CN–WP Interactions

Only a few studies have been published regarding the effect of the shear rate or the mixing speed on the milk protein interactions in milk products, with most of them relating to WP aggregation, but none concerning the CN–WP interactions (Table 8.5). Lee et al. (2003) studied the change in the apparent viscosity of two model-processed cheeses prepared from rennet CN, with and without fat, as a function of the cooking time (80°C) and at a constant shear rate of 15 sec^{-1}. Although

Table 8.5   Effect of Shear Rate on the Milk Protein Interactions

Equipment Used	Test Conditions	Interactions Observed	References
Bohlin CVO rheometer. Cup C25 concentric cylinder	Cooking at 80°C at a constant shear rate of 15 sec^{-1}; pH 5.7	The apparent viscosity of cheese prepared with or without fat increased with increasing cooking (shearing) time, reaching a peak followed by decreasing values. The increase–decrease in viscosity was attributed to a strong protein interaction, followed by a collapse of the protein gel structure due to further cooking and shearing	Lee et al. (2003)
Mixer rheometer	55°C; 600 rpm. Use of disodium orthophosphate as an emulsifying salt	A change in viscosity during cooking with a maximum peak was observed. The maximum viscosity index reached in the experiment was related to the emulsifying salt concentration, and it was found to be 80 at 0.35% w/w	Ennis, O'Sullivan, and Mulvihill (1998)
Couette apparatus	70°C–90°C; 111–625 (sec^{-1}); 1.5% WPC35 solution	An increase in the shear rate before total whey denaturation has proceeded enables the particles to collide and agglomerate, with the maximum size dependent on the shear rate. Once a significant amount of WP has denatured, the rate of aggregate size increase is almost independent of the shear rate	Simmons, Jayaraman, and Fryer (2007)

the viscosity values reached were different for both samples (full-fat cheese exhibiting considerably higher maximum viscosities compared with fat-free cheese, ~18 vs. ~6 Pa sec^{-1}, respectively), they observed a similar pattern of increase in the apparent viscosity over time and the existence of a peak, apparently due to the reassociation of the protein network and a strong protein interaction, followed by a decrease in the apparent viscosity as a result of the collapse of the protein network structure upon further cooking and shearing—a phenomenon called "creaming." A similar pattern-related change in viscosity was observed some years ago (Ennis, O'Sullivan, and Mulvihill 1998). They measured the viscosity index of the rennet CN solutions with emulsifying salts (disodium orthophosphate) using a mixer rheometer stirred at constant speed and temperature (600 rpm, 55°C). The maximum viscosity reached in the experiment was related to the emulsifying salt concentration. The salt concentrations used were 0%, 0.3%, 0.35%, 0.5%, 0.7%, and 1% w/w, and the highest viscosity indexes were roughly 5, 15, 30, 80, 40, and 5, respectively.

An interesting work regarding the effect of temperature and shear rate on the aggregation of WPs has recently been published (Simmons, Jayaraman, and Fryer 2007). The authors used a Couette apparatus to control the temperature (70°C–90°C) and shear rate (111–625 sec^{-1}) in 1.5% WPC35 solutions. They observed that whey denaturation is a strong function of temperature, while WP aggregation is a function of the applied shear field. An increase in the shear rate before the total whey denaturation has proceeded enables the particles to collide and agglomerate, generating more dense particles, causing acceleration in the aggregate growth rate with a maximum size dependent on the shear rate in a complex relationship. Once a significant amount of WP has been denatured, the rate of the aggregate size increase is almost independent of the shear rate. It was observed that at the initial stages of heating, aggregates produced at a shear rate of 305 sec^{-1} grew at a higher rate and produced larger particles than those created at 222 or 111 sec^{-1}; however, after some critical time (20 min), the final aggregate size showed an inverse relationship with the final shear rate. Although the dependence of aggregation on the shear rate was demonstrated, the relationships among all the factors and the observed responses were not clearly established.

8.7 Conclusion

The heat-induced interactions between the CN and the WP in milk and model systems are strongly influenced by factors such as the time–temperature and the pH of heating as well as the protein concentration and, most importantly, the CN:WP ratio. CN–WP interactions are strongly influenced by the pH, and at a pH less than 6.7, a greater quantity of denatured WP associates with the CN micelles; whereas at a pH higher than 6.7, the β-LG/κ-CN complexes dissociate from the micelle surface due to the dissociation of κ-CN. The properties of milk gels are strongly influenced by the heat treatment and the pH of milk before gelation, and a higher modulus and gelation pH as well as a decrease in the gelation time have been observed in the gels prepared from heated milk compared with those prepared from unheated milk, possibly due to more interconnectivity and cross-links between the proteins. The same effect has been observed as the pH of milk increased from 6.5 to 7.1 during the heat treatment. The rate of shear applied during the heating of milk has been a factor in possibly influencing the aggregation rate and final particle size of the CN–WP aggregates, but its effect and magnitude have not been well established.

References

Alexander, M. and Dalgleish, D. G. 2005. Interactions between denatured milk serum proteins and casein micelles studied by diffusing wave spectroscopy. *Langmuir* 21: 11380–86.

Alting, A. C., Hamer, R. J., de Kruif, C. G., Paques, M., and Visschers, R. W. 2003. Number of thiol groups rather than the size of the aggregates determines the hardness of cold set whey protein gels. *Food Hydrocolloids* 17: 469–79.

Alting, A. C., Hamer, R. J., de Kruif, G. G., and Visschers, R. W. 2000. Formation of disulfide bonds in acid-induced gels of preheated whey protein isolate. *Journal of Agricultural and Food Chemistry* 48: 5001–7.

Alting, A. C., Weijers, M., De Hoog, E. H. A., et al. 2004. Acid-induced cold gelation of globular proteins: Effects of protein aggregate characteristics and disulfide bonding on rheological properties. *Journal of Agricultural and Food Chemistry* 52: 623–31.

Anema, S. G. 1998. Effect of milk concentration on heat-induced, pH-dependent dissociation of casein from micelles in reconstituted skim milk at temperatures between 20 and 120 degrees C. *Journal of Agricultural and Food Chemistry* 46: 2299–2305.

Anema, S. G. 2000. Effect of milk concentration on the irreversible thermal denaturation and disulfide aggregation of beta-lactoglobulin. *Journal of Agricultural and Food Chemistry* 48: 4168–75.

Anema, S. G. and Klostermeyer, H. 1997. Heat-induced, pH-dependent dissociation of casein micelles on heating reconstituted skim milk at temperatures below 100 degrees C. *Journal of Agricultural and Food Chemistry* 45: 1108–15.

Anema, S. G., Lee, S. K., Lowe, E. K., and Klostermeyer, H. 2004. Rheological properties of acid gels prepared from heated pH-adjusted skim milk. *Journal of Agricultural and Food Chemistry* 52: 337–43.

Anema, S. G. and Li, Y. M. 2003a. Association of denatured whey proteins with casein micelles in heated reconstituted skim milk and its effect on casein micelle size. *Journal of Dairy Research* 70: 73–83.

Anema, S. G. and Li, Y. M. 2003b. Effect of pH on the association of denatured whey proteins with casein micelles in heated reconstituted skim milk. *Journal of Agricultural and Food Chemistry* 51: 1640–46.

Beaulieu, M., Pouliot, Y., and Pouliot, M. 1999. Thermal aggregation of whey proteins in model solutions as affected by casein/whey protein ratios. *Journal of Food Science* 64: 776–80.

Bikker, J. F., Anema, S. G., Li, Y. M., and Hill, J. P. 2000. Rheological properties of acid gels prepared from heated milk fortified with whey protein mixtures containing the A, B and C variants of beta-lactoglobulin. *International Dairy Journal* 10: 723–32.

Cayot, P., Fairise, J. F., Colas, B., Lorient, D., and Brule, G. 2003. Improvement of rheological properties of firm acid gels by skim milk heating is conserved after stirring. *Journal of Dairy Research* 70: 423–31.

Cho, Y. H., Singh, H., and Creamer, L. K. 2003. Heat-induced interactions of beta-lactoglobulin A and kappa-casein B in a model system. *Journal of Dairy Research* 70: 61–71.

Considine, T., Patel, H. A., Anema, S. G., Singh, H., and Creamer, L. K. 2007. Interactions of milk proteins during heat and high hydrostatic pressure treatments – A review. *Innovative Food Science & Emerging Technologies* 8: 1–23.

Corredig, M. and Dalgleish, D. G. 1996. Effect of temperature and pH on the interactions of whey proteins with casein micelles in skim milk. *Food Research International* 29: 49–55.

Corredig, M. and Dalgleish, D. G. 1999. The mechanisms of the heat-induced interaction of whey proteins with casein micelles in milk. *International Dairy Journal* 9: 233–36.

Creamer, L. K., Bienvenue, A., Nilsson, H., et al. 2004. Heat-induced redistribution of disulfide bonds in milk proteins. 1. Bovine beta-lactoglobulin. *Journal of Agricultural and Food Chemistry* 52: 7660–68.

Dalgleish, D., Alexander, M., and Corredig, M. 2004. Studies of the acid gelation of milk using ultrasonic spectroscopy and diffusing wave spectroscopy. *Food Hydrocolloids* 18: 747–55.

Dalgleish, D. G., van Mourik, L., and Corredig, M. 1997. Heat-induced interactions of whey proteins and casein micelles with different concentrations of alpha-lactalbumin and beta-lactoglobulin. *Journal of Agricultural and Food Chemistry* 45: 4806–13.

de Kruif, C. G. and May, R. P. 1991. Kappa-casein micelles: Structure, interaction and gelling studied by small-angle neutron scattering. *European Journal of Biochemistry* 200: 431–36.

Ennis, M. P., O'Sullivan, M. M., and Mulvihill, D. M. 1998. The hydration behaviour of rennet caseins in calcium chelating salt solution as determined using a rheological approach. *Food Hydrocolloids* 12: 451–57.

Graveland-Bikker, J. F. and Anema, S. G. 2003. Effect of individual whey proteins on the rheological properties of acid gels prepared from heated skim milk. *International Dairy Journal* 13: 401–8.

Guyomarc'h, F. 2006. Formation of heat-induced protein aggregates in milk as a means to recover the whey protein fraction in cheese manufacture, and potential of heat-treating milk at alkaline pH values in order to keep its rennet coagulation properties. A review. *Le Lait* 86: 1–20.

Guyomarc'h, F., Law, A. J. R., and Dalgleish, D. G. 2003a. Formation of soluble and micelle-bound protein aggregates in heated milk. *Journal of Agricultural and Food Chemistry* 51: 4652–60.

Guyomarc'h, F., Queguiner, C., Law, A. J. R., Horne, D. S., and Dalgleish, D. G. 2003b. Role of the soluble and micelle-bound heat-induced protein aggregates on network formation in acid skim milk gels. *Journal of Agricultural and Food Chemistry* 51: 7743–50.

Haugland, R. P. 2002. *Handbook of Fluorescent Probes and Research Chemicals.* Eugene, OR: Molecular Probes Inc.

Havea, P. 2006. Protein interactions in milk protein concentrate powders. *International Dairy Journal* 16: 415–22.

Havea, P., Carr, A. J., and Creamer, L. K. 2004. The roles of disulphide and non-covalent bonding in the functional properties of heat-induced whey protein gels. *Journal of Dairy Research* 71: 330–39.

Havea, P., Singh, H., and Creamer, L. K. 2000. Formation of new protein structures in heated mixtures of BSA and alpha-lactalbumin. *Journal of Agricultural and Food Chemistry* 48: 1548–56.

Havea, P., Singh, H., and Creamer, L. K. 2001. Characterization of heat-induced aggregates of beta-lactoglobulin, alpha-lactalbumin and bovine serum albumin in a whey protein concentrate environment. *Journal of Dairy Research* 68: 483–97.

Havea, P., Singh, H., and Creamer, L. K. 2002. Heat-induced aggregation of whey proteins: Comparison of cheese WPC with acid WPC and relevance of mineral composition. *Journal of Agricultural and Food Chemistry* 50: 4674–81.

Havea, P., Singh, H., Creamer, L. K., and Campanella, O. H. 1998. Electrophoretic characterization of the protein products formed during heat treatment of whey protein concentrate solutions. *Journal of Dairy Research* 65: 79–91.

Hinrichs, J. 2001. Incorporation of whey proteins in cheese. *International Dairy Journal* 11: 495–503.

Hoffmann, M. A. M. and van Mil, P. J. J. M. 1997. Heat-induced aggregation of beta-lactoglobulin: Role of the free thiol group and disulfide bonds. *Journal of Agricultural and Food Chemistry* 45: 2942–48.

Hoffmann, M. A. M. and van Mil, P. J. J. M. 1999. Heat-induced aggregation of beta-lactoglobulin as a function of pH. *Journal of Agricultural and Food Chemistry* 47: 1898–1905.

Holt, C. 1992. Structure and stability of bovine casein micelles. *Advances in Protein Chemistry* 43: 63–151.

Holt, C. and Horne, D. S. 1996. The hairy casein micelle: Evolution of the concept and its implications for dairy technology. *Netherlands Milk and Dairy Journal* 50: 85–111.

Horne, D. S. 1998. Casein interactions: Casting light on the black boxes, the structure in dairy products. *International Dairy Journal* 8: 171–77.

Horne, D. S. 2006. Casein micelle structure: Models and muddles. *Current Opinion in Colloid & Interface Science* 11: 148–53.

Kontopidis, G., Holt, C., and Sawyer, L. 2004. Invited review: Beta-lactoglobulin: Binding properties, structure, and function. *Journal of Dairy Science* 87: 785–96.

Lee, S. K., Buwalda, R. J., Euston, S. R., Foegeding, E. A., and McKenna, A. B. 2003. Changes in the rheology and microstructure of processed cheese during cooking. *Lebensmittel-Wissenschaft Und-Technologie-Food Science and Technology* 36: 339–45.

Lowe, E. K., Anema, S. G., Bienvenue, A., Boland, M. J., Creamer, L. K., and Jimenez-Flores, R. 2004. Heat-induced redistribution of disulfide bonds in milk proteins. 2. Disulfide bonding patterns between bovine beta-lactoglobulin and kappa-casein. *Journal of Agricultural and Food Chemistry* 52: 7669–80.

Lucey, J. A. 2002. Formation and physical properties of milk protein gels. *Journal of Dairy Science* 85: 281–94.

Lucey, J. A., Munro, P. A., and Singh, H. 1999. Effects of heat treatment and whey protein addition on the rheological properties and structure of acid skim milk gels. *International Dairy Journal* 9: 275–79.

Lucey, J. A., Tamehana, M., Singh, H., and Munro, P. A. 1998a. Effect of interactions between denatured whey proteins and casein micelles on the formation and rheological properties of acid skim milk gels. *Journal of Dairy Research* 65: 555–67.

Lucey, J. A., Teo, C. T., Munro, P. A., and Singh, H. 1997. Rheological properties at small (dynamic) and large (yield) deformations of acid gels made from heated milk. *Journal of Dairy Research* 64: 591–600.

Lucey, J. A., Teo, C. T., Munro, P. A., and Singh, H. 1998b. Microstructure, permeability and appearance of acid gels made from heated skim milk. *Food Hydrocolloids* 12: 159–65.

Oldfield, D. J., Singh, H., and Taylor, M. W. 2005a. Kinetics of heat-induced whey protein denaturation and aggregation in skim milks with adjusted whey protein concentration. *Journal of Dairy Research* 72: 369–78.

Oldfield, D. J., Singh, H., Taylor, M. W., and Pearce, K. N. 1998. Kinetics of denaturation and aggregation of whey proteins in skim milk heated in an ultra-high temperature (UHT) pilot plant. *International Dairy Journal* 8: 311–18.

Oldfield, D. J., Singh, H., Taylor, M. W., and Pearce, K. N. 2000. Heat-induced interactions of beta-lactoglobulin and alpha-lactalbumin with the casein micelle in pH-adjusted skim milk. *International Dairy Journal* 10: 509–18.

Oldfield, D. J., Taylor, M. W., and Singh, H. 2005b. Effect of preheating and other process parameters on whey protein reactions during skim milk powder manufacture. *International Dairy Journal* 15: 501–11.

Parris, N. and Baginski, M. A. 1991. A rapid method for the determination of whey-protein denaturation. *Journal of Dairy Science* 74: 58–64.

Patel, H. A., Singh, H., Anema, S. G., and Creamer, L. K. 2006. Effects of heat and high hydrostatic pressure treatments on disulfide bonding interchanges among the proteins in skim milk. *Journal of Agricultural and Food Chemistry* 54: 3409–20.

Schokker, E. P., Singh, H., Pinder, D. N., Norris, G. E., and Creamer, L. K. 1999. Characterization of intermediates formed during heat-induced aggregation of beta-lactoglobulin AB at neutral pH. *International Dairy Journal* 9: 791–800.

Schorsch, C., Wilkins, D. K., Jones, M. G., and Norton, I. T. 2001. Gelation of casein-whey mixtures: Effects of heating whey proteins alone or in the presence of casein micelles. *Journal of Dairy Research* 68: 471–81.

Simmons, M. J. H., Jayaraman, P., and Fryer, P. J. 2007. The effect of temperature and shear rate upon the aggregation of whey protein and its implications for milk fouling. *Journal of Food Engineering* 79: 517–28.

Singh, H. and Waungana, A. 2001. Influence of heat treatment of milk on cheesemaking properties. *International Dairy Journal* 11: 543–51.

Vasbinder, A. J., Alting, A. C., and de Kruif, K. G. 2003a. Quantification of heat-induced casein-whey protein interactions in milk and its relation to gelation kinetics. *Colloids and Surfaces B: Biointerfaces* 31: 115–23.

Vasbinder, A. J., Alting, A. C., Visschers, R. W., and de Kruif, C. G. 2003b. Texture of acid milk gels: Formation of disulfide cross-links during acidification. *International Dairy Journal* 13: 29–38.

Vasbinder, A. J. and de Kruif, C. G. 2003. Casein–whey protein interactions in heated milk: The influence of pH. *International Dairy Journal* 13: 669–77.

Vasbinder, A. J., de Velde, F. V., and de Kruif, C. G. 2004. Gelation of casein-whey protein mixtures. *Journal of Dairy Science* 87: 1167–76.

Vasbinder, A. J., van Mil, P., Bot, A., and de Kruif, K. G. 2001. Acid-induced gelation of heat-treated milk studied by diffusing wave spectroscopy. *Colloids and Surfaces B: Biointerfaces* 21: 245–50.

Walstra, P. 1999. Casein sub-micelles: Do they exist? *International Dairy Journal* 9: 189–92.

Walstra, P., Geurts, T. J., Noomen, A., Jelema, A., and van Boekel, M. A. J. S. 1999. *Dairy Technology: Principles of Milk Properties and Processes*. New York: Marcel Dekker.

Waungana, A., Singh, H., and Bennett, R. J. 1996. Influence of denaturation and aggregation of beta-lactoglobulin on rennet coagulation properties of skim milk and ultrafiltered milk. *Food Research International* 29: 715–21.

Yalcin, A. S. 2006. Emerging therapeutic potential of whey proteins and peptides. *Current Pharmaceutical Design* 12: 1637–43.

Chapter 9

Protein–Saccharide Interaction

Hiroki Saeki

Contents

9.1 Introduction

Interactions among food components, such as proteins, saccharides, lipids, and nucleic acids, occur constantly during the storage and processing of foods. Various phenomena in the food processing system involving the quality improvement or quality loss of processed foods result from these interactions. For example, the Maillard reaction (nonenzymatic browning reaction) between proteins/amino acids and reducing saccharides affects the color, flavor/odor, and texture of processed foods. The reactive oxygen groups generated from lipid peroxidation accelerate protein oxidation, and oxidation of the myofibrillar proteins (Mf) often affects the physical properties of muscle foods. Furthermore, enzymes such as proteases and lipases degrade the food components during storage and such enzymatic reactions cause a loss of food quality.

The first objective of this chapter is to review the effects of protein–saccharide interactions on the physicochemical properties of food components and the quality of processed foods. In food storage or processing systems with coexisting proteins and saccharides, the functionality of the food proteins is often altered by the saccharides. The effects of saccharides on the physicochemical characteristics of proteins are introduced in Section 9.2. The stabilization of proteins, which occurs without covalent bonding, is one of the most important positive effects of saccharides. Examples of the application of protein stability in food processing and preservation are described in Section 9.3.

The second objective of this chapter is to explain the attempts to improve the functionality of the food proteins and saccharides based on the covalent binding reactions among them. In the past 20 years, many synthetic glycoproteins (neoglycoproteins) have been developed as novel functional food materials. Several combinations of protein–saccharides have been attempted using cross-linkers and modified reagents, as described in Section 9.4. Recently, the Maillard reaction has been used as a key reaction in the binding of proteins with saccharides by covalent attachment between the proteins/peptides and the reducing saccharides.

This protein glycosylation system is superior to other synthetic systems because it can proceed under mild conditions without any chemicals. The availability of Maillard-type neoglycoprotein and neoglycopeptide as novel functional materials is discussed in Section 9.4. Additionally, in Section 9.5, the fish protein–saccharide conjugates developed using the Maillard reaction are introduced as examples of neoglycoprotein synthesis from thermally and chemically unstable proteins.

9.2 Physicochemical Characteristics of the Proteins Affected by Saccharides

9.2.1 Stabilization of Proteins in the Presence of Saccharides

The physicochemical properties of proteins in aqueous conditions are often affected by saccharides. The phenomena can be classified as interactions with or without covalent binding, and the conformational stabilization of proteins by saccharides results from the noncovalent binding interaction between proteins and saccharides (Lee and Timasheff 1981; Arakawa and Timasheff 1982; Baier and McClements 2001). Attempts to understand the influences of saccharides on the thermal stability of the food proteins in aqueous solutions have been carried out through thermodynamic studies using differential scanning calorimetry (DSC). The results have shown that the thermal denaturation temperature (Td) of the proteins increases in the presence of saccharides. Monosaccharides and disaccharides have a positive effect on various food proteins. Glucose, sucrose, and lactose increased the Td and denaturation enthalpy of the food proteins such as bovine serum albumin (BSA) (Boye, Alli, and Ismail 1996a), β-lactoglobulin (Boye, Ismail, and Alli 1996b), and egg proteins (ovalubumin and ovotransferrin) (Christ, Takeuchi, and Cunha 2005). Figure 9.1 shows the effect of sucrose on the thermal denaturation of BSA measured by ultrasensitive DSC. The Td was increased with an increase in the sucrose concentration.

The mechanism of conformational stabilization of the globular proteins has been discussed in terms of preferential hydration. Gekko and Ito (1990) summarized the stabilization mechanism by polyols as follows: "Globular proteins are preferentially hydrated in polyol–water mixtures as well as in sugar–water mixtures. The main origin of such preferential hydration would be local exclusion of these polyhydric compounds from the nonpolar surface of proteins, although solvent surface tension partially contributes in sugar–water systems. The resultant increase in the chemical potentials of proteins should be the driving force minimizing the protein-solvent interface, leading to stabilization of the native conformations of proteins." This opinion is widely supported by scientists.

The degree of the stabilizing effect varies with the sugars. Ooizumi et al. (1981), investigating the protective effect of saccharides (sugars and polyoles) on the heat denaturation of the myosin Ca-ATPase activity of the fish Mf, demonstrated that

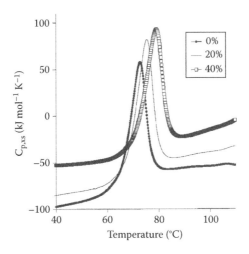

Figure 9.1 **The heat capacity profiles of 1 wt% BSA in aqueous sucrose solutions (pH 6.9) scanned at 90° Ch⁻¹. (From Baier, S. and McClements, D. J., *Journal of Agricultural and Food Chemistry* 49, 2600–8, 2001. With permission.)**

the effect was lower in the order of maltotriose > lactitol > sucrose > sorbitol = glucose = maltose > mannitol > mannose = fructose = xylitol, xylose, sorbitan, glycerol, and sorbite. In addition, the degree of stabilization is closely related to the number of OH groups in a molecule (Figure 9.2). A similar trend has been reported based on important research concerning the relative stabilizing effects of polysaccharides investigated by DSC using ovalubumin (Back, Oakenfull, and Smith 1979).

9.2.2 Cryoprotective Effect of Saccharides

Various saccharides suppress protein denaturation during the freezing and freeze-drying processes (Matsumoto et al. 1992; Izutsu, Aoyagi, and Kojima 2004), which are indispensable operations in the modern food industry. Allison et al. (1999) reported that the hydrogen bonding between saccharides and proteins suppressed protein unfolding induced by dehydration. A series of research investigating the cryoprotective effect of saccharides on the fish muscle protein demonstrated that hexoses (glucose and fructose) and disaccharides (sucrose and lactose) have superior cryoprotective properties compared with pentoses (xylose and ribose) (Noguchi, Oosawa, and Matsumoto 1976; Matsumoto and Noguchi 1992). They also examined the sugar alcohols and found that glycerol, xylitol, and sorbitol were good cryoprotectant additives. Sorbitol is now widely used as a commercial cryoprotectant in frozen fish and as a stabilizer in dried food processes, as described in Section 9.3.

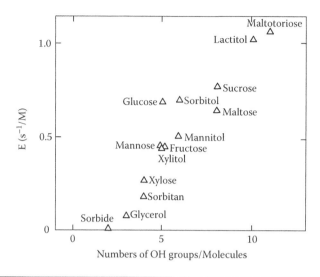

Figure 9.2 **The relationship between the number of OH groups of sugars and of sugar alcohols and their protective effects against the inactivation of mackerel myofibrillar Ca-ATPase. The protective effect of sugar and sugar alcohol was defined by the relation E = 1/a(log kD₀ – log kDₐ), where kD₀ and kDₐ, respectively, are the first-order rate constants for the inactivation of myofibrillar Ca-ATPase in the absence and presence of an M sugar or sugar alcohol. (From Ooizumi, T., Hashimoto, K., Ogura, J., and Arai, K.,** *Bulletin of the Japanese Society of Scientific Fisheries* **47, 901–8, 1981. With permission.)**

9.2.3 Maillard Reaction and Its Effect on Food Quality

The Maillard reaction (nonenzymatic browning), a common interaction between proteins and saccharides with covalent binding, occurs in various unit operations, such as heating (roasting, baking, frying, and boiling), dehydrating, drying, and concentrating processes. Understanding the Maillard reaction is important in producing high-quality processed foods. The Maillard reaction is affected by the temperature, humidity, pH, metal ions, and oxygen in the food system. High temperature, low moisture, and alkaline conditions all promote the Maillard reaction. Although the Maillard reaction occurs in aqueous conditions, 60%–80% of relative humidity is the optimum condition for the Maillard reaction.

The first step in the Maillard reaction is the covalent attachment of the reducing sugars to the reactive amino groups (ε- or α-amino group) of the proteins, peptides, and amino acids. The reaction mechanism of the Maillard reaction is well recognized as the Hodge scheme (Hodge 1953, 1967) (Figure 9.3). At an early stage, the condensation product of a protein and a reducing sugar forms a Schiff base and an Amadori rearrangement (ketoamine). There are two covalent

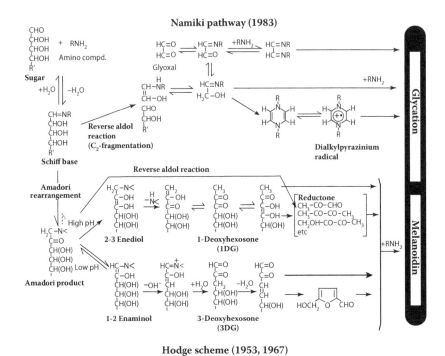

Figure 9.3 The main pathway of the Maillard reaction. (From Namiki, M., *JMARS News Letter* (April 22, 2006): 1–3. http://www.maillard.umin.jp/. With permission.)

reaction schemes between the amino groups and the reducing sugars. In the advanced stage, the Maillard reaction goes through two pathways, depending on the pH, and stable brown melanoidin pigments are produced. Additionally, the Amadori products react with the cross-linkages between the proteins or other amino compounds in foods (Friedman 1996). The polymeric aggregate products are called advanced glycation end-products (AGE). The Maillard reaction in foods is generally evaluated by measuring the color change. Additionally, changes in the available lysine, Amadori compounds, or 3-deoxyglucosone are examined to evaluate the Maillard reaction.

Another mechanism of the Maillard reaction, proposed by Namiki and Hayashi (1983), involves sugar fragmentation and free radical (pyrazinium radical) formation prior to the Amadori rearrangement (Figure 9.3). Glomb and Monnier (1995) estimated that approximately 50% of the lysine residues formed in a glucose/lysine system originate from the oxidation of an Amadori product. The result indicates the contribution of the Namiki pathway in the Maillard reaction. Hofmann, Bors, and Stettmaier (1999) also proved the progress of the Namiki pathway in a model food system. Cammer and Kroh (1996) showed that the radical formation pathway (Namiki pathway) is predominant in the Maillard reaction at a pH higher than 7.0. The Namiki pathway seems to participate in protein glycation in a living body (Hofmann, Bors, and Stettmaier 1999; Wondrak et al. 2000). Namiki (2006) indicated that the reaction mechanism of Hodge may be dominant in the food system and the Namiki pathway in the living system.

9.3 Food Properties Affected by Protein–Saccharide Interactions

9.3.1 Effects of Saccharides on the Gel Formation of Proteins

The gel-forming ability of the food proteins is important in obtaining the significant texture of processed foods. The formation of a heat-induced gel involves denaturation, aggregation/coagulation, and cross-linking of proteins. A conformational change in the protein structure induced by heating is the first step in the formation of a gel matrix. The improved thermal stability of proteins in the presence of saccharides leads to an increase in the gel transition temperature (Figure 9.4) and a delay in the gel-forming process. The gelation temperature of BSA (Baier and McClements 2001), whey protein (Dierckx and Huyghebaert 2002), and hen egg albumin (Christ, Takeuchi, and Cunha 2005) sols increased in the presence of low-molecular-weight saccharides, such as sucrose and sorbitol.

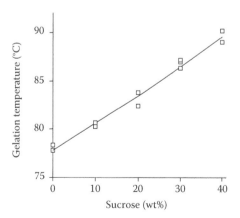

Figure 9.4 The dependence of the gelation temperature on the sucrose concentration for 2 wt% BSA in aqueous sucrose solutions at neutral pH. (From Baier, S. and McClements, D. J., *Journal of Agricultural and Food Chemistry* 49, 2600–8, 2001. With permission.)

Saccharides often inhibit the heat-induced gel formation of proteins and alter the gel properties. The stabilization of proteins occurs with saccharides, which interfere with the gel formation of proteins because protein unfolding is the first trigger of gel formation (Kulmyrzaev, Cancelliere, and McClements 2000). Indeed, sucrose diminished the rigidity of the egg albumin gel (Christ, Takeuchi, and Cunha 2005), and a BSA solution with a high concentration of sucrose required a prolonged heating time to obtain a heat-induced gel with a high rigidity (Baier and McClements 2001). Chanasattru, Decker, and McClements (2007) explained the mechanism of inhibition for the gelation of β-lactoglobulin induced by sorbitol as follows: sorbitol increased the thermal Td of the protein and diminished the active protein molecules to participate in gel formation.

Protein aggregation and the subsequent gel formation induced by high hydrostatic pressure are suppressed by low-molecular-weight saccharides (Dumay, Kalichevsky, and Cheftel 1994). He et al. (2006) reported that monosaccharides and disaccharides, such as xylose, glucose, sucrose, and lactose, decreased the rigidity of the pressure-induced whey protein gel at 800 MPa. They concluded that sugars influenced the characteristics of the pressure-induced gels by inhibiting the intermolecular cross-linking of the proteins and restricting the phase separation of change in the microstructure of the gel from a honeycomb to a thin-stranded structure.

Some researchers have reported that saccharides enhance the gel formation of proteins. Kumeno et al. (1996) reported that the gel strength and viscoelasticity of a pressure-induced gel (300–600 MPa) from a concentrated milk protein were improved by the addition of 10% (w/w) saccharides. Dierckx and Huyghebaert (2002) showed that a high concentration of sucrose or sorbitol increased the gel rigidity of the whey protein at a pH of 6.0, a level at which noncovalent interactions predominate to form a heat-induced gel. No effect was observed in the whey sol at a pH of 8.0, a level at which covalent S–S bonding is important for gel formation. Thus, they attributed the gel rigidity to the enhancement of the hydrophilic protein–protein interaction. This case indicates that the influence of saccharides on the protein gel formation changes according to the mechanism of gel formation.

9.3.2 Effects of Protein–Saccharide Interactions on Food Emulsion

The stabilization of the oil-in-water emulsion in food systems is strongly influenced by the interaction of proteins and saccharides. Most proteins act as emulsifiers in food systems and are adsorbed on the surface of the oil droplets, resulting in the stabilization of the oil droplets and the suppression of the aggregation of each oil droplet. The hydrophobicity of the proteins plays an important role in adsorbing the proteins on the surface of the oil droplets, and a sufficient surface coverage by the proteins in the oil droplets contributes to the stabilization of the emulsion. On the other hand, polysaccharides stabilize the emulsion or accelerate the flocculation of the oil droplets (e.g., depletion flocculation or bridge flocculation). These influences depend on the type of polysaccharide and its concentration in the emulsion system.

The combination of proteins and polysaccharides in a suitable condition contributes to the stabilization of a food emulsion system. For example, a mixture of sodium caseinate and polysaccharide (high-molecular-weight dextrose) can stabilize a soya oil emulsion during spray drying and encapsulate the soya oil in the powder particles (Hogan et al. 2001). Additionally, various combinations, such as soy protein–arabic gum (Kim, Morr, and Schenz 1996), whey protein isolate–arabic gum (Ibanogle 2002), β-lactoglobulin–alginate or –carrageenan (Harnsilawat, Pongsawatmanit, and McClements 2006), and whey protein–pectin (Neirynck et al. 2007), have been proposed to stabilize the emulsions.

Figure 9.5 illustrates the relationship between the polysaccharides and the oil droplets in stabilizing a protein layer. Charged polysaccharides can interact with the proteins that saturate on the surface of the oil droplets, resulting in a strong stabilization of the emulsion (Figure 9.5a). When the adsorbed proteins have a large positive net charge, anionic polysaccharides, such as anionic dextran sulfate (Dickinson 1996) and carrageenan (Dickinson and Pawlowsky 1997), can associate electrostatically with the proteins to stabilize the emulsion. Furthermore, an excess

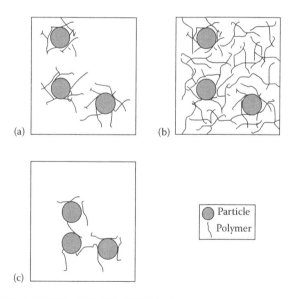

Figure 9.5 A schematic representation of three alternative effects of the adsorption of stiff hydrocolloid polymers on the surface of the spherical emulsion droplets depending on the hydrocolloid concentration and the nature of the hydrocolloid–protein interactions: (a) a sterically stabilized system, (b) an emulsion gel, and (c) a system flocculated by macromolecular bridging. (From Dickinson, E., *Food Hydrocolloids* 17, 25–39, 2003. With permission.)

concentration of polysaccharides can convert an emulsion solid to an emulsion gel (Figure 9.5b).

On the other hand, when a protein–polysaccharide interaction is weak or when there is insufficient surface coverage by the proteins or polysaccharides, the emulsion may destabilize. The stability of the emulsion with the whey protein is severely reduced by heat treatment in the presence of xanthan and carrageenan gums (Euston, Finnigan, and Hirst 2002), which could be related to the weakening of the protein–saccharide interactions with a conformational change of the protein. In another case, the size of the polysaccharides could affect the emulsion stability. Herceg et al. (2007) reported that glucose and sucrose improved the emulsifying property of the whey protein isolate and β-lactoglobulin, but the emulsions were destabilized by the addition of inulin or starch. Leroux et al. (2003) also showed that low-molecular-weight and high-acetylated pectins suppressed the aggregation of the oil droplets through electrostatically associated adsorbed proteins, but high-molecular-weight and low-acetylated pectins destabilized the emulsion because of calcium-binding flocculation and pectin–pectin interactions. Figure 9.5c illustrates the destabilization of an emulsion by "bridge flocculation." Dickinson and Pawlowsky (1997) showed the formation of bridge flocculation for a BSA-stabilized emulsion by the addition of 0.01–0.1 wt% iota-carrageenan. This result indicates that when the proteins have a positive charge on the surface, the concentration of the anionic polysaccharides is one of the important factors affecting the emulsion stability.

Many researchers have reported that the emulsifying property of a protein can be improved when the protein forms a complex with the saccharides. The improved food functionality of the synthetic glycoprotein is described in Section 9.4.

9.3.3 Effects of Protein–Saccharide Interactions on Food-Drying Processes

Food proteins often undergo severe physicochemical stress during dehydration (water removal) because water is necessary to maintain the conformational structure of the native protein. Saccharides can suppress protein denaturation during freeze drying and spray drying and, as a result, food proteins maintain good functionality, such as gel forming, emulsion forming, and water-holding ability (Matsumoto et al. 1992; Yoo and Lee 1993). Saccharides can also contribute to the stabilization of the dry-state proteins (Izutsu, Yoshioka, and Takeda 1991; Allison et al. 1999). The protective effect of saccharides in the dry state has been explained in several theories. One theory states that proteins can be stabilized by forming hydrogen bonds between proteins and saccharides in a dried food state (Crowe et al. 1987; Arakawa, Kita, and Carpenter 1991). According to another theory, saccharides form amorphous matrices and embed the protein molecules through the progressive removal of water, and the structural change in the proteins can be suppressed by embedding the proteins in the amorphous matrices (Costantino et al.

1998; Imamura et al. 2003). Buera, Schebor, and Elizalde (2005) reported that the formation and maintenance of the amorphous sugar matrix are important in stabilizing the proteins and delaying the interactions among the food components (Figure 9.6).

Several researchers have compared the degrees to which saccharides exhibit a protective action. Murray and Liang (1999) reported that trehalose is the best stabilizer to prevent the loss of the foaming properties of spray-dried whey protein and β-lactoglobulin. Matsumoto et al. (1992) compared the preventive effects of nine saccharides on fish protein denaturation during freeze drying by measuring the myofibrillar Ca-ATPase activity, and they demonstrated that the change in

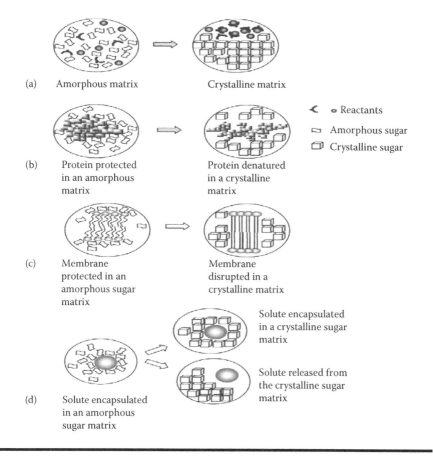

(a) Amorphous matrix Crystalline matrix

≺ ● Reactants
   ↩ Amorphous sugar
   ▱ Crystalline sugar

(b) Protein protected in an amorphous matrix Protein denatured in a crystalline matrix

(c) Membrane protected in an amorphous sugar matrix Membrane disrupted in a crystalline matrix

Solute encapsulated in a crystalline sugar matrix

Solute released from the crystalline sugar matrix

(d) Solute encapsulated in an amorphous sugar matrix

Figure 9.6 **Schematic interpretation of the effect of sugar crystallization on (a) the acceleration of the chemical reactions, (b) the protein denaturation, (c) the membrane integrity, and (d) the release of the encapsulated compounds. (From Buera, P., Schebor, C., and Elizalde, B.,** *Journal of Food Engineering.* **67, 157–65, 2005. With permission.)**

activity coincided with the trend of the thermoprotective effects of all saccharides (Figure 9.2), except glucose. Izutsu, Aoyagi, and Kojima (2004) investigated the protective effect of the saccharides with different molecular weights on the protein structural changes induced by freeze drying, and they found that the structure-stabilizing effects of malto-oligosaccharides diminished with an increase in the molecular weight of the saccharides.

9.3.4 Protective Effect of Saccharides on Protein Denaturation During Frozen Storage

Saccharides effectively suppress protein denaturation during frozen storage. Frozen surimi is one example of the successful application of saccharides as a cryoprotectant. The excellent gel-forming ability of the fish Mf is preserved during the extended frozen storage of surimi when cryoprotectants, 6%–8% (w/w) sorbitol, and/or sucrose are added to the washed fish meat (Toyoda et al. 1992). The pH of the washed fish meat is often adjusted to 6.8–7.5 because protein denaturation is effectively suppressed under a neutral pH (Matsumoto, Ooizumi, and Aral 1985).

Sorbitol is one of the most popular cryoprotectants because no Maillard reaction occurs. Additionally, other saccharides, such as sucrose, lactitol (Sultanbawa and Li-Chan 2001), branched oligosaccharide (Auh et al. 1999), maltodextrin (Carvajal, MacDonald, and Lanier 1999), and trehalose (Osako et al. 2005), have been reported as cryoprotectants in frozen surimi. Sucrose and sorbitol also suppressed protein denaturation in freeze-dried broiler muscle (Kijowski and Richardson 1996).

Fish myosin, a major component of the fish Mf, is thermally less stable than other vertebrates, and it is easily denatured by freezing. Saccharides suppress the structural and conformational changes of the protein, such as the loss of salt solubility, myosin Ca-ATPase activity, and surface hydrophobicity, as well as an increase in the S–S bonding, resulting in the preservation of such food functions as the water-holding capacity, solubility, emulsifying property, and gel-forming ability (Zhou et al. 2006; Sultanbawa and Li-Chan 2001).

9.4 Designing Glycosylated Proteins for Food Functionality

9.4.1 Chemical Synthesis of Glycosylated Proteins

Synthetic glycoproteins are called *neoglycoproteins*. Various glycosylation techniques have been developed (Christopher, Stowell, and Lee 1980) and applied to research in biochemistry, medical science, and pharmaceutical chemistry. In the field of food science, attempts have been made to improve the physicochemical and functional properties of the food proteins by covalent attachment of the glycosyl units. Table 9.1 contains a list of examples of the food proteins and enzymes that have been conjugated with saccharides using chemical reagents or genetic engineering. These physicochemical properties of the food proteins have been improved by the introduction of the glycosyl units into the protein molecules.

The use of cross-linking reagents is a popular strategy to synthesize neoglycoproteins. For example, β-lactoglobulin showed improved emulsifying properties (Nagasawa et al. 1996) and retinol-binding activity (Hattori et al. 1994) when conjugated with carboxymethyl dextran using water-soluble carbodiimide (EDC: 1-ethyl-3-(3-dimethylaminopropyl) carbodiimide hydrochloride) as a cross-linker (Kobayashi et al. 1989). Conjugation with glucosamine using EDC improved the solubility and thermal stability of lysozyme and casein, and their emulsifying properties and foaming ability were also enhanced (Ramezani, Esmailpour, and Aminlari 2008).

Reductive alkylation is another popular method used to prepare neoglycoproteins. The introduction of monosaccharides to legumin and vicilin (pea storage proteins) using reductive alkylation (cyanoborohydride is required) improved their solubility, emulsifying properties, and foaming ability (Baniel et al. 1992; Pedrosa, Trisciuzzi, and Ferreira 1997). A trypsin–carboxymethyl dextran conjugate prepared by reductive alkylation obtained high thermal and chemical stability: the trypsin activity was stabilized in denaturants such as urea and guanidinium chloride (Villalonga, Villalonga, and Gomez 2000). The stabilization of trypsin was also performed by conjugation with a dextran derivative containing two reactive aldehyde groups (Kobayashi and Takatsu 1994).

9.4.2 Protein–Saccharide Conjugate Prepared Using the Maillard Reaction and Improved Functional Properties

In 1990, Kato et al. (1990) proposed a simple glycosylation method using the Maillard reaction, in which the ε-amino groups in a lyophilized protein react with the reducing end in the saccharides. In that study, the ovalbumin–dextran conjugate that was prepared using the Maillard reaction under dry conditions showed excellent emulsifying properties. The general preparation procedure for neoglycoproteins by Maillard reaction is as follows: the proteins are mixed with the reducing sugars, lyophilized, and then heated under controlled temperature–humidity conditions in which covalent bonding between the proteins and the glycosyl units is produced. This is the most popular method used for introducing the glycosyl units into the food proteins because it can progress under relatively mild conditions (generally 40°C–60°C and 40%–70% relative humidity) without chemical reagents and special equipment. Various kinds of food proteins, such as lysozyme, ovalbumin, β-lactoglobulin, BSA, gluten, protamine, and phosvitin, have been conjugated with monosaccharides, oligosaccharides, and polysaccharides using the Maillard reaction. These Maillard-type neoglycoproteins are superior to the

Table 9.1 Methods for Introducing the Glycosyl Units into Proteins

Method	Protein	Saccharide	Improved Property	References
Cyclic carbonate method	β-Lactoglobulin	Maltose and glucosamine	Digestibility	Wanisca and Kinsella (1984)
Cross-linking reagent (EDC)	β-Lactoglobulin	Melibionic acid	Thermal stability and solubility	Kitabatake, Cuq, and Cheftel (1985)
Cross-linking reagent (EDC)	β-Lactoglobulin	Carboxymethyl dextran	Emulsifying property and retinol-binding ability	Hattori et al. (1994); Nagasawa et al. (1996)
Cross-linking reagent (EDC)	Whey protein	Carboxymethyl starch	Digestion resistance of starch	Hattori, Yang, and Takahashi (1995)
Cross-linking reagent (EDC)	β-Lactoglobulin	Alginic acid	Emulsifying property, retinol-binding activity, and stability	Hattori et al. (1996)
Cross-linking enzyme (transglutaminase)	Legumin and gliadin	Galactose	Solubility	Colas, Caer, and Fournier (1993)
Cross-linking reagent (EDC)	Lysozyme and casein	Glucosamine	Thermal stability, emulsifying property, and foaming capacity	Ramezani, Esmailpour, and Aminlari (2008)
Reductive alkylation	Superoxide dismutase	Mannan and carboxymethyl cellulose	Stability (resistance to inactivation with H_2O_2)	Valdivia et al. (2006); Dominguez et al. (2005)

Reaction between the amino group and the reactive dialdehyde	Trypsin	Dextran-N-hydroxy-succinimide ester	Thermal stability	Yamasaki and Ikebe (1992); Kobayashi and Takatsu (1994)
Reductive alkylation	Trypsin and α-amylase	Carboxymethyl dextran and sodium alginate	Thermal stability, pH stability, and resistance to denaturants	Villalonga, Villalonga, and Gomez (2000); Gomez, Ramirez, and Villalonga (2001)
Reductive alkylation	Pea vicilin	Lactose, galactose, glucose, and galacturonic acid	Emulsifying property	Pedrosa, Trisciuzzi, and Ferreira (1997)
Reductive alkylation	Pea legumin	Lactose, galactose, and galacturonic acid	Solubility, emulsifying property, and foaming ability	Baniel et al. (1992)
Genetic engineering	Pepsin	N-linked oligomannosyl chain	Stability (heating, pH, and denaturant)	Yoshimasu et al. (2004)
Genetic engineering	Cystatin	N-linked oligomannosyl chain	Thermal stability, digestion resistance, and antirotavirus action	Tzeng and Jiang (2004); Nakamura et al. (2004)
Genetic engineering	Lysozyme	N-linked oligomannosyl chain	Thermal stability, pH stability, and emulsifying property	Nakamura et al. (1993); Saito et al. (2003)

EDC: N-(dimethylaminopropyl)-N'-ethyl carbodiimide hydrochloride.

native proteins in terms of their physicochemical and functional properties, such as thermal stability, solubility, emulsifying property, antimicrobial action, and antioxidative activity. Additionally, the solubility of the insoluble saccharides and the physicochemical stability of starch can be improved by conjugation with proteins or peptides. Table 9.2 lists examples of Maillard-type glycosylation, improved food functionality, type of proteins, and type of saccharides.

9.4.3 Effect of Maillard-Type Glycosylation on Protein Stability

Various kinds of proteins can be stabilized by conjugation with polysaccharides, such as dextran, chitosan, and galactomannan. The following protein–saccharide conjugates showed higher thermal stability than the native proteins: lysozyme–galactomannan (Shu et al. 1996), lysozyme–dextran or casein–dextran (Aminlari, Ramezani, and Jadidi 2005), β-lactoglobulin–dextran (Jimenez-Castano et al. 2005), ovoinhibitor–galactomannan (Begum et al. 2003), and fish Mf–dextran (Fujiwara, Oosawa, and Saeki 1998). The stabilization effect was enhanced by an increase in the length of the attached polysaccharides. For example, excellent heat stability was achieved when 3.5–6 kDa of galactomannan was attached to the lysozyme (Shu et al. 1996). Therefore, Kato (2002) suggested that protein stabilization is likely due to the suppression of the unfolded protein–protein aggregation by the glycosyl units attached to the proteins.

On the other hand, Broersen et al. (2004) reported that β-lactoglobulin reacted with glucose or fructose is more thermally stable than the native form and that modification suppresses the protein–protein aggregation by heating. In this case, glycosylation has a significant impact on the heat stability of the structure without affecting the structural integrity of β-lactoglobulin. This finding suggests that the glycosyl units introduced into the protein molecules can stabilize the structure of the protein.

Proteins tend to increase their solubility under a wide pH range by conjugation with saccharides. Examples of protein–saccharide combinations are as follows: β-lactoglobulin–alginic acid (Hattori et al. 1996), acid-precipitated soy protein–galactomannan (Babiker et al. 1998), fish myosin–alginate oligosaccharide (AO) (Maitena et al. 2004), β-lactoglobulin–dextran (Jimenez-Castano, Villamiel, and Lopez-Fandino 2007), and soy protein isolate–porphyran (Takano et al. 2007). Since the positively charged lysine residue was modified by the reaction with the glycosyl units, the solubility improvement at a pH range of 4–6 was related to the shift in the isoelectric point of the protein to the acidic side.

9.4.4 Improved Emulsifying Property of the Maillard-Type Protein–Polysaccharide Conjugate

As described in Section 9.3.2, anionic polysaccharides can associate electrostatically with the proteins that cover the surface of the oil droplets and the polysaccharides

Table 9.2 Examples of Improved Functional Properties of Protein/Peptide–Saccharide Conjugates Prepared Using the Maillard Reaction

Improved Functionality	Protein/Peptide	Saccharides	References
Thermal stability	β-Lactoglobulin	Dextran	Jimenez-Castano et al. (2005)
	Fish myofibrillar protein	Dextran	Fujiwara, Oosawa, and Saeki (1998)
	Lysozyme	Galactomannan	Shu et al. (1996)
	Lysozyme and casein	Dextran	Aminlari, Ramezani, and Jadidi (2005)
	Ovoinhibitor	Galactomannan	Begum et al. (2003)
Thermal stability and enzyme activity	Trypsin (bovine)	Glucose	Kato, Minaki, and Kobayashi (1993a)
Thermal stability and solubility	Fish myofibrillar protein	Alginate oligosaccharide	Sato et al. (2003)
Solubility	β-Lactoglobulin	Alginic acid	Hattori et al. (1996)
Solubility	Soy protein	Galactomannan	Babiker et al. (1998)
Solubility	Soy protein	Porphyran	Takano et al. (2007)
Emulsifying property	Acid-precipitated soy protein	Galactomannan	Babiker et al. (1998)
	β-Lactoglobulin	Dextran	Jimenez-Castano et al. (2007)
	β-Lactoglobulin	Dextran	Wooster and Augustin (2007)

(continued)

Table 9.2 (Continued)   Examples of Improved Functional Properties of Protein/Peptide–Saccharide Conjugates Prepared Using the Maillard Reaction

Improved Functionality	Protein/Peptide	Saccharides	References
	β-Lactoglobulin	Chitosan	Hattori et al. (2000)
	Bovine serum albumin	Galactomannan	Kim et al. (2003)
	Buckwheat protein (Fag e 1)	Arabinogalactan	Nakamura et al. (2008)
	Casein	Maltodextrin	Shepherd, Robertson, and Ofman (2000)
	Casein	Pectin	Einhorn-Stoll et al. (2005)
	Egg white protein	Galactomannan	Kato, Matsuda, and Nakamura (1993b)
	Fish myofibrillar protein	Dextran	Fujiwara, Oosawa, and Saeki (1998)
	Fish myofibrillar protein	Alginate oligosaccharide	Sato et al. (2003)
	Fish myosin	Alginate oligosaccharide	Maitena et al. (2004)
	Lysozyme	Chitosan	Song et al. (2002)
	Lysozyme	Galactomannan	Shu et al. (1996)
	Lysozyme	Galactomannan	Nakamura, Kato, and Kobayashi (1992)
	Lysozyme	Chitosan	Song et al. (2002)

	Protein	Saccharide	Reference
	Lysozyme	Dextran	Nakamura, Kato, and Kobayashi (1991)
	Ovalbumin	Dextran	Kato et al. (1990)
	Ovoinhibitor	Galactomannan	Begum et al. (2003)
	Sorghum seed protein	Dextran	Babiker et al. (1998)
	Soy protein	Dextran	Usui et al. (2004)
	Soybean protein	Porphyran	Takano et al. (2007)
	Whey protein	Maltodextrin	Akhtar and Dickinson (2007)
	Whey protein and casein	Pectin	Einhorn-Stoll et al. (2005)
	Protamine	Dextran	Tanaka, Kunisaki, and Ishizaki (1999)
Antimicrobial activity	Gluten peptides	Chitosan	Babiker (2002)
	Lysozyme	Dextran	Nakamura, Kato, and Kobayashi (1991); Aminlari, Ramezani, and Jadidi (2005)
	Protamine	Dextran	Tanaka, Kunisaki, and Ishizaki (1999)
	Lysozyme	Glucose stearic acid monoester	Takahashi et al. (2000)
	Lysozyme	Chitosan	Song et al. (2002)
	Soy protein	Chitosan	Usui et al. (2004)

(continued)

Table 9.2 (Continued) Examples of Improved Functional Properties of Protein/Peptide–Saccharide Conjugates Prepared Using the Maillard Reaction

Improved Functionality	Protein/Peptide	Saccharides	References
Reduction of allergenicity	Soy protein	Chitosan	Usui et al. (2004)
	Squid tropomyosin	Glucose	Nakamura et al. (2006)
	Buckwheat allergen (Fag e 1)	Arabinogalactan	Nakamura et al. (2008)
	Soy protein 34 kDa-allergen	Galactomannan	Babiker et al. (1998)
Antioxidative activity	Lysozyme	Galactomannan	Nakamura and Kato (2000)
	Phosvitin	Galactomannan	Nakamura et al. (1998)
Immunoreactive effect	Pea albumin	Glucose	Mierzejewska et al. (2008)
Antioxidative activity, gel-forming activity, emulsifying property, radical-scavenging activity, and solubility at neutral pH of curdlan	Soy protein hydrolysate	Curdlan	Fan et al. (2006)
Antioxidative activity	α-Lactalbumin hydrolysate	Allose and psicose	Sun et al. (2005)
Reduction of swelling, solubility, retrogradation, and antimicrobial activity of starch	Polylysine	Potato starch	Yang et al. (1995)

can suppress the coalescence of the oil droplets. Therefore, the protein–saccharide conjugates by the Maillard reaction are also expected to be strong emulsifiers. Indeed, the improvement in the emulsifying properties of the protein is the significant benefit of protein glycosylation. Examples of protein–saccharide conjugates that can promote the stabilization of the oil-in-water emulsion are listed in Table 9.2. As listed here, the obtained protein–neutral saccharide conjugates improved the emulsifying property.

Kato (2002) summarized the effect of the molecular weight of the polysaccharides on the emulsifying properties of the protein–saccharide conjugates (Figure 9.7). Figure 9.7 indicates that greater than 10 kDa saccharides are effective for obtaining better emulsifying properties. Hiller and Lorenzen (2008) observed a marked reduction in the surface hydrophobicity of the proteins with the progression of the conjugation with the polysaccharides, and they also found a negative correlation between the surface hydrophobicity and the emulsion stability in the protein–saccharide conjugates. Wooster and Augustin (2007) also showed that the β-lactoglobulin–maltodextrin conjugate exists as a diblock copolymer in the oil-in-water emulsion, with the protein anchored at the surface of an oil droplet and the polysaccharide protruding into the aqueous continuous phase. In addition, the covalently bound polysaccharides with the protein would protect the absorbed protein layer from unfavorable conditions, such as high heat, low pH, and

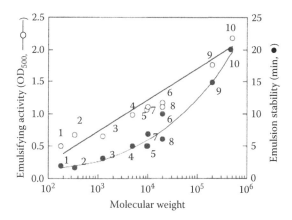

Figure 9.7 The relationships between the emulsifying properties and the molecular weights of the saccharides in conjugation with the soy protein. The numbers 1–10 indicate the saccharides conjugated with the soy protein: 1: glucose; 2: lactose; 3: xyloglucan (Mw 1400); 4: galactomannan (Mw 3500–6000); 5: galactomannan (Mw 10,000); 6: galactomannan (Mw 20,000); 7: dextran (Mw 9300); 8: dextran (Mw 19,600); 9: dextran (Mw 200,000–300,000); 10: xyloglucan (Mw 470,000). (From Kato, A., *Food Science and Technology Research* 8, 193–99, 2002. With permission.)

high electrolyte concentrations (Dickinson 2009). Recently, the water-containing oil droplets of multiple emulsions (W/O/W) were successfully prepared by using the sodium caseinate–dextran conjugates (Fechner et al. 2007). In the future, it is expected that the protein–saccharide conjugates can be used as a strong edible emulsifier in the food industry.

9.4.5 Reduction of Allergenicity of Protein Caused by Glycosylation

The Maillard reaction often affects the allergenicity (specific immunoglobulin E [IgE] binding ability) of proteins. The Maillard reaction with glucose decreased the allergenicity of squid tropomyosin (Nakamura et al. 2006), which is a major allergen in invertebrates. However, scallop tropomyosin (Nakamura et al. 2005), the major peanut allergen (Ara h 2) (Gruber, Becker, and Hofmann 2005), and the major cherry allergen (Pru av 1) (Gruber et al. 2004) increased in allergenicity with the progression of the Maillard reaction with glucose or ribose. The effect of the Maillard reaction with monosaccharides on allergenicity is obviously different for each allergenic protein. There is no information about the reason why some proteins increased their allergenicity with the progress of the Maillard reaction. The structural change of the allergenic proteins by the Maillard reaction may generate exposure of the epitope (IgE-binding site) or increase in resistance to protease/peptidase digestion. On the other hand, the attachment of the polysaccharides tends to decrease the allergenicity of the allergenic proteins. For example, the allergenicity of the major buckwheat allergen (Fag e 1) and that of the soy protein 34 kDa allergen were effectively reduced by conjugation with arabinogalactan (Nakamura et al. 2008) and galactomannan (Babiker et al. 1998), respectively. The decrease in allergenicity could have been caused by the masking of the epitope by the attached glycosyl unit, the conformational changes near the epitope through glycosylation, or the modification of the lysine residues belonging to the epitope site.

9.4.6 Peptide–Saccharide Conjugates for the Preparation of New Functional Materials

Natural polysaccharides containing covalently attached peptides, such as gum arabic and soybean polysaccharide (Maeda 2000), have good emulsifying activities and inhibit lipid oxidation in an oil-in-water emulsion (Matsumura et al. 2003). These findings suggest that a high functional material based on saccharides can be prepared by conjugating polysaccharides with low-molecular-weight peptides. In fact, the emulsifying properties and the antioxidative activity of the soy protein hydrolysate, which has a low emulsion-forming ability, were improved by conjugation with curdlan through the Maillard reaction (Fan et al. 2006). Additionally, the solubility of curdlan, which is an insoluble saccharide at a neutral pH, is highly improved by the attachment of a peptide. The enhancement of the antioxidative activity of a protein hydrolysate also occurred in an α-lactoalbumin hydrolysate by conjugation with allose or piscose (Sun et al. 2005). Furthermore, Yang et al. (1995) developed an acid-treated potato starch conjugated with ε-poly(L-lysine) that obtained a high antibacterial activity and improved the physicochemical properties, including the reduction of swelling, solubility, gelatinization, retrogradation, and digestibility with amylase. These findings indicate that the attachment of low-molecular-weight peptides to a saccharide also contributes to the development of new functional food materials.

9.5 Improved Physicochemical Properties of the Fish Myofibrillar Proteins Using the Maillard Reaction

9.5.1 Importance of Suppression of Protein Denaturation During Glycosylation

The fish Mf–saccharide conjugates can be prepared using the Maillard reaction in a dry state in the same manner as that used for globular food proteins. Fish Mf or myosin can be conjugated with reducing sugars (monosaccharides, oligosaccharides, and polysaccharides) when lyophilized protein–saccharide mixtures are incubated at 30°C–60°C under a controlled humidity. Despite having excellent functional characteristics, such as emulsifying properties, gel-forming ability, and a water-holding capacity, the fish Mf is thermally and chemically less stable than that of other vertebrates (Hashimoto, Kobayashi, and Arai 1982). Since the protein solubility and other functional properties are easily lowered as denaturation progresses, the fish protein–saccharide conjugates should be prepared while an attempt is made to suppress the protein denaturation during the preparation step. The following procedures are required in the preparation of the fish Mf–saccharide conjugates: (a) maintaining control of the temperature and keeping the humidity at the glycosylation step as low as possible, (b) regulating the Maillard reaction in the early stages, and (c) using saccharides with protective effects on the protein denaturation, if necessary.

The temperature and humidity are positive factors affecting the glycosylation through the Maillard reaction in a dry state. On the other hand, increases in both of these factors accelerate the protein denaturation (Katayama, Haga, and Saeki 2004; Takeda et al. 2007). Thus, the glycosylation of the fish Mf should be performed at low temperatures and low humidity. However, an excess decrease in temperature and humidity in the reaction system suppresses the Maillard reaction. Generally, conducting the reaction at 40°C–60°C and less than 35% relative humidity is recommended in order to prepare the Mf–saccharide conjugates with high solubility.

Many reducing sugars have protective effects on protein denaturation, as described in Section 9.2. The denaturation of the fish Mf during freeze drying and the Maillard reaction is effectively suppressed when the reducing sugars, such as glucose, ribose, and dextran, are attached to the fish Mf at the appropriate concentrations. For example, when the Mf was suspended in a 2%–5% (w/w) dextran solution, the protein denaturation during freeze drying and the Maillard reaction were effectively suppressed, and the prepared Mf–dextran conjugate obtained a high thermal stability (Fujiwara, Oosawa, and Saeki 1998). The Mf–monosaccharide conjugates with high solubility can also be prepared by suspending the Mf in a 0.3 M glucose or ribose solution before the freeze drying and the Maillard reaction steps (Saeki and Inoue 1997; Tanabe and Saeki 2001). When the conjugating glycosyl unit has no effect or a protective effect on protein denaturation, the addition of other edible protectants with no responsiveness to protein (e.g., sorbitol) is recommended to suppress protein denaturation (Sato et al. 2000). However, excess progress of the Maillard reaction should be avoided because the solubility loss of the Mf–saccharide conjugate occurs as glycosylation progresses.

9.5.2 Water Solubilization of the Fish Myofibrillar Protein by Conjugation with Saccharides and Its Molecular Mechanism

Myosin, the major component of muscle protein, is insoluble in a physiological condition. The solubility of myosin and actin in low-salt media is significantly improved by conjugation with low-molecular-weight saccharides, such as glucose, ribose, maltose, maltotriose, and AO (Saeki and Inoue 1997; Tanabe and Saeki 2001; Sato et al. 2003; Katayama and Saeki 2004). Since the progress of the Maillard reaction impairs improvement of the solubility, regulation of the Maillard reaction at an early stage and suppression of the protein denaturation are required to obtain a water-soluble Mf–saccharide conjugate.

A myosin molecule can be separated into two portions by trypsin digestion—the water-soluble head portion containing two ATPase active sites and the water-insoluble tail portion called the myosin rod. The myosin molecules assemble into insoluble filaments in a low-salt medium, and the rod region of the myosin plays a significant role in the formation of the filaments. Katayama, Haga, and Saeki (2004) indicated that the water solubilization of the Mf–saccharide conjugate is caused by the loss of the filament-forming ability of the myosin rod region, and they proposed the following water-soluble mechanism: briefly, saccharides are covalently attached to the lysine residues located on the surface of the α-helix in the myosin rod region. The attached glycosyl units act as a physical barrier against the self-assembly of the rod regions and trigger the degradation of the assembled myosin filaments. Finally, the glycosylated myosin molecules are solubilized in water and stabilized with the help of the hydrophilic glycosyl units introduced into the molecules.

9.5.3 Improved Food Functionality of the Fish Myofibrillar Protein by Conjugation with Saccharides

The thermal stability of the Mf is effectively improved by conjugation with oligosaccharides and polysaccharides. When the fish Mf is conjugated with AO (the degree of average polymerization is 6.1), a water-soluble Mf–AO conjugate with excellent heat tolerance can be prepared. Heat treatment at 80°C for 2 h had no effect on the solubility of the Mf–AO conjugate (Sato et al. 2003). Although the Mf–dextran conjugate was water-insoluble, it also showed excellent thermal stability (Fujiwara, Oosawa, and Saeki 1998). The emulsion-forming ability of the fish Mf is enhanced by conjugation with monosaccharides (Saeki and Inoue 1997), oligosaccharides (Sato et al. 2003), and polysaccharides (Fujiwara, Oosawa, and Saeki 1998). The improved emulsion-forming ability of the Mf was not impaired by a high NaCl concentration (0.5–0.5 M) or an acidic pH condition near the isoelectric point of myosin.

9.6 Concluding Remarks

As described in this chapter, the protein–saccharide interaction is an important factor affecting the physicochemical characteristics of food materials and the quality of processed foods. The current knowledge of the protein–sugar interaction has been utilized in the production of food materials and processed foods. A greater understanding of the protein–sugar interaction and its positive application will contribute to the effective utilization of bioresources and the development of novel processing and preserving technologies.

References

Akhtar, M. and Dickinson, E. 2007. Whey protein–maltodextrin conjugates as emulsifying agents: An alternative to gum arabic. *Food Hydrocolloids* 21: 607–16.

Allison, S. D., Chang, B., Randolph, T. W., and Carpenter, J. F. 1999. Hydrogen bonding between sugar and protein is responsible for inhibition of dehydration-induced protein unfolding. *Archives of Biochemistry and Biophysics* 365: 289–98.

Aminlari, M., Ramezani, R., and Jadidi, F. 2005. Effect of Maillard-based conjugation with dextran on the functional properties of lysozyme and casein. *Journal of the Science of Food and Agriculture* 85: 2617–24.

Arakawa, T., Kita, Y., and Carpenter, J. F. 1991. Protein–solvent interaction in pharmaceutical formulations. *Pharmaceutical Research* 8: 285–91.

Arakawa, T. and Timasheff, S. N. 1982. Stabilization of protein structure by sugars. *Biochemistry (Wash.)* 21: 6536–44.

Auh, J. H., Lee, H. G., Kim, J. W., Kim, J. C., Yoon, H. S., and Park, K. H. 1999. Highly concentrated branched oligosaccharides as cryoprotectant for surimi. *Journal of Food Science* 64: 418–22.

Babiker, E. E. 2002. Effect of chitosan conjugation on the functional properties and bactericidal activity of gluten peptides. *Food Chemistry* 79: 367–72.

Babiker, E. E., Hiroyuki, A., Matsudomi, N., Iwata, H., Ogawa, T., Bando, N., and Kato, A. 1998. Effect of polysaccharide conjugation or transglutaminase treatment on the allergenicity and functional properties of soy protein. *Journal of Agricultural and Food Chemistry* 46: 866–71.

Back, J. F., Oakenfull, D., and Smith, M. B. 1979. Increased thermal stability of proteins in the presence of sugars and polyols. *Biochemistry (Wash.)* 18: 5191–96.

Baier, S. and McClements, D. J. 2001. Impact of preferential interactions on thermal stability and gelation of bovine serum albumin in aqueous sucrose solutions. *Journal of Agricultural and Food Chemistry* 49: 2600–8.

Baniel, A., Caer, D., Colas, B., and Gueguen, J. 1992. Functional properties of glycosylated derivatives of the 11S storage protein from pea (Pisum sativum L.). *Journal of Agricultural and Food Chemistry* 40: 200–205.

Begum, S., Saito, A., Xu, X. H., and Kato, A. 2003. Improved functional properties of the ovoinhibitor by conjugating with galactomannan. *Bioscience, Biotechnology, and Biochemistry* 67: 1897–1902.

Boye, J. I., Alli, I., and Ismail, A. A. 1996a. Interactions involved in the gelation of bovine serum albumin. *Journal of Agricultural and Food Chemistry* 44: 996–1004.

Boye, J. I., Ismail, A. A., and Alli, I. 1996b. Effects of physicochemical factors on the secondary structure of beta-lactoglobulin. *The Journal of Dairy Research* 63: 97–109.

Broersen, K., A. Voragen, G. J., Hamer, R. J., and de Jongh, H. H. J. 2004. Glycoforms of beta-lactoglobulin with improved thermostability and preserved structural packing. *Biotechnology and Bioengineering* 86: 76–87.

Buera, P., Schebor, C., and Elizalde, B. 2005. Effects of carbohydrate crystallization on stability of dehydrated foods and ingredient formulations. *Journal of Food Engineering* 67: 157–65.

Cammerer, B. and Kroh, L. W. 1996. Investigation of the contribution of radicals to the mechanism of the early stage of the Maillard reaction. *Food Chemistry* 57: 217–21.

Carvajal, P. A., MacDonald, G. A., and Lanier, T. C. 1999. Cryostabilization mechanism of fish muscle proteins by maltodextrins. *Cryobiology* 38: 16–26.

Chanasattru, W., Decker, E. A., and McClements, D. J. 2007. Modulation of thermal stability and heat-induced gelation of b-lactoglobulin by high glycerol and sorbitol levels. *Food Chemistry* 103: 512–20.

Christ, D., Takeuchi, K. P., and Cunha, R. L. 2005. Effect of sucrose addition and heat treatment on egg albumen protein gelation. *Journal of Food Science* 70: E230–38.

Christopher, P., Stowell, P., and Lee, Y. C. 1980. Neoglycoproteins. The preparation and application of synthetic glycoproteins. *Advances in Carbohydrate Chemistry and Biochemistry* 37: 225–81.

Colas, B., Caer, D., and Fournier, E. 1993. Transglutaminase-catalyzed glycosylation of vegetable Proteins. Effect on solubility of pea legumin and wheat gliadines. *Journal of Agricultural and Food Chemistry* 41: 1811–15.

Costantino, H. R., Curley, J. G., Wu, S., and Hsu, C. C. 1998. Water sorption behavior of lyophilized protein–sugar systems and implications for solid-state interactions. *International Journal of Pharmaceutics* 166: 211–21.

Crowe, J. H., Crowe, L. M., Carpenter, J. F., and Wistrom, C. A. 1987. Stabilization of dry phospholipid bilayers and proteins by sugars. *Biochemical Journal* 242: 1–10.

Dickinson, E. 1996. Biopolymer interactions in emulsion systems: Influences on creaming, flocculation, and rheology. In *Macromolecular Interactions in Food Technology, ACS Symposium Series*, eds. N. Parris, A. Kato, L. K. Creamer, and J. Pearce, pp. 197–207. Washington, DC: American Chemical Society.

Dickinson, E. 2003. Hydrocolloids at interfaces and the influence on the properties of dispersed system. *Food Hydrocolloids* 17: 25–39.

Dickinson, E. 2009. Hydrocolloids as emulsifiers and emulsion stabilizers. *Food Hydrocolloids* 23: 1473–82.

Dickinson, E. and Pawlowsky, K. 1997. Effect of ι-carrageenan on flocculation, creaming, and rheology of a protein-stabilized emulsion. *Journal of Agricultural and Food Chemistry* 45: 3799–3806.

Dierckx, S. and Huyghebaert, A. 2002. Effect of sucrose and sorbitol on the gel formation of a whey protein isolate. *Food Hydrocolloids* 16: 489–97.

Dominguez, A., Valdivia, A., Caballero, J., Villalonga, R., Martinez, G., and Schacht, E. H. 2005. Improved pharmacological properties for superoxide dismutase modified with carboxymethycellulose. *Journal of Bioactive and Compatible Polymers* 20: 557–70.

Dumay, E. M., Kalichevsky, M. T., and Cheftel, J. C. 1994. High-pressure unfolding and aggregation of b-Lactoglobulin and the baroprotective effects of sucrose. *Journal of Agricultural and Food Chemistry* 42: 1861–68.

Einhorn-Stoll, U., Ulbrich, M., Sever, S., and Kunzek, H. 2005. Formation of milk protein–pectin conjugates with improved emulsifying properties by controlled dry heating. *Food Hydrocolloids* 19: 329–40.

Euston, S. R., Finnigan, S. R., and Hirst, R. L. 2002. Kinetics of droplet aggregation in heated whey protein-stabilized emulsions: Effect of polysaccharides. *Food Hydrocolloids* 16: 499–505.

Fan, J. F., Zhang, Y. Y., Szesze, T., et al. 2006. Improving functional properties of soy protein hydrolysate by conjugation with curdlan. *Journal of Food Science* 71: C285–91.

Fechner, A., Knoth, A., Scherze, I., and Muschiolik, G. 2007. Stability and release properties of double-emulsions stabilised by caseinate-dextran conjugates. *Food Hydrocolloids* 21: 943–52.

Friedman, M. 1996. Food browning and its prevention: An overview. *Journal of Agricultural and Food Chemistry* 44: 631–53.

Fujiwara, K., Oosawa, T., and Saeki, H. 1998. Improved thermal stability and emulsifying properties of carp myofibrillar proteins by conjugation with dextran. *Journal of Agricultural and Food Chemistry* 46: 1257–61.

Gekko, K. and Ito, H. 1990. Competing solvent effects of polyols and guanidine hydrochloride on protein stability. *Journal of Biochemistry (Tokyo)* 107: 572–77.

Glomb, M. A. and Monnier, V. M. 1995. Mechanism of protein modification by glyoxal and glycolalgehyde, reactive intermediates of the Maillard reaction. *Journal of Biological Chemistry* 270: 10017–26.

Gomez, L., Ramirez, H., and Villalonga, R. 2001. Modification of alpha-amylase by sodium alginate. *Acta Biotechnologica* 21: 265–73.

Gruber, P., Becker, W. M., and Hofmann, T. 2005. Influence of the Maillard reaction on the allergenicity of rAra h 2, a recombinant major allergen from peanut (Arachis hypogaea), its major epitopes, and peanut agglutinin. *Journal of Agricultural and Food Chemistry* 53: 2289–96.

Gruber, P., Vieths, S., Wangorsch, A., Nerkamp, J., and Hofmann, T. 2004. Maillard reaction and enzymatic browning affect the allergenicity of Pru av 1, the major allergen from cherry (Prunus avium). *Journal of Agricultural and Food Chemistry* 52: 4002–7.

Harnsilawat, T., Pongsawatmanit, R., and McClements, D. J. 2006. Stabilization of model beverage cloud emulsions using protein–polysaccharide electrostatic complexes formed at the oil–water interface. *Journal of Agricultural and Food Chemistry* 54: 5540–47.

Hashimoto, A., Kobayashi, A., and Arai, K. 1982. Thermostability of fish myofibrillar Ca-ATPase and adaptation to environmental temperature. *Bulletin of the Japanese Society of Scientific Fisheries* 48: 671–84.

Hattori, M., Aiba, Y., Nagasawa, K., and Takahashi, K. 1996. Functional improvement of alginic acid by conjugating with beta-lactoglobulin. *Journal of Food Science* 61: 1171–76.

Hattori, M., Nagasawa, K., Ametani, A., Kaminogawa, S., and Takahashi, K. 1994. Functional changes in beta-lactoglobulin by conjugation with carboxymethyl dextran. *Journal of Agricultural and Food Chemistry* 42: 2120–25.

Hattori, M., Numamoto, K., Kobayashi, K., and Takahashi, K. 2000. Functional changes in beta-lactoglobulin by conjugation with cationic saccharides. *Journal of Agricultural and Food Chemistry* 48: 2050–56.

Hattori, M., Yang, W. H., and Takahashi, K. 1995. Functional-changes of carboxymethyl potato starch by conjugation with whey proteins. *Journal of Agricultural and Food Chemistry* 43: 2007–11.

He, J. S., Azuma, N., Hagiwara, T., and Kanno, C. 2006. Effects of sugars on the cross-linking formation and phase separation of high-pressure induced gel of whey protein from bovine milk. *Bioscience, Biotechnology, and Biochemistry* 70: 615–25.

Herceg, Z., Režek, A., Lelas, V., Krešic, G., and Franetovic, M. 2007. Effect of carbohydrates on the emulsifying, foaming and freezing properties of whey protein suspensions. *Journal of Food Engineering* 79: 279–86.

Hiller, B. and Lorenzen, P. C. 2008. Surface hydrophobicity of physicochemically and enzymatically treated milk proteins in relation to techno-functional properties. *Journal of Agricultural and Food Chemistry* 56: 461–68.

Hodge, J. E. 1953. Chemistry of browning reactions in model systems. *Journal of Agricultural and Food Chemistry* 1: 928–43.

Hodge, J. E. 1967. Origin of flavor in foods nonenzymatic browning. In *Chemistry and Physiology of Flavors*, eds. H. W. Schultz, E. A. Day, and L. M. Libbey, pp. 465–91. Westport, CT: AVI Publishing.

Hofmann, T., Bors, W., and Stettmaier, K. 1999. Radical-assisted melanoidin formation during thermal processing of foods as well as under physiological conditions. *Journal of Agricultural and Food Chemistry* 47: 391–95.

Hogan, S. A., McNamee, B. F., O'Riordan, E. D., and O'Sullivan, M. 2001. Emulsification and microencapsulation properties of sodium caseinate/carbohydrate blends. *International Dairy Journal* 11: 137–44.

Ibanogle, E. 2002. Rheological behaviour of whey protein stabilized emulsions in the presence of gum arabic. *Journal of Food Engineering* 52: 273–77.

Imamura, K., Ogawa, T., Sakiyama, T., and Nakanishi, K. 2003. Effects of types of sugar on the stabilization of protein in the dried state. *Journal of Pharmaceutical Sciences* 92: 266–74.

Izutsu, K., Aoyagi, N., and Kojima, S. 2004. Protection of protein secondary structure by saccharides of different molecular weights during freeze-drying. *Chemical & Pharmaceutical Bulletin* 52: 199–203.

Izutsu, K., Yoshioka, S., and Takeda, Y. 1991. The effects of additives on the stability of freeze-dried beta-galactosidase stored at elevated temperature. *International Journal of Pharmaceutics* 71: 137–46.

Jimenez-Castano, L., Lopez-Fandino, R., Olano, A., and Villamiel, M. 2005. Study on beta-lactoglobulin glycosylation with dextran: Effect on solubility and heat stability. *Food Chemistry* 93: 689–95.

Jimenez-Castano, L., Villamiel, M., and Lopez-Fandino, R. 2007. Glycosylation of individual whey proteins by Maillard reaction using dextran of different molecular mass. *Food Hydrocolloids* 21: 433–43.

Katayama, S., Haga, Y., and Saeki, H. 2004. Loss of filament formation of myosin by non-enzymatic glycosylation and its molecular mechanism. *FEBS Letters* 575: 9–13.

Katayama, S. and Saeki, H. 2004. Cooperative effect of relative humidity and glucose concentration on improved solubility of shellfish muscle protein by the Maillard reaction. *Fisheries Science* 70: 158–65.

Kato, A. 2002. Industrial applications of Maillard-type protein-polysaccharide conjugates. *Food Science and Technology Research* 8: 193–99.

Kato, Y., Matsuda, T., and Nakamura, R. 1993b. Improvement of physicochemical and enzymatic properties of bovine trypsin by non-enzymatic glycation. *Bioscience, Biotechnology, and Biochemistry* 57: 1–5.

Kato, A., Minaki, K., and Kobayashi, K. 1993a. Improvement of emulsifying properties of egg white proteins by the attachment of polysaccharide through Maillard reaction in dry state. *Journal of Agriculture and Food Chemistry* 41: 540–43.

Kato, A., Sasaki, Y., Furuta, R., and Kobayashi, K. 1990. Functional protein-polysaccharide conjugate prepared by controlled dry-heating of ovalbumin-dextran mixtures. *Agricultural and Biological Chemistry* 54: 107–12.

Kijowski, J. and Richardson, R. I. 1996. The effect of cryoprotectants during freezing or freeze drying upon properties of washed mechanically recovered broiler meat. *International Journal of Food Science Technology* 31: 41–55.

Kim, H. J., Choi, S. J., Shin, W. S., and Moon, T. W. 2003. Emulsifying properties of bovine serum albumin-galactomannan conjugates. *Journal of Agricultural and Food Chemistry* 51: 1049–56.

Kim, Y. D., Morr, C. V., and Schenz, T. W. 1996. Microencapsulation properties of gum arabic and several food proteins: Liquid orange oil emulsion particles. *Journal of Agricultural and Food Chemistry* 44: 1308–13.

Kitabatake, N., Cuq, J. L., and Cheftel, J. C. 1985. Covalent binding of glycosyl residues to beta-lactoglobulin: Effects on solubility and heat stability. *Journal of Agricultural and Food Chemistry* 33: 125–30.

Kobayashi, M. and Takatsu, K. 1994. Cross-linked stabilization of trypsin with dextran-dialdehyde. *Bioscience, Biotechnology, and Biochemistry* 58: 275–78.

Kobayashi, M., Yanagihara, S., Kitae, T., and Ichishima, E. 1989. Use of water-soluble carbodiimide (EDC) for immobilization of EDC-sensitive dextranase. *Agricultural and Biological Chemistry* 53: 2211–16.

Kulmyrzaev, A., Cancelliere, C., and McClements, D. J. 2000. Influence of sucrose on cold gelation of heat-denatured whey protein isolate. *Journal of the Science of Food and Agriculture* 80: 1314–18.

Kumeno, K., Nakahama, N., Honma, K., Makino, T., and Watanabe, M. 1996. Production and characterization of pressure-induced gel from freeze-concentrated milk. *Bioscience, Biotechnology, and Biochemistry* 57: 750–52.

Lee, J. C. and Timasheff, S. N. 1981. Stabilization of proteins by sucrose. *Journal of Biological Chemistry* 256: 7193–7201.

Leroux, J., Langendorff, V., Schick, G., Vaishnav, V., and Mazoyer, J. 2003. Emulsion stabilizing properties of pectin. *Food Hydrocolloids* 17: 455–62.

Maeda, H. 2000. Soluble soybean polysaccharide. In *Handbook of Hydrocolloids*, eds. G. O. Phillips and P. A. Williams, pp. 309–20. Cambridge: Woodhead Publishing.

Maitena, U., Katayama, S., Sato, R., and Saeki, H. 2004. Improved solubility and stability of carp myosin by conjugation with alginate oligosaccharide. *Fisheries Science* 70: 896–902.

Matsumoto, I., Nakakumi, T., Ito, Y., and Arai, K. 1992. Preventive effect of various sugars against denaturation of carp myofibrillar protein caused by freeze-drying. *Nippon Suisan Gakkaishi* 58: 1913–18.

Matsumura, Y., Egami, M., Satake, C., et al. 2003. Inhibitory effects of peptide-bound polysaccharides on lipid oxidation in emulsions. *Food Chemistry* 83: 107–19.

Matsumoto, J. J. and Noguchi, S. 1992. Cryostabilization of protein in surimi. In *Surimi Technology*, eds. T. C. Lanier and C. M. Lee, pp. 364–66. New York: Marcel Dekker.

Matsumoto, I., Ooizumi, T., and Arai, K. 1985. Protective effect of sugar on freeze-denaturation of carp myofibrillar protein. *Bulletin of the Japanese Society of Scientific Fisheries* 51: 833–39.

Mierzeiewska, D., Mitrowska, P., Rudnicka, B., Kubicka, E., and Kostyra, H. 2008. Effect of non-enzymatic glycosylation of pea albumins on their immunoreactive properties. *Food Chemistry* 111: 127–31.

Murray, B. S. and Liang, H. 1999. Enhancement of the foaming properties of protein dried in the presence of trehalose. *Journal of Agricultural and Food Chemistry* 47: 4984–91.

Nagasawa, K., Ohgata, K., Takahashi, K., and Hattori, M. 1996. Role of the polysaccharide content and net charge on the emulsifying properties of beta-lactoglobulin-carboxymethyldextran conjugates. *Journal of Agricultural and Food Chemistry* 44: 2538–43.

Nakamura, S., Hata, J., Kawamukai, M., et al. 2004. Enhanced anti-rotavirus action of human cystatin C by site-specific glycosylation in yeast. *Bioconjugate Chemistry* 15: 1289–96.

Nakamura, S. and Kato, A. 2000. Multi-functional biopolymer prepared by covalent attachment of galactomannan to egg-white proteins through naturally occurring Maillard reaction. *Nahrung-Food* 44: 201–6.

Nakamura, S., Kato, A., and Kobayashi, K. 1991. New antimicrobial characteristics of lysozyme dextran conjugate. *Journal of Agricultural and Food Chemistry* 39: 647–50.

Nakamura, S., Kato, A., and Kobayashi, K. 1992. Bifunctional lysozyme-galactomannan conjugate having excellent emulsifying properties and bactericidal effect. *Journal of Agricultural and Food Chemistry* 40: 735–39.

Nakamura, S., Ogawa, M., Nakai, S., Kato, A., and Kitts, D. D. 1998. Antioxidant activity of a Maillard-type phosvitin-galactomannan conjugate with emulsifying activity and heat stability. *Journal of Agricultural and Food Chemistry* 46: 3958–63.

Nakamura, A., Sasaki, F., Watanabe, K., Ojima, T., Ahn, D. H., and Saeki, H. 2006. Changes in allergenicity and digestibility of squid tropomyosin during the Maillard reaction with ribose. *Journal of Agricultural and Food Chemistry* 54: 9529–34.

Nakamura, S., Suzuki, Y., Ishikawa, E., et al. 2008. Reduction of in vitro allergenicity of buckwheat Fag e 1 through the Maillard-type glycosylation with polysaccharides. *Food Chemistry* 109: 538–45.

Nakamura, S., Takasaki, H., Kobayashi, K., and Kato, A. 1993. Hyperglycation of hen egg white lysozyme in yeast. *Journal of Biological Chemistry* 268: 12706–12.

Nakamura, A., Watanabe, K., Ojima, T., Ahn, D. H., and Saeki, H. 2005. Effect of Maillard reaction on allergenicity of scallop tropomyosin. *Journal of Agricultural and Food Chemistry* 53: 7559–64.

Namiki, M. 2006. The Maillard reaction and free radicals: Discovery of the Namiki pathway. *JMARS News Letter* (April 22, 2006): 1–3. http://www.maillard.umin.jp/.

Namiki, M. and Hayashi, T. 1983. A new mechanism of the Maillard reaction involving sugar fragmentation and free-radical formation. In *The Maillard Reaction in Foods and Nutrition*, eds. G. R. Waller and M. S. Feather; *ACS Symposium Series*, pp. 21–46. Washington, DC: American Chemical Society.

Neirynck, N., Van der Meeren, P., Lukaszewicz-Lausecker, M., Cocquyt, J., Verbeken, D., and Dewettinck, K. 2007. Influence of pH and biopolymer ratio on whey protein–pectin interactions in aqueous solutions and in O/W emulsion. *Colloids and Surfaces A: Physicochemical and Engineering Aspects* 298: 99–107.

Noguchi, S., Oosawa, K., and Matsumoto, J. J. 1976. Studies on the control of denaturation of fish muscle proteins during frozen storage-Vi. Preventive effect of carbohydrates. *Bulletin of the Japanese Society of Scientific Fisheries* 42: 77–82.

Ooizumi, T., Hashimoto, K., Ogura, J., and Arai, K. 1981. Quantitative aspect for protective effect of sugar and sugar alcohol against denaturation of fish myofibrils. *Bulletin of the Japanese Society of Scientific Fisheries* 47: 901–8.

Osako, K., Hossain, M. A., Kuwahara, K., and Nozaki, Y. 2005. Effect of trehalose on the gel-forming ability, state of water and myofibril denaturation of horse mackerel Trachurus japonicus surimi during frozen storage. *Fisheries Science* 71: 367–73.

Pedrosa, C., Trisciuzzi, C., and Ferreira, S. T. 1997. Effects of glycosylation on functional properties of vicilin, the 7S storage globulin from pea (Pisum sativum). *Journal of Agricultural and Food Chemistry* 45: 2025–30.

Ramezani, R., Esmailpour, M., and Aminlari, M. 2008. Effect of conjugation with glucosamine on the functional properties of lysozyme and casein. *Journal of the Science of Food and Agriculture* 88: 2730–37.

Saeki, H. and Inoue, K. 1997. Improved solubility of carp myofibrillar proteins in low ionic strength medium by glycosylation. *Journal of Agricultural and Food Chemistry* 45: 3419–22.

Saito, A., Sako, Y., Usui, M., Azakami, H., and Kato, A. 2003. Functional properties of glycosylated lysozyme secreted in Pichia pastoris. *Bioscience, Biotechnology, and Biochemistry* 67: 2334–43.

Sato, R., Katayama, S., Sawabe, T., and Saeki, H. 2003. Stability and emulsion-forming ability of water-soluble fish myofibrillar protein prepared by conjugation with alginate oligosaccharide. *Journal of Agricultural and Food Chemistry* 51: 4376–81.

Sato, R., Sawabe, T., Kishimura, H., Hayashi, K., and Saeki, H. 2000. Preparation of neoglycoprotein from carp myofibrillar protein and alginate oligosaccharide: Improved solubility in low ionic strength medium. *Journal of Agricultural and Food Chemistry* 48: 17–22.

Shepherd, R., Robertson, A., and Ofman, D. 2000. Dairy glycoconjugate emulsifiers: Casein-maltodextrins. *Food Hydrocolloids* 14: 281–86.

Shu, Y. W., Sahara, S., Nakamura, S., and Kato, A. 1996. Effects of the length of polysaccharide chains on the functional properties of the Maillard-type lysozyme-polysaccharide conjugate. *Journal of Agricultural and Food Chemistry* 44: 2544–48.

Song, Y., Babiker, E. E., Usui, M., Saito, A., and Kato, A. 2002. Emulsifying properties and bactericidal action of chitosan-lysozyme conjugates. *Food Research International* 35: 459–66.

Sultanbawa, Y. and Li-Chan, E. C. Y. 2001. Structural changes in natural actomyosin and surimi from ling cod (Ophiodon elongatus) during frozen storage in the absence or presence of cryoprotectants. *Journal of Agricultural and Food Chemistry* 49: 4716–25.

Sun, Y. X., Hayakawa, S., Ogawa, M., and Izumori, K. 2005. Evaluation of the site specific protein glycation and antioxidant capacity of rare sugar-protein/peptide conjugates. *Journal of Agricultural and Food Chemistry* 53: 10205–12.

Takahashi, K., Lou, X. F., Ishii, Y., and Hattori, M. 2000. Lysozyme-glucose stearic acid monoester conjugate formed through the Maillard reaction as an antibacterial emulsifier. *Journal of Agricultural and Food Chemistry* 48: 2044–49.

Takano, K., Hattori, M., Yoshida, T., Kanuma, S., and Takahashi, K. 2007. Porphyran as a functional modifier of a soybean protein isolate through conjugation by the Maillard reaction. *Journal of Agricultural and Food Chemistry* 55: 5796–5802.

Takeda, H., Iida, T., Okada, A., et al. 2007. Feasibility study on water solubilization of spawned out salmon meat by conjugation with alginate oligosaccharide. *Fisheries Science* 73: 924–34.

Tanabe, M. and Saeki, H. 2001. Effect of Maillard reaction with glucose and ribose on solubility at low ionic strength and filament-forming ability of fish myosin. *Journal of Agricultural and Food Chemistry* 49: 3403–7.

Tanaka, M., Kunisaki, N., and Ishizaki, S. 1999. Improvement of emulsifying and antibacterial properties of salmine by the Maillard reaction with dextran. *Fisheries Science* 65: 623–28.

Toyoda, K., Kimura, I., Noguchi, S., and Lee, C. M. 1992. The surimi manufacturing process. In *Surimi Technology*, eds. T. C. Lanier and C. M. Lee, pp. 79–112. New York: Marcel Dekker.

Tzeng, S. S. and Jiang, S. T. 2004. Glycosylation modification improved the characteristics of recombinant chicken cystatin and its application on mackerel surimi. *Journal of Agricultural and Food Chemistry* 52: 3612–16.

Usui, M., Tamura, H., Nakamura, K., et al. 2004. Enhanced bactericidal action and masking of allergen structure of soy protein by attachment of chitosan through Maillard-type protein-polysaccharide conjugation. *Nahrung-Food* 48: 69–72.

Valdivia, A., Perez, Y., Dominguez, A., Caballero, J., Hernandez, Y., and Villalonga, R. 2006. Improved pharmacological properties for superoxide dismutase modified with mannan. *Biotechnology and Applied Biochemistry* 44: 159–65.

Villalonga, R., Villalonga, M. L., and Gomez, L. 2000. Preparation and functional properties of trypsin modified by carboxymethylcellulose. *Journal of Molecular Catalysis B: Enzymatic* 10: 483–90.

Wanisca, R. D. and Kinsella, J. E. 1984. Enzymatic hydrolysis of maltosyl and glucosaminyl derivatives of b-lactoglobulin. *Journal of Agricultural and Food Chemistry* 32: 1042–44.

Wondrak, G. T., Varadarajan, S., Butterfield, D. A., and Jacobson, M. 2000. Formation of a protein-bound pyrazinium free radical cation during glycation of histone H1. *Free Radical Biology & Medicine* 29: 557–67.

Wooster, T. J. and Augustin, M. A. 2007. The emulsion flocculation stability of protein-carbohydrate diblock copolymers. *Journal of Colloid and Interface Science* 313: 665–75.

Yamasaki, N. and Ikebe, K. 1992. A new amylose derivative for the preparation of protein-carbohydrate conjugates. *Bioscience, Biotechnology, and Biochemistry* 56: 2091–92.

Yang, W., Komine, N., Hattori, M., Ishii, Y., and Takahashi, K. 1995. Improvement of potato starch by conjugation with ε-poly(L-lysine) through the Maillard reaction. *Journal of Applied Glycoscience* 52: 253–59.

Yoo, B. and Lee, C. M. 1993. Thermoprotective effect of sorbitol on proteins during dehydration. *Journal of Agricultural and Food Chemistry* 41(2): 190–92.

Yoshimasu, M. A., Tanaka, T., Ahn, J. K., and Yada, R. Y. 2004. Effect of N-linked glycosylation on the aspartic proteinase porcine pepsin expressed from Pichia pastoris. *Glycobiology* 14: 417–29.

Zhou, A., Benjakul, S., Pan, K., Gong, J., and Liu, X. 2006. Cryoprotective effects of trehalose and sodium lactate on tilapia (Sarotherodon nilotica) surimi during frozen storage. *Food Chemistry* 96: 96–103.

Chapter 10

Peptide–Lipid Interactions and Functionalities

Ameliorative Action of Peptides on Cholesterol and Lipid Metabolism from the Viewpoint of Peptide–Lipid Interactions

Satoshi Nagaoka

Contents

10.1 Introduction

Hyperlipidemia, especially hypercholesterolemia, is one of the most important risk factors of heart disease and ischemic heart disease (Kannel et al. 1971; Martin et al. 1986; Law 1999). The increased mortality rate from ischemic heart disease, such as angina pectoris, due to coronary arteriosclerosis and myocardial infarction is remarkable in Japan and other countries. The prevention and improvement of hypercholesterolemia by dietary regulation are considered important (Grundy and Denke 1990; Ginsberg, Barr, and Gilbert 1990). Dietary protein is considerably useful as a regulator of serum cholesterol concentration (Anderson, Johnstone, and Cook-Newell 1995). In fact, diets low in saturated fat and cholesterol that include 25 g of soy protein per day may reduce the risk of heart disease. A health claim that US Food and Drug Administration has made to lower the serum cholesterol concentration to 5%–10% by an intake of 25 g soybean protein per day was approved in 1999 (Food and Drug Administration Soy Proteins and Coronary Heart Disease 1999). In Japan, "Food for specified health use" has been produced for the prevention and treatment of lifestyle-related diseases, and also "Food for specified health use" for preventing and improving hypercholesterolemia has been created.

A couple of approaches have been proposed to control the cholesterol and lipid levels in the blood, liver, and other parts of the body. This chapter focuses on the issue of the ameliorative action of peptides on cholesterol and lipid metabolism from the viewpoint of peptide–lipid interactions.

10.2 Soy Protein Hydrolysate with Bound Phospholipids and Wheat Gluten Hydrolysate with Bound Phospholipids

Global efforts to make a more active soy peptide than the soybean protein are ongoing. One of these trials is the development of soy peptides with bound phospholipids (Nagaoka et al. 1999, 2002; Hori et al. 2001). It is expected that the hypocholesterolemic action of soy peptides with bound phospholipids is more active than that of soybean proteins (Nagaoka et al. 1999).

Sirtori et al. (1985) reported the effects of textured SP containing 6% lecithin in type-II hyperlipidemic patients, among whom they noted an increase in HDL cholesterol concentration. This textured soy protein containing 6% lecithin was made with 6% lecithin by mixing. Nagaoka et al. (1999) reported that the effects of a crude type of soy protein hydrolysate with bound phospholipids (CSPHP) on cholesterol metabolism were compared with the effects of soy protein peptic hydrolysate with bound phospholipids (SPHP), casein, soy protein peptic hydrolysate (SPH), and chitosan (Nagaoka et al. 2006). In this study, enzyme-modified soy phospholipids were used as a phospholipid source. Enzyme-modified soy phospholipids were prepared from soy phospholipids by phospholipase A_2 hydrolysis. For example, SPHP was prepared as follows: first, the soy protein was dispersed in water and stirred using an homogenizer at 2800× g for 5 min. Enzyme-modified soy phospholipids were then added to the solution and stirred as described above. The soy protein and enzyme-modified soy phospholipids were added in the ratio of 4:1 (wt/wt). The resulting mixture was freeze-dried and identified as SP (soyprotein with bound phospholipids), which was hydrolyzed by porcine pepsin. The digest was centrifuged at 4500× g for 20 min. The sediment was washed with water three times and centrifuged at 4500× g for 20 min. The sediment was freeze dried and identified as SPHP.

The serum and liver cholesterol levels were significantly decreased with CSPHP, SPHP, SPH, and chitosan feeding, compared with casein feeding. The cholesterol level was most significantly lowered by SPHP. In vivo radioisotope studies suggested that SPHP or chitosan feeding inhibited the absorption of cholesterol and the reabsorption of taurocholate in rats (Nagaoka et al. 2006). A Caco-2 cell culture study indicated that cholesterol uptake from micelles containing SPHP or CSPHP was significantly lower than from micelles containing casein tryptic hydrolysate (CTH) or casein (Nagaoka et al. 1999). The taurocholate-binding capacities of CTH, SPH, and SPHP were measured in vitro. The bile acid–binding capacity of SPHP was significantly higher than that of the CTH or SPH, indicating that the SPHP has the highest bile acid–binding capacity among the protein hydrolysates tested (Nagaoka et al. 1999). The micellar solubility of cholesterol was significantly lower in the presence of SPH or SPHP than CTH or soy protein (Nagaoka et al. 1999). Our results suggested that the inhibition of the micellar solubility of cholesterol by direct interaction between the bile acid in cholesterol-mixed micelles and the SPHP causes the suppression of cholesterol absorption. Thus, the direct interaction between the peptide and the bile acid or the micelle is important for the inhibition of cholesterol absorption. The screening of the bile acid–binding peptide derived from food protein is important for the screening of the hypocholesterolemic peptide.

Clinical trials with CSPHP have been done. Human studies suggested that serum total cholesterol and low-density lipoprotein (LDL) cholesterol levels were decreased by CSPHP feeding. CSPHP is useful for normalizing the high cholesterol level of hypercholesterolemic patients whose serum total cholesterol levels are >220 mg/dL. A daily intake of only 3 g of CSPHP for 3 months reduced both the serum total cholesterol and the LDL cholesterol levels of hypercholesterolemic patients (Hori et al. 2001). The US Food and Drug Administration suggested that a daily intake of 25 g of SP might reduce the risk of heart disease. Therefore, because less CSPHP than soy protein is needed for the same effect, CSPHP has a remarkable effect on the reduction of the risk of heart disease. "Food for specified health use" using soy peptides with bound soy phospholipids (CSPHP) for the prevention and improvement of hypercholesterolemia has been created (Nagaoka et al. 1999, 2002; Hori et al. 2001).

On the other hand, wheat gluten also induces a hypocholesterolemic action in rats (Mokady and Liener 1982; McGregor 1971). Thus, similar to the trials of soybean protein, a wheat gluten with bound phospholipids (WGP) and a wheat gluten hydrolysate with bound phospholipids (WGHP) were developed, as shown in Table 10.1.

Table 10.1 Effects of Dietary Casein, Wheat Gluten, Wheat Gluten with Bound Phospholipids (WGP), Wheat Gluten Peptic Hydrolysate (WGH), and Wheat Gluten Peptic Hydrolysate with Bound Phospholipids (WGHP) on Body and Liver Weights, Food Intake, Serum and Liver Lipids, and Fecal Steroid Excretion in Rats[a]

	Diet Group				
	Casein	Wheat Gluten	WGP	WGH	WGHP
Body weight gain (g/10 day)	28.9 ± 1.3 a	27.3 ± 2.1 a	26.7 ± 1.1 a	26.1 ± 2.9 a	28.7 ± 1.3 a
Liver weight (g/100 g body weight)	4.32 ± 0.04 d	4.07 ± 0.09 c	4.07 ± 0.04 c	3.87 ± 0.07 b	3.69 ± 0.04 a
Food intake (d 8, g/d)	13.8 ± 0.5ab	15.4 ± 0.9 b	14.3 ± 0.34 ab	14.5 ± 0.34 ab	13.2 ± 0.5 a
Serum (mg/dL)					
Total cholesterol (a)	231.7 ± 14.7 b	102.1 ± 5.5 a	101.9 ± 7.5 a	88.3 ± 3.7 a	78.2 ± 3.2 a
HDL cholesterol (b)	27.2 ± 1.7 a	30.3 ± 2.1 ab	30.0 ± 1.7 ab	36.5 ± 2.7 bc	38.3 ± 2.7 c
LDL + VLDL cholesterol[b]	204.5 ± 15.7 c	71.7 ± 6.4 b	71.9 ± 7.6 b	51.8 ± 5.5 ab	39.9 ± 0.8 a
(b)/(a)	0.12 ± 0.02 a	0.30 ± 0.03 b	0.30 ± 0.03 b	0.42 ± 0.04 c	0.49 ± 0.02 c
Triglyceride	55.5 ± 4.3 b	47.3 ± 5.1 ab	50.1 ± 2.6 ab	35.9 ± 3.9 a	48.3 ± 8.1 ab

Liver					
Total lipids (mg/g liver)	146.8 ± 4.4 d	132.6 ± 5.4 c	118.1 ± 3.7 b	106.0 ± 3.9 b	90.8 ± 2.2 a
Cholesterol (mg/g liver)	32.8 ± 1.2 e	28.8 ± 0.9 cd	26.1 ± 1.7 c	21.3 ± 0.8 b	13.5 ± 0.5 a
Triglyceride (mg/g liver)	29.7 ± 2.5 b	28.9 ± 3.2 b	22.0 ± 2.0 ab	17.3 ± 1.9 a	17.3 ± 1.3 a
Fecal					
Dry weight (g/3 day)	3.10 ± 0.09 a	3.54 ± 0.2 ab	3.37 ± 0.22 a	5.69 ± 0.07 c	3.87 ± 0.11 b
Neutral steroids (mg/3 day)					
Cholesterol	59.0 ± 2.8 a	71.1 ± 3.8 b	59.1 ± 3.5 a	52.8 ± 3.4 a	82.3 ± 4.3 c
Coprostanol	11.3 ± 1.4 a	6.1 ± 0.5 a	13.9 ± 2.1 a	32.8 ± 2.0 b	43.3 ± 8.2 b
Total	70.3 ± 3.9 a	77.2 ± 3.4 ab	73.0 ± 4.8 ab	85.6 ± 1.8 b	125.6 ± 6.0 c
Acidic steroids (mg/3 day)	86.9 ± 7.7 a	115.3 ± 6.1 b	90.5 ± 5.9 a	108.0 ± 7.9 ab	95.9 ± 8.2 ab
Total steroids[c] (mg/3 day)	157.2 ± 10.0 a	192.5 ± 8.1 bc	163.5 ± 10.3 ab	193.6 ± 9.2 bc	221.5 ± 12.6 c

[a] Values are means (n = 6) and pooled SEM. Within a row, means with different superscript letters are significantly different (p < .05) by Duncan's multiple range test.
[b] Values were calculated as follows: LDL + VLDL cholesterol = total cholesterol – HDL cholesterol.
[c] Total steroids = neutral steroids + acidic steroids.

The preparation procedure of WGHP was almost the same as that of SPHP. Serum and liver cholesterol levels were significantly decreased with WGHP feeding, compared with casein feeding. The bile acid–binding capacity of WGHP was significantly higher than that of WGH. The micellar solubility of cholesterol was significantly lower in the presence of WGH or WGHP than CTH or wheat gluten. Fecal steroid excretion was significantly increased in rats fed WGH or WGHP, compared with rats fed casein, as shown in Table 10.1. Our results suggest that the inhibition of the micellar solubility of cholesterol by direct interaction between the bile acid in cholesterol-mixed micelles and WGHP causes the suppression of cholesterol absorption.

SPHP originating from soybean protein can induce cholesterol 7α-hydroxylase (CYP7A1) mRNA and enzyme activity (Nagaoka et al. unpublished result) in rats with hypocholesterolemia that are fed a cholesterol-enriched diet. Currently, the mechanism by which SPHP induces the stimulation of CYP7A1 mRNA in rats in this study is unknown. Fecal neutral and acidic steroid excretion was stimulated by SPHP feeding (Nagaoka et al. 1999). We hypothesized that the increased fecal steroid excretion with an SPHP diet may induce a shift in the liver cholesterol metabolism, thereby providing cholesterol for the enhancement of bile acid synthesis related to the stimulation of CYP7A1 gene expression. Realization of this possibility is currently underway.

10.3 Cholesterol-Lowering Peptides and Bile Acid–Binding Peptides

10.3.1 Lactostatin

A previous paper by Sugano, Goto, and Yamada (1990) demonstrated the hypothesis that hypocholesterolemic peptides derived from soybean protein might exist and influence the serum cholesterol level. However, to date, no one has found the hypocholesterolemic peptide in soybean protein or in any protein. Sugano, Goto, and Yamada (1990) suggested that the hypocholesterolemic peptides derived from soybean protein consisted of those peptides of molecular weights between 1K and 10K. However, the hypocholesterolemic peptide of soybean protein remains unidentified. Although a previous study (Morimatsu et al. 1996) demonstrated that the hypocholesterolemic effects of papain-hydrolyzed pork meat contained peptides with molecular weights of 3K or less, the hypocholesterolemic peptide of pork protein is also unidentified.

We reported that the hypocholesterolemic action of β-lactoglobulin tryptic hydrolysate (LTH) was induced by the suppression of cholesterol absorption evidenced by an in vivo cholesterol absorption study and an in vitro Caco-2 cell study (Nagaoka et al. 2001). We tried to identify the active component related to the hypocholesterolemic action of LTH. We hypothesized that the peptide derived from bovine milk β-lactoglobulin might induce a hypocholesterolemic action. Our experimental system to evaluate the cholesterol uptake in Caco-2 cells, which we called "Caco-2 cell screening," is useful for clarifying the active component underlying the inhibitory effect of a peptide on cholesterol absorption in the small intestine (Nagaoka et al. 1997, 1999). Thus, by using Caco-2 cell screening, for the first time, we identified four kinds of novel peptide sequences that inhibit cholesterol absorption in vitro (Nagaoka et al. 2001).

Subsequently, we tried to evaluate the hypocholesterolemic activity of new peptides in animal studies in vivo. Although no one could find a hypocholesterolemic peptide from any protein origin, we identified, for the first time, a new hypocholesterolemic peptide, Ile-Ile-Ala-Glu-Lys (IIAEK: we designated as "lactostatin"). We found that the peptide derived from β-lactoglobulin can greatly influence the serum cholesterol level and that the hypocholesterolemic activity of IIAEK exhibited a greater hypocholesterolemic activity in comparison with the β-sitosterol in rats (Nagaoka et al. 2001).

Our results demonstrated that the ERK pathway and the calcium channel were involved in CYP7A1 transactivation induced by lactostatin in HepG2 cells (Morikawa et al. 2007). We also reported that lactostatin induces the suppression of CYP7A1 mRNA in mice (Nagaoka et al. 2006) and rats (Nagaoka et al. unpublished result). CYP7A1 is the rate-limiting enzyme for cholesterol degradation (bile acid synthesis). The overexpression of CYP7A1 ameliorates hypercholesterolemia and atherosclerosis in animal models (Spady et al. 1995, 1998). Thus, regulation of the CYP7A1 gene expression is very important in the strategy of prevention and improvement of hypercholesterolemia and atherosclerosis. Currently, the mechanism by which lactostatin induces the suppression of CYP7A1 mRNA in rodents is unknown. In primary cultured rat hepatocytes, bile acids inhibit the bile acid synthesis via activation of the stress kinases (Gupta et al. 2001; Wang et al. 2002; Fabiani et al. 2001). Wang et al. (2002) examined the status of the c-Jun N-terminal kinase (JNK) as a favored candidate in this process.

It is well known that the CYP7A1 gene expression has species differences (Repa et al. 2000; Davis et al. 2002; Goodwin et al. 2003; Agellon et al. 2002; Chiang, Kimmel, and Stroup 2001). In rodents, such as mice, the LXRα agonist (T0901317) activates the CYP7A1 gene expression (Repa et al. 2000), but in primary cultures of human hepatocytes, T0901317 cannot activate the CYP7A1 gene expression (Goodwin et al. 2003). The species differences in the CYP7A1 gene expression between rodents and humans may explain the differences in LXRα responsive elements in the CYP7A1 gene promoter (Davis et al. 2002; Agellon et al. 2002; Chiang, Kimmel, and Stroup 2001). Thus, the effect on the CYP7A1 gene expression induced by lactostatin may change among species. Realization of this possibility is currently under investigation. Taken together, our results suggest the involvement of a new regulatory pathway in the calcium channel–related mitogen-activated protein kinase (MAPK) signaling pathway of lactostatin-mediated cholesterol degradation (Morikawa et al. 2007).

10.3.2 Soystatin

In a previous paper, Sugano et al. (1990) have hypothesized that soybean peptides that induce the inhibition of cholesterol absorption exist and influence the serum cholesterol concentration. Yet, none of the earlier investigations have actually found a peptide derived from soybean protein that inhibits cholesterol absorption. Many studies have been conducted to identify the hypocholesterolemic peptide derived from soybean protein with screening systems in vitro. One of the in vitro screening systems to identify the hypocholesterolemic peptide is the evaluation system based on the changes in LDL-receptor mRNA in HepG2 cells. In this system, a candidate of the hypocholesterolemic peptide FVVNATSN was found to increase the LDL-receptor mRNA without corroborative evidence from animal experiments in vivo (Cho, Juillerat, and Lee 2008). Another in vitro screening to identify the hypocholesterolemic peptide evaluates the increase of CYP7A1 mRNA in HepG2 cells in vitro. In this system, a mixture of milk casein peptides was found to increase CYP7A1 mRNA without corroborative evidence from animal experiments in vivo (Nass et al. 2008). Another in vitro screening to identify the hypocholesterolemic peptide evaluates the micellar solubility of cholesterol and the bile acid–binding ability in vitro; that is, the same parameters examined in this study. Based on earlier animal experiments in vivo and assays on the micellar solubility of cholesterol and the level of bile acid–binding ability in the presence of SPH in vitro, our group has already proposed that the suppression of cholesterol absorption by direct interaction between the cholesterol-mixed micelles and the SPH in the intestinal epithelia contributes to the hypocholesterolemic action induced by SPH (Nagaoka et al. 1999). Convinced that SPH must contain a yet-to-be-identified hypocholesterolemic peptide that inhibits cholesterol absorption, we hypothesized that VAWWMY per se may induce the inhibition of cholesterol absorption in vivo by inhibiting the micellar solubility of cholesterol via its high bile acid–binding capacity in vitro, as proposed by our previous report (Nagaoka et al. 1999). We reported that VAWWMY (Val-Ala-Trp-Trp-Met-Tyr; designated soystatin) had a significantly greater ability to bind bile acid than GGGGGG (Gly-Gly-Gly-Gly-Gly-Gly), SPH, and CTH (Nagaoka et al. 2010). Surprisingly, the bile acid–binding ability of VAWWMY is almost as strong as that of the hypocholesterolemic medicine cholestyramine. The micellar solubility of cholesterol was significantly lower in the presence of VAWWMY than that of GGGGGG, SPH, or CTH. We found that soystatin derived from soybean glycinin acted as an inhibitor of cholesterol absorption in vivo (Nagaoka et al. 2010). Now that the present study has found the bile acid–binding ability of soystatin per se in vitro, the molecular mechanism of the binding of the taurocholic acid molecule to the soystatin molecule is likely to be clarified in the near future. By elucidating the molecular mechanisms underlying the binding of bile acid to soystatin, we will be able to improve the soystatin activity and create a novel bile acid–binding peptide.

10.4 The Ameliorative Action of Peptides on Lipid Metabolism

10.4.1 Hypotriglyceridemic Action of Peptides

In recent years, it has been reported that the oligopeptides have a hypotriglyceridemic property by suitable protease digestion of various edible proteins, such as globin, soy protein, and casein. Globin digest (GD), prepared from globin by acidic protease treatment, suppressed the elevation of the serum triglyceride level in not only total but also chylomicron fraction after oral administration of olive oil. By screening with this lowering activity, it was reported that Val-Val-Tyr-Pro (VVYP) would be the most effective constituent having a hypotriglyceridemic action in GD (Kagawa et al. 1996). Furthermore, the administration of GD caused a more prominent activation of the hepatic triglyceride lipase (HTGL) and an increase in hepatic free fatty acid (FFA) concentration in the early phase after the administration of fat. From these results, it could be elucidated that GD, and also VVYP, inhibited fat absorption from the digestive tract and enhanced the activity of HTGL, thereby causing a more rapid clearance of dietary hypertriglyceridemia. They also reported the suppressive effect of GD on postprandial hyperlipidemia in male volunteers (Kagawa et al. 1998).

10.4.2 Antiobesity Action of Peptides

An antiobesity therapy based on targeted induction of apoptosis in the vasculature of adipose tissue has been reported (Kolonin et al. 2004). They used an in vivo phage display to isolate a peptide motif (sequence CKGGRAKDC) that homes white fat vasculature. They also showed that the CKGGRAKDC peptide associates with prohibitin, a multifunctional membrane protein, and establishes prohibitin as a vascular marker of adipose tissue. Targeting a proapoptotic peptide, CKGGRAKDC-GG-KLAKLAKKLAKLAK, to establish prohibitin in the adipose vasculature caused ablation of white fat.

10.4.3 Antiatherogenic Action of Peptides

Apolipoprotein A-I (apoA-I) is the main protein component in high-density lipoprotein (HDL). The beneficial effects of HDL have largely been attributed to its major protein, apoA-I. Decades of research have gone into the synthesis of apoA-I mimetic peptides. There has been a long search for peptides smaller than apoA-I, but with many of the properties of apoA-I. The structural requirements for the antioxidative and anti-inflammatory properties of apoA-I mimetic peptides have recently been reviewed (Van Lenten et al. 2009). The 4F (DWFKAFYDKVAEKFKEAF) peptides, L-4F and D-4F, are apoA-I mimetics that demonstrate prominent anti-inflammatory properties in vitro and in animal models (Van Lenten et al. 2009).

The 4F peptide is an anti-inflammatory, apoA-I mimetic peptide that is active in vivo at nanomolar concentrations in the presence of a large molar excess of apoA-I. The physiological concentrations (~35 µM) of human apoA-I did not inhibit the production of LDL-induced monocyte chemotactic activity by human aortic endothelial cell cultures, but adding nanomolar concentrations of 4F in the presence of ~35 µM apoA-I significantly reduced this inflammatory response. The anti-inflammatory 4F peptide bound the oxidized lipids with much higher affinity than did apoA-I. Initially, they examined the binding of PAPC (1-palmitoyl-2-arachidonoyl-sn-glycero-3-phosphatidylcholine) and observed that its oxidized products bound the 4F peptide with an affinity that was ~4–6 orders of magnitude higher than that of apoA-I.

Meanwhile, a peptide containing only four amino acid residues (KRES), which is too small to form an amphipathic helix, reduced lipoprotein lipid hydroperoxides (LOOH), increased paraoxonase activity, increased plasma HDL-cholesterol levels, rendered HDL anti-inflammatory, and reduced atherosclerosis in apoE null mice (Navab et al. 2006). Changing the order of two amino acids (from KRES to KERS) resulted in the loss of all biological activity. The solubility in ethyl acetate and the interaction with lipids indicated significant differences between KRES and KERS. Negative stain electron microscopy showed that KRES formed organized peptide–lipid structures, whereas KERS did not. After oral administration, KRES and FREL were found to be associated with HDL, whereas KERS was not. They concluded that the ability of peptides to interact with lipids, remove LOOH, and activate antioxidant enzymes associated with HDL determines their anti-inflammatory and antiatherogenic properties regardless of their ability to form amphipathic helixes.

10.4.4 Inhibitory Action of Fatty Acid Synthase by Peptides

Fatty acid synthase (FAS, EC 3.2.1.85) is a multicomponent enzyme that catalyzes the biosynthesis of long-chain fatty acids through an NADPH-dependent cyclic reaction (Chakravarty et al. 2004). FAS is homodimeric, and each polypeptide chain carries seven catalytic domains integrating all the steps needed for fatty acid synthesis (Smith and Tsai 2007; Maier, Jenni, and Ben 2006). The discovery and development of pharmacological FAS inhibitors promise the prevention of obesity, related metabolic disorders, and cancer (Sheng, Niu, and Sun 2009; Ronnett et al. 2005). The inhibition of FAS in the central nervous system markedly reduces food intake and body weight in animals (Loftus et al. 2000). The inhibition of FAS in the hypothalamus and pancreatic b cells protects mice against high-fat diet–induced metabolic syndrome (Chakravarthy, Zhu, and Yin 2009). Martinez-Villaluenga et al. (2010) reported that three peptides derived from β-conglycinin (KNPQLR, EITPEKNPQLR, and RKQEEDEDEEQQRE) inhibited FAS. The biological activity of these peptides was confirmed by their inhibitory activity against purified chicken FAS and a high correlation ($r = -0.7$) with lipid accumulation in the

3T3-L1 adipocytes. The FAS inhibitory potency of soy peptides also correlated with their molecular mass, their pI value, and the number of negatively charged and hydrophilic residues. Molecular modeling predicted that the large FAS inhibitory peptides (EITPEKNPQLR and RKQEEDEDEEQQRE) bond to the thioesterase domain of human FAS with lower interaction energies than classical thioesterase inhibitors (Orlistat). Docking studies suggested that soy peptides blocked the active site through interactions within the catalytic triad, the interface cavity, and the hydrophobic groove in the human FAS thioesterase domain. FAS thioesterase inhibitory activities displayed by the synthetic soy peptides EITPEKNPQLR and RKQEEDEDEEQQRE were higher than C75 but lower than Orlistat.

10.5 Conclusion

This chapter deals with the issue of the ameliorative action of peptides on cholesterol and lipid metabolism from the viewpoint of peptide–lipid interactions. The SPHP are useful materials for the ameliorative action of cholesterol metabolism in humans. The cholesterol-lowering peptide (lactostatin: IIAEK) involves a new regulatory pathway in the calcium channel–related MAPK signaling pathway of cholesterol degradation. The bile acid–binding peptide (soystatin: VAWWMY) binding to taurocholate inhibits the micellar solubility of cholesterol in vitro. Soystatin exhibits the inhibitory action of intestinal cholesterol absorption in vivo.

In Section 10.4 on the ameliorative action of peptides on lipid metabolism, VVYP is the most effective constituent having hypotriglyceridemic action in GDs. The suppressive effect of GD on postprandial hyperlipidemia in humans has been reported. Targeting a proapoptotic peptide (CKGGRAKDC-GG-KLAKLAKKLAKLAK) to establish prohibitin in the adipose vasculature causes ablation of white fat. It is concluded that the ability of peptides (KRES and FREL: apoA-I mimetic peptides) to interact with lipids, remove LOOH, and activate antioxidant enzymes associated with HDL determines their anti-inflammatory and antiatherogenic properties. It is reported that three peptides derived from β-conglycinin, KNPQLR, EITPEKNPQLR, and RKQEEDEDEEQQRE, inhibit FAS.

Taken together, a higher level of molecular, cellular, genetic, and in vivo studies is needed to create the innovative new peptides that ameliorate cholesterol and lipid metabolism.

References

Agellon, L. B., Drover, V. A., Cheema, S. K., Gbaguidi, G. F., and Walsh, A. 2002. Dietary cholesterol fails to stimulate the human cholesterol 7α-hydroxylase gene (CYP7A1) in transgenic mice. *Journal of Biological Chemistry* 277: 20131–34.

Anderson, J. W., Johnstone, B. M., and Cook-Newell, M. E. 1995. Meta-analysis of the effects of soy protein intake on serum lipids. *New England Journal of Medicine* 333: 276–82.

Chakravarty, B., Gu, Z. W., Chirala, S. S., Wakil, S. J., and Quiocho, F. A. 2004. Human fatty acid synthase: Structure and substrate selectivity of the thioesterase domain. *Proceedings of the National Academy of Sciences of United States of America* 101: 15567–72.

Chakravarthy, M. V., Zhu, Y., and Yin, L. 2009. Inactivation of hypothalamic FAS protects mice from diet-induced obesity and inflammation. *Journal of Lipid Research* 50: 630–40.

Chiang, J. Y. L., Kimmel, R., and Stroup, D. 2001. Regulation of cholesterol 7α-hydroxylase gene (CYP7A1) transcription by the liver orphan receptor (LXRα). *Gene* 262: 257–65.

Cho, S. J., Juillerat, M. A., and Lee, C. H. 2008. Identification of LDL-receptor transcription stimulating peptides from soybean hydrolysate in human hepatocytes. *Journal of Agricultural and Food Chemistry* 56: 4372–76.

Davis, R. A., Miyake, J. H., Hui, T. Y., and Spann, N. J. 2002. Regulation of cholesterol 7α-hydroxylase: BAREly missing a SHP. *Journal of Lipid Research* 43: 533–43.

Fabiani, E. D., Mitro, N., Anzulovich, A. C., Pinelli, A., Galli, G., and Crestani, M. 2001. The negative effects of bile acids and tumor necrosis factor-α on the transcription of cholesterol 7-hydroxylase gene (CYP7A1) converge to hepatic nuclear factor-4: A novel mechanism of feedback regulation of bile acid synthesis mediated by nuclear receptors. *Journal of Biological Chemistry* 276: 30708–16.

Food and Drug Administration Soy Proteins and Coronary Heart Disease. 1999. *Federal Register* 64: 57688–733.

Ginsberg, H. N., Barr, S. L., and Gilbert, A. 1990. Reduction of plasma cholesterol levels in normal men on an American Heart Association step I diet or a step I diet with added monounsaturated fat. *New England Journal of Medicine* 322: 574–79.

Goodwin, B., Watson, M. A., Kim, H., Miao, J., Kemper, J. K., and Kliewer, S. A. 2003. Differential regulation of rat and human CYP7A1 by the nuclear oxysterol receptor liver X receptor-α. *Molecular Endocrinology* 17: 386–94.

Grundy, S. M. and Denke, M. A. 1990. Dietary influences on serum lipids and lipoproteins. *Journal of Lipid Research* 31: 1149–72.

Gupta, S., Stravitz, R. T., Dent, P., and Hylemon, P. B. 2001. Down-regulation of cholesterol 7-hydroxylase (CYP7A1) gene expression by bile acids in primary rat hepatocytes is mediated by the c-jun N-terminal kinase pathway. *Journal of Biological Chemistry* 276: 15816–22.

Hori, G., Wang, M. F., Chan, Y. C., et al. 2001. Soy protein hydrolyzate with bound phospholipids reduces serum cholesterol levels in hypercholesterolemic adult male volunteers. *Bioscience, Biotechnology, and Biochemistry* 65: 72–78.

Kagawa, K., Matsutaka, H., Fukuhama, C., Fujino, H., and Okuda, H. 1998. Suppressive effect of globin digest on postprandial hyperlipidemia in male volunteers. *Journal of Nutrition* 128: 56–60.

Kagawa, K., Matsutaka, H., Fukuhama, C., Watanabe, Y., and Fujino, H. 1996. Globin digest, acidic protease hydrolysate, inhibits dietary hypertriglyceridemia and val-val-tyr-pro, one of its constituents, possesses most superior effect. *Life Sciences* 58: 1745–55.

Kannel, W. B., Castelli, W. P., Gordon, T., and McNamara, P. M. 1971. Serum cholesterol, lipoproteins, and the risk of coronary heart disease. The Framingham Study. *Annals of Internal Medicine* 74: 1–12.

Kolonin, M. G., Saha, P. K., Chan, L., Pasqualini, R., and Arap, W. 2004. Reversal of obesity by targeted ablation of adipose tissue. *Nature Medicine* 10: 625–32.

Law, M. R. 1999. Lowering heart disease risk with cholesterol reduction: Evidence from observational studies and clinical trials. *European Heart Journal* (Suppl. 1): S3–8.

Loftus, T. M., Jaworsky, D. E., Frehywot, G. L., et al. 2000. Reduced food intake and body weight in mice treated with fatty acid synthase inhibitors. *Science* 288: 2379–81.

Maier, T., Jenni, S., and Ben, N. 2006. Architecture of mammalian fatty acid synthase at 4.5 Å resolution. *Science* 311: 1258–62.

Martin, M. J., Hully, S. B., Browner, W. S., Kuller, L. H., and Wentworth, D. 1986. Serum cholesterol, blood pressure, and mortality: Implications from a cohort of 361,662 men. *Lancet* 2: 933–36.

Martinez-Villaluenga, C., Rupasinghe, S. G., Schuler, M. A., and Gonzalez, E. 2010. Peptides from purified soybean β-conglycinin inhibit fatty acid syntase by interaction with the thioesterase catalytic domain. *FEBS Journal* 277: 1481–93.

McGregor, D. 1971. The effects of some dietary changes upon the concentrations of serum lipids in rats. *British Journal of Nutrition* 25: 213–24.

Mokady, S. and Liener, I. E. 1982. Effect of plant proteins on cholesterol metabolism in growing rats fed atherogenic diets. *Annals of Nutrition and Metabolism* 26: 138–44.

Morikawa, K., Kondo, I., Kanamaru, Y., and Nagaoka, S. 2007. A novel regulatory pathway for cholesterol degradation via lactostatin. *Biochemical and Biophysical Research Communications* 352: 697–702.

Morimatsu, F., Ito, M., Budijanto, S., Watanabe, I., Furukawa, Y., and Kimura, S. 1996. Plasma cholesterol-supressing effect of papain-hydrolyzed pork meat in rats fed hypercholesterolemic diet. *Journal of Nutritional Science and Vitaminology* 42: 145–53.

Nagaoka, S., Awano, T., Nagata, N., Masaoka, M., Hori, G., and Hashimoto, K. 1997. Soyprotein peptic hydrolyzate lowers serum cholesterol level and inhibits cholesterol absorption in CaCo-2 cells. *Bioscience, Biotechnology, and Biochemistry* 61: 354–56.

Nagaoka, S., Fujimura, W., Morikawa, K., et al. 2006. Lactostatin (IIAEK) and SPHP: New cholesterol-lowering peptides derived from food proteins. In *Dietary Fat and Risk of Common Diseases*, ed. Y.-S. Huang, 168–85. Champaign, IL: American Oil Chemist's Society (AOCS) Press.

Nagaoka, S., Futamura, Y., Miwa, K., et al. 2001. Identification of novel hypocholesterolemic peptides derived from bovine milk β-lactoglobulin. *Biochemical, Biophysical and Research Communications* 281: 11–17.

Nagaoka, S., Hori, G., Yamamoto, K., Yamamoto, K., and Yamamoto, S. 2002. Improvements in cholesterol metabolism induced by soypeptides with bound phospholipids. *Journal of Nutrition* 132: 604S.

Nagaoka, S., Miwa, K., Eto, M., Kuzuya, Y., Hori, G., and Yamamoto, K. 1999. Soyprotein peptic hydrolyzate with bound phospholipids decrease micellar solubility and cholesterol absorption in rats and Caco-2 cells. *Journal of Nutrition* 129: 1725–30.

Nagaoka, S., Nakamura, A., Shibata, H., and Kanamaru, Y. 2010. Soystatin (VAWWMY), a novel bile acid-binding peptide, decreases micellar solubility and inhibits cholesterol absorption in rats. *Bioscience, Biotechnology, and Biochemistry* 74: 1738–41.

Nass, N., Schoeps, R., Ulbrich-Hofmann, R., et al. 2008. Screening for nutritive peptides that modify cholesterol 7 alpha-hydroxylase expression. *Journal of Agricultural and Food Chemistry* 56: 4987–94.

Navab, M., Anantharamaiah, G. M., Reddy, S. T., et al. 2006. Oral small peptides render HDL anti-inflammatory in mice and monkeys and reduce atherosclerosis in apoE null mice. *Circulation Research* 97: 524–32.

Repa, J. J., Turley, S. D., Lobaccaro, J. M. A., et al. 2000. Regulation of absorption and ABC1-mediated efflux of cholesterol by RXR heterodimers. *Science* 289: 1524–29.

Ronnett, G. V., Kim, E.-K., Landree, L. E., and Tu, Y. 2005. Fatty acid metabolism as a target for obesity treatment. *Physiology and Behavior* 85: 25–35.

Sheng, H., Niu, B., and Sun, H. 2009. Metabolic targeting of cancers: From molecular mechanisms to therapeutic strategies. *Current Medicinal Chemistry* 16: 1561–87.

Sirtori, C. R., Zucchi-Dentone, C., Sirtori, M., et al. 1985. Cholesterol-lowering and HDL-raising properties of lecithinated soy proteins in type II hyperlipidemic patients. *Annals of Nutrition and Metabolism* 29: 348–57.

Smith, S. and Tsai, S. C. 2007. The type I fatty acid and polyketide synthases: A tale of two megasynthases. *Natural Product Reporta* 24: 1041–72.

Spady, D. K., Cuthbert, J. A., Willard, M. N., and Meidell, R. S. 1995. Adenovirus-mediated transfer of a gene encoding cholesterol 7α-hydroxylase into hamsters increases hepatic enzyme activity and reduces plasma total and low density lipoprotein cholesterol. *Journal of Clinical Investigation* 96: 700–709.

Spady, D. K., Cuthbert, J. A., Willard, M. N., and Meidell, R. S. 1998. Overexpression of cholesterol 7α-hydroxylase (CYP7A1) in mice lacking the low density lipoprotein (LDL) receptor gene. *Journal of Biological Chemistry* 273: 126–32.

Sugano, M., Goto, S., Yamada, Y., et al. 1990. Cholesterol-lowering activity of various undigested fractions of soybean protein in rats. *Journal of Nutrition* 120: 977–85.

Van Lenten, B. J., Wagner, A. C., Anantharamaiah, G. M., et al. 2009. Apolipoprotein A-I mimetic peptides. *Current Atherosclerosis Report* 11: 52–57.

Wang, L., Lee, Y. K., Bundman, D., et al. 2002. Redundant pathways for negative feedback regulation of bile acid production. *Developmental Cell* 2: 721–31.

Chapter 11

Proteins and Peptides with Taste

Motonaka Kuroda

Contents

11.1 Introduction

The palatability of a food is determined by various factors, as illustrated in Figure 11.1. Among these factors, the basic taste, that is, sweetness, sourness, saltiness, bitterness, and umami, is one of the most important factors affecting an individual's preference for a given food. In addition, recent studies have characterized several peptides with a "kokumi" taste. The kokumi taste is designated as an increase in the thickness (intensity), continuity, and mouthfulness of a food. The kokumi taste is also considered to be an important factor in the palatability of a food.

Recent progress in molecular biology and analysis using a specific gene knockout technique on mice have enabled the identification and characterization of the taste receptors. The G-protein–coupled receptor (GPCR) T1R2/T1R3 was identified as a sweetness receptor (Nelson et al. 2001), and T1R1/T1R3 was also identified as an umami receptor (Brand 2000). In addition, a group of GPCRs designated as T2Rs were identified as the receptors for various bitter compounds (Mueller et al. 2005). These receptors are distributed in the taste buds on the tongue. Two groups independently revealed that the receptor for sourness consists of PKD2L1 molecules (Huang et al. 2006; Ishimaru et al. 2006). On the other hand, although the receptor for saltiness has not yet been clarified, epithelium sodium channels (ENaCs) have been considered the primary candidate (Heck et al. 1984).

It is well known that several kinds of proteins have a strong sweetness, and the binding site of such proteins to the sweetness receptor (T1R2/T1R3) has been investigated. In addition, some types of peptides have bitterness and some have

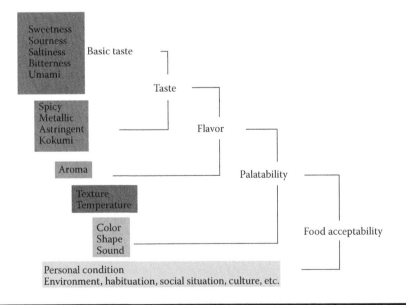

Figure 11.1   Factors in the palatability of a food.

been reported to have umami. Recent studies have revealed that some γ-glutamyl peptides have a kokumi taste and that the kokumi taste is perceived through a calcium-sensing receptor (CaSR) in humans.

In this chapter, proteins and peptides with tastes such as sweetness, kokumi, umami, and bitterness are described. It is expected that the recent progress in the study of receptors will contribute to a full understanding of the relationship between the taste sensation and the chemical structure.

11.2 Proteins and Peptides with Taste

11.2.1 Proteins and Peptides Associated with Sweetness

Sweetness is a desirable taste for humans, who have enjoyed it in fruits and honey since ancient times. A sweet taste is presumed to be a signal that the food will provide energy. The major sweet compounds are sugars, such as sucrose, glucose, fructose, and maltose. Also, in some geographical areas, people use plants containing sweet substances in addition to sugars, such as licorice (*Glycyrrhiza glabra*) containing glycyrrhizin; stevia (*Stevia rebaudiana*) containing stevioside; *Siraitia grosvenorii* containing mogroside; hydrangea tea (*Hydrangea macrophylla*) containing phyllodulcin; *Thaumatococcus daniellii* containing thaumatin; serendipity berry (also called "Nigeria berry") (*Dioscoreophyllum cumminsii*) containing monellin; mabinlang (*Capparis masaikai* Levl.) containing mabinlins; Oubli (*Pentadiplandra brazzeana*) containing brazzein and pentadin; miracle fruit (*Synsepalum dulcificum*) containing miraculin; and lumbah (*Curculigo latifolia*) containing curculin.

In 2000, the receptor for sweet substances was clarified using the techniques of molecular biology and specific gene knockout mice. Three members of a novel subfamily of GPCRs, T1R1, T1R2, and T1R3, have been proposed to function as taste receptors based on their expression in the taste cells. This subfamily, which is found in humans and mice, contains a long extracellular region composed of a highly conserved amino acid sequence with approximately 570 residues and is called the T1R family. While each receptor expressed by the cultured cells does not react with the sweet substances by itself, the cells that coexpress T1R2 and T1R3 show this reactivity. These results suggest that a heterodimer of T1R2 and T1R3 acts as a sweet receptor (Morini, Bassoli, and Temussi 2005). In addition, recent studies have clarified the binding sites of several sweet proteins in T1R2/T1R3. Specifically, brazzein and monellin were reported to bind to the cysteine-rich region of T1R3 in the T1R2/T1R3 heterodimer (Jiang et al. 2004).

As mentioned previously, various substances are associated with extreme sweetness. Among them, several proteins have been shown to elicit highly intense sweetness. Many studies have been done to characterize these proteins, which are summarized in Table 11.1.

This section describes the characteristics of the proteins associated with sweetness. In particular, the chemical nature and the amino acid residues that contribute to the function of these proteins are discussed.

Table 11.1 Characteristics of Proteins Associated with Sweetness

	Thaumatin	Monellin	Mabinlin II	Brazzein	Lysozyme	Miraculin	Curculin
Origin	Thaumatococcus daniellii	Dioscoreophyllum comminsii	Capparis masaikai	Pentadiplandra brazzeana	Hen egg-white	Synsepalum dulcificum	Curculigo latifolia
Molecular weight[a]	22,206	11,086	12,441	6,473	13,556	24,600	12,491
Isoelectric point	12.0	9.3	11.3	5.0	11.2	9.1	7.1
Heat stability	+++	−	+++	+++	−	−	+
Intensity of sweetness[b]	2,000	3,000	10	2,000	710	0[c]	550

[a] Calculated molecular weight.
[b] Intensity of sweetness was determined by sensory evaluation by comparing with sucrose and expressed as the times compared with sucrose on a weight basis.
[c] Miraculin itself cannot elicit sweetness.

11.2.1.1 *Thaumatin*

Thaumatin is a sweet protein that was isolated from the arils in the fruit of *T. daniellii*. The molecular weight of the representative homolog, thaumatin I, is 22,206. It is a basic protein with an isoelectric point of approximately 12.0. This protein is 1600 times sweeter than sucrose, on a weight basis. Thaumatin is cultivated on a commercial scale and is used as a sweetener, flavor enhancer, and flavor modifier. An aqueous solution of commercially available thaumatin is stable under conditions of pH 2–10. There are two main forms: thaumatin I and thaumatin II. Thaumatin I and II are each composed of 207 amino acids with 8 intramolecular disulfide bonds, as shown in Figure 11.2. Thaumatin I and II differ in eight amino acids, which suggests that the two proteins are 98% homologous (Iyengar et al. 1979; Edens et al. 1982). The tertiary structure of thaumatin I was analyzed by x-ray at resolutions of 3.1 Å (de Vos et al. 1985)

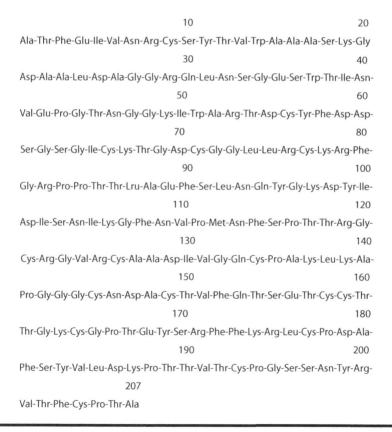

```
                   10                              20
Ala-Thr-Phe-Glu-Ile-Val-Asn-Arg-Cys-Ser-Tyr-Thr-Val-Trp-Ala-Ala-Ala-Ser-Lys-Gly
                   30                              40
Asp-Ala-Ala-Leu-Asp-Ala-Gly-Gly-Arg-Gln-Leu-Asn-Ser-Gly-Glu-Ser-Trp-Thr-Ile-Asn-
                   50                              60
Val-Glu-Pro-Gly-Thr-Asn-Gly-Gly-Lys-Ile-Trp-Ala-Arg-Thr-Asp-Cys-Tyr-Phe-Asp-Asp-
                   70                              80
Ser-Gly-Ser-Gly-Ile-Cys-Lys-Thr-Gly-Asp-Cys-Gly-Gly-Leu-Leu-Arg-Cys-Lys-Arg-Phe-
                   90                              100
Gly-Arg-Pro-Pro-Thr-Thr-Lru-Ala-Glu-Phe-Ser-Leu-Asn-Gln-Tyr-Gly-Lys-Asp-Tyr-Ile-
                   110                             120
Asp-Ile-Ser-Asn-Ile-Lys-Gly-Phe-Asn-Val-Pro-Met-Asn-Phe-Ser-Pro-Thr-Thr-Arg-Gly-
                   130                             140
Cys-Arg-Gly-Val-Arg-Cys-Ala-Ala-Asp-Ile-Val-Gly-Gln-Cys-Pro-Ala-Lys-Leu-Lys-Ala-
                   150                             160
Pro-Gly-Gly-Gly-Cys-Asn-Asp-Ala-Cys-Thr-Val-Phe-Gln-Thr-Ser-Glu-Thr-Cys-Cys-Thr-
                   170                             180
Thr-Gly-Lys-Cys-Gly-Pro-Thr-Glu-Tyr-Ser-Arg-Phe-Phe-Lys-Arg-Leu-Cys-Pro-Asp-Ala-
                   190                             200
Phe-Ser-Tyr-Val-Leu-Asp-Lys-Pro-Thr-Thr-Val-Thr-Cys-Pro-Gly-Ser-Ser-Asn-Tyr-Arg-
                   207
Val-Thr-Phe-Cys-Pro-Thr-Ala
```

Figure 11.2   Amino acid sequence of thaumatin I. There are eight disulfide bonds between Cys9–Cys204, Cys56–Cys66, Cys71–Cys77, Cys121–Cys193, Cys126–Cys177, Cys134–Cys145, Cys149–Cys158, and Cys159–Cys164.

and 1.65 Å (Ogata et al. 1992). An investigation of the relationship between the chemical structure and the sweetness elucidated that four lysine residues (Lys49, Lys67, Lys106, and Lys163) and three arginine residues (Arg76, Arg79, and Arg82) play significant roles in the sweetness of thaumatin I (Ohta et al. 2008). It has been reported that thaumatin elicits a sweet taste in humans and causes a significant electrophysiological response in the chorda tympani and glossopharyngeal nerves in the Old World monkey, but not in the guinea pig or rat (van der Wel 1980). However, it was revealed that in Slc:ICR mice, the chorda tympani, the taste receptor cell response profiles, and the behavioral results for monellin and thaumatin are similar to the response profiles for sucrose (Tonosaki, Miwa, and Kanemura 1997). Thaumatin has been approved as a sweetener in Israel and Japan. In the United Nations, it is listed in Table III of the Codex General Standard for Food Additives (GSFA), which means that it is permitted for general use in food. Commercialized thaumatin is produced after purification of a water extract from the seeds of *T. daniellii*.

11.2.1.2 Monellin

Monellin is a sweet protein that was isolated from the fruit of *D. cumminsii*, which is known as the serendipity berry or Nigeria berry. The molecular weight of this protein is 11,086. It is a basic protein with an isoelectric point of approximately 9.3. Monellin is 3000 times sweeter than sucrose, on a weight basis (Morris and Cagan 1972). Monellin has two noncovalently associated polypeptide chains: chain A contains 44 amino acid residues and chain B has 50 residues. Figure 11.3 shows the

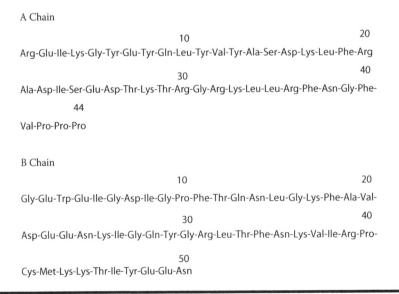

A Chain

                 10   20

Arg-Glu-Ile-Lys-Gly-Tyr-Glu-Tyr-Gln-Leu-Tyr-Val-Tyr-Ala-Ser-Asp-Lys-Leu-Phe-Arg

                 30   40

Ala-Asp-Ile-Ser-Glu-Asp-Thr-Lys-Thr-Arg-Gly-Arg-Lys-Leu-Leu-Arg-Phe-Asn-Gly-Phe-

          44

Val-Pro-Pro-Pro

B Chain

                 10   20

Gly-Glu-Trp-Glu-Ile-Gly-Asp-Ile-Gly-Pro-Phe-Thr-Gln-Asn-Leu-Gly-Lys-Phe-Ala-Val-

                 30   40

Asp-Glu-Glu-Asn-Lys-Ile-Gly-Gln-Tyr-Gly-Arg-Leu-Thr-Phe-Asn-Lys-Val-Ile-Arg-Pro-

                 50

Cys-Met-Lys-Lys-Thr-Ile-Tyr-Glu-Glu-Asn

Figure 11.3 **Amino acid sequences of the A chain and the B chain of monellin.**

amino acid sequences of the A and B chains of monellin (Bohak and Li 1976; Frank and Zuber 1976; Hudson and Biemann 1976; Kohmura, Nio, and Ariyoshi 1990a). Since chains A and B are not individually sweet (Bohak and Li 1976; Kohmura, Nio, and Ariyoshi 1990b), it is believed that the expression of a sweet taste requires a natural three-dimensional structure. X-ray structural analysis of monellin was carried out at resolutions of 3 Å (Ogata et al. 1987) and 2.75 Å (Somoza et al. 1993). A structure–sweetness relationship study of monellin analogs suggested that the Asp residue at position 7 of chain B (AspB7) plays an important role in eliciting a sweet taste (Kohmura, Nio, and Ariyoshi 1990c, 1991). In addition, recent studies have clarified the binding sites of monellin in T1R2/T1R3. Specifically, monellin was reported to bind to the cysteine-rich region of T1R3 in the T1R2/T1R3 heterodimer (Jiang et al. 2004). As mentioned previously, it was revealed that in Slc:ICR mice, the chorda tympani, the taste receptor cell response profiles, and the behavioral results for monellin and thaumatin are similar to the response profiles for sucrose (Tonosaki, Miwa, and Kanemura 1997). Although monellin has not been approved as a sweetener in the European Union or in the United States, the Nigeria berry extract containing monellin is approved as a food additive in Japan. However, since there are no past records of production, the Nigeria berry extract was omitted from the list of food additives in 2004.

11.2.1.3 Mabinlin

Mabinlins are sweet proteins extracted from the seed of Mabinlang (*C. masaikai* Levl.). There are at least five homologs. The molecular weight of the representative homolog, mabinlin II, is 12,441. It is also a basic protein with an isoelectric point of 11.3. Mabinlin II is 10 times sweeter than sucrose, on a weight basis. Mabinlin II is a heterodimer consisting of two different chains, A and B, similar to monellin. Chain A is composed of 33 amino acid residues, and chain B is composed of 72 amino acid residues. Chain B contains two intramolecular disulfide bonds and is connected to chain A through two intermolecular disulfide bridges, as shown in Figure 11.4 (Nirasawa et al. 1993). It has also been suggested that the difference in the heat stability of the different mabinlin homologs is due to the presence of either an arginine (heat-stable homolog) or glutamine (heat-unstable homolog) residue at position 47 of chain B (Nirasawa et al. 1994). The tertiary structure of mabinlin II was analyzed by x-ray at a resolution of 1.7 Å (Li et al. 2008). Although the amino acid residues that contribute to the sweet taste have not been clarified, an investigation of the interaction between the hT1R2/hT1R3 expressed on the HEK293 cells and mabinlin II and its subunits has revealed that the B chain of mabinlin II is responsible for binding to the receptor.

11.2.1.4 Brazzein

Brazzein is a sweet protein that was isolated from the fruit of Oubli (*Pentadiplandra brazzeana* Baillon). The molecular weight of this protein is 6473. It is an acidic

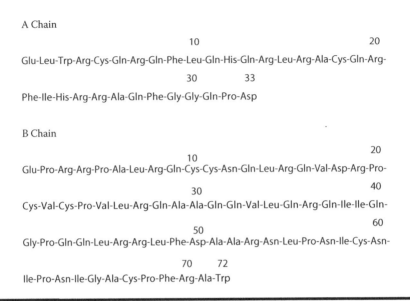

A Chain

                              10   20
Glu-Leu-Trp-Arg-Cys-Gln-Arg-Gln-Phe-Leu-Gln-His-Gln-Arg-Leu-Arg-Ala-Cys-Gln-Arg-

                    30   33
Phe-Ile-His-Arg-Arg-Ala-Gln-Phe-Gly-Gly-Gln-Pro-Asp

B Chain

                                                               20
                              10
Glu-Pro-Arg-Arg-Pro-Ala-Leu-Arg-Gln-Cys-Cys-Asn-Gln-Leu-Arg-Gln-Val-Asp-Arg-Pro-

                    30   40
Cys-Val-Cys-Pro-Val-Leu-Arg-Gln-Ala-Ala-Gln-Gln-Val-Leu-Gln-Arg-Gln-Ile-Ile-Gln-

                                                               60
                          50
Gly-Pro-Gln-Gln-Leu-Arg-Arg-Leu-Phe-Asp-Ala-Ala-Arg-Asn-Leu-Pro-Asn-Ile-Cys-Asn-

                      70   72
Ile-Pro-Asn-Ile-Gly-Ala-Cys-Pro-Phe-Arg-Ala-Trp

Figure 11.4   Amino acid sequences of the A chain and the B chain of mabinlin II. There are four disulfide bonds between CysA5–CysB21, CysA18–CysB10, CysB11–CysB59, and CysB23–CysB67.

protein with an isoelectric point of 5.0. Brazzein is 2000 times sweeter than sucrose, on a weight basis. Brazzein is stable over a broad pH range from 2.5 to 8 and is heat-stable at 80°C for 4 h (Ming and Hellekant 1994). Brazzein is a monomer protein consisting of 54 amino acid residues with 8 disulfide bonds, as shown in Figure 11.5 (Ming and Hellekant 1994; Kohmura et al. 1996). The chemical synthesis (Izawa et al. 1996) of brazzein was performed, and the recombinant proteins were successfully produced by *Escherichia coli* (Assadi-Porter et al. 2000a). X-ray analysis (Ishikawa et al. 1996) and nuclear magnetic resonance (NMR) studies of brazzein (Caldwell et al. 1998) revealed that residues 29–33, residues 39–43, and residue 36, as well as the C-terminus, were found to contribute to the sweet taste of the protein. The charge on Arg43 in the protein also plays an important role in its

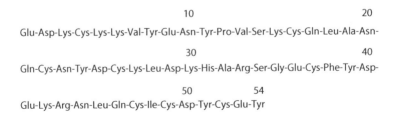

                          10   20
Glu-Asp-Lys-Cys-Lys-Lys-Val-Tyr-Glu-Asn-Tyr-Pro-Val-Ser-Lys-Cys-Gln-Leu-Ala-Asn-

                          30   40
Gln-Cys-Asn-Tyr-Asp-Cys-Lys-Leu-Asp-Lys-His-Ala-Arg-Ser-Gly-Glu-Cys-Phe-Tyr-Asp-

                          50   54
Glu-Lys-Arg-Asn-Leu-Gln-Cys-Ile-Cys-Asp-Tyr-Cys-Glu-Tyr

Figure 11.5   Amino acid sequence of brazzein. There are four disulfide bonds between Cys4–Cys52, Cys16–Cys37, Cys22–Cys47, and Cys26–Cys49.

interaction with the sweet taste receptor (Assadi-Porter et al. 2000b). In addition, recent studies have clarified the binding sites of brazzein in T1R2/T1R3. Brazzein was reported to specifically bind to the cysteine-rich region of T1R3 in the T1R2/T1R3 heterodimer (Jiang et al. 2004).

11.2.1.5 Lysozyme

Lysozyme is a bacteriolytic enzyme that catalyzes the hydrolysis of xc, $cx\beta$-1,4-glycosidic bonds between the C1 of N-acetyl muramic acids and the C4 of N-acetyl glucosamine of peptidoglycans in the bacterial cell walls. It is one of the most thoroughly characterized enzymes. In 1998, Maehashi and Udaka (1998) reported that chicken-type (c-type) lysozymes from hen egg-white, turkey egg-white, quail egg-white, guinea fowl egg-white, and soft-shelled turtle egg-white have a sweet taste. In addition, it was reported that the lysozyme from goose (g-type lysozyme), which is quite different from the c-type lysozyme on the basis of its structural, immunological, and enzymatical properties, also elicits a sweet taste (Masuda, Ueno, and Kitabatake 2001). In this section, the chemical nature of the hen egg-white lysozyme, the representative lysozyme with sweetness, is described. The molecular weight of the hen egg-white lysozyme is 13,566. It is a basic protein with an isoelectric point of 11.2. This protein is 710 times sweeter than sucrose, on a weight basis (Masuda, Ide, and Kitabatake 2005). The hen egg-white lysozyme is composed of 129 amino acids with 4 intramolecular disulfide bonds, as shown in Figure 11.6. An investigation of the relationship between its

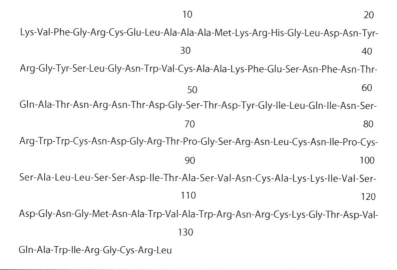

<pre>
              10   20
Lys-Val-Phe-Gly-Arg-Cys-Glu-Leu-Ala-Ala-Ala-Met-Lys-Arg-His-Gly-Leu-Asp-Asn-Tyr-
              30   40
Arg-Gly-Tyr-Ser-Leu-Gly-Asn-Trp-Val-Cys-Ala-Ala-Lys-Phe-Glu-Ser-Asn-Phe-Asn-Thr-
              50   60
Gln-Ala-Thr-Asn-Arg-Asn-Thr-Asp-Gly-Ser-Thr-Asp-Tyr-Gly-Ile-Leu-Gln-Ile-Asn-Ser-
              70   80
Arg-Trp-Trp-Cys-Asn-Asp-Gly-Arg-Thr-Pro-Gly-Ser-Arg-Asn-Leu-Cys-Asn-Ile-Pro-Cys-
              90   100
Ser-Ala-Leu-Leu-Ser-Ser-Asp-Ile-Thr-Ala-Ser-Val-Asn-Cys-Ala-Lys-Lys-Ile-Val-Ser-
             110   120
Asp-Gly-Asn-Gly-Met-Asn-Ala-Trp-Val-Ala-Trp-Arg-Asn-Arg-Cys-Lys-Gly-Thr-Asp-Val-
             130
Gln-Ala-Trp-Ile-Arg-Gly-Cys-Arg-Leu
</pre>

Figure 11.6 **Amino acid sequence of hen egg-white lysozyme. There are four disulfide bonds between Cys6–Cys127, Cys30–Cys115, Cys64–Cys80, and Cys76–Cys94.**

chemical structure and its sweetness elucidated that two lysine residues (Lys13 and Lys96) and three arginine residues (Arg14, Arg21, and Arg73) play significant roles in the sweetness of the hen egg-white lysozyme (Masuda, Ide, and Kitabatake 2005).

11.2.1.6 Miraculin

Miraculin is a taste-modifying, that is, sweetness-inducing, glycoprotein that was extracted from miracle fruit (*S. dulcificum* or *Richardella dulcifica*). Miraculin itself is not sweet; however, when one perceives ordinarily sour foods such as citrus after tasting this protein, one senses a strong sweetness that lasts for up to approximately 2 h. The perceived taste of 0.1 M citrate after tasting 1 μM of miraculin corresponds to the sweetness of 0.4 M sucrose, which means that miraculin is 400,000 times sweeter than sucrose on a molar basis and approximately 5600 times sweeter than sucrose on a weight basis. Miraculin was found to be a glycoprotein consisting of 191 amino acids (Figure 11.7) and some sugar

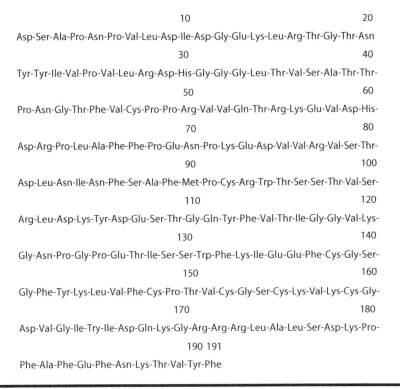

```
                          10                          20
Asp-Ser-Ala-Pro-Asn-Pro-Val-Leu-Asp-Ile-Asp-Gly-Glu-Lys-Leu-Arg-Thr-Gly-Thr-Asn
                          30                          40
Tyr-Tyr-Ile-Val-Pro-Val-Leu-Arg-Asp-His-Gly-Gly-Gly-Leu-Thr-Val-Ser-Ala-Thr-Thr-
                          50                          60
Pro-Asn-Gly-Thr-Phe-Val-Cys-Pro-Pro-Arg-Val-Val-Gln-Thr-Arg-Lys-Glu-Val-Asp-His-
                          70                          80
Asp-Arg-Pro-Leu-Ala-Phe-Phe-Pro-Glu-Asn-Pro-Lys-Glu-Asp-Val-Val-Arg-Val-Ser-Thr-
                          90                         100
Asp-Leu-Asn-Ile-Asn-Phe-Ser-Ala-Phe-Met-Pro-Cys-Arg-Trp-Thr-Ser-Ser-Thr-Val-Ser-
                         110                         120
Arg-Leu-Asp-Lys-Tyr-Asp-Glu-Ser-Thr-Gly-Gln-Tyr-Phe-Val-Thr-Ile-Gly-Gly-Val-Lys-
                         130                         140
Gly-Asn-Pro-Gly-Pro-Glu-Thr-Ile-Ser-Ser-Trp-Phe-Lys-Ile-Glu-Glu-Phe-Cys-Gly-Ser-
                         150                         160
Gly-Phe-Tyr-Lys-Leu-Val-Phe-Cys-Pro-Thr-Val-Cys-Gly-Ser-Cys-Lys-Val-Lys-Cys-Gly-
                         170                         180
Asp-Val-Gly-Ile-Try-Ile-Asp-Gln-Lys-Gly-Arg-Arg-Arg-Leu-Ala-Leu-Ser-Asp-Lys-Pro-
                        190 191
Phe-Ala-Phe-Glu-Phe-Asn-Lys-Thr-Val-Tyr-Phe
```

Figure 11.7 Amino acid sequence of miraculin. There are three intrachain disulfide bonds at Cys47–Cys92, Cys148–Cys159, and Cys152–Cys155 and an interchain disulfide bond at Cys138.

chains. The molecular weight of miraculin is 24.6 kDa, including 3.4 kDa (13.9% of the weight) of sugar composed (on a molar basis) of glucosamine (31%), mannose (30%), fucose (22%), xylose (10%), and galactose (7%). The sugar is linked to Asn42 and Asn186. It is a basic glycoprotein with an isoelectric point of 9.1. Miraculin occurs as a tetramer (98.4 kDa), a combination of four monomers grouped into dimers. Within each dimer, two miraculin glycoproteins are linked by an intermolecular disulfide bridge (Theerasilp and Kurihara 1988; Theerasilp et al. 1989; Igeta et al. 1991). It was concluded that native miraculin in pure form is a tetramer of a 25-kDa peptide and that native miraculin in a crude state in pure form is a dimer of the peptide. Both tetramer miraculin and native dimer miraculin in a crude state have a taste-modifying activity. The study using the point mutation technique (Ito et al. 2007) and molecular modeling suggests that two histidine residues, His30 and His60, are responsible for the taste-modifying activity of miraculin. Although miraculin has not been approved as a food additive in the European Union or in the United States, a miracle fruit extract containing miraculin was approved as a food additive in Japan in 1996. However, since there are no past records of production, the miracle fruit extract was omitted from the list of food additives in 2004.

11.2.1.7 Curculin

Curculin is a protein with a sweet taste and a taste-modifying activity, that is, sweetness-inducing activity, which was isolated from the fruit of lumbah (*C. latifolia*) (Hypoxidaceae). Like miraculin, when one perceives sour foods such as citrus fruit after tasting curculin, one senses a strong sweetness. The molecular weight of this protein is 12,491. Its isoelectric point is appoximately 7.1. Curculin is approximately 550 times sweeter than sucrose, on a weight basis. Curculin was reported to be a homodimer of two proteins connected through two disulfide bridges (Yamashita et al. 1990). It contains a sequence of 114 amino acids, as shown in Figure 11.8. Curculin was crystallized, and an x-ray structural analysis was performed (Harada et al. 1994). Curculin is rather stable in an acidic solution, and heating it at 55°C for 1 h does not reduce its potency. After curculin is held in the mouth for a short time, the sweet taste diminishes. However, the sweet taste is elicited again on exposure to clear water. Yamashita, Akabane, and Kurihara (1995) explained this phenomenon as follows. The sweet taste is suppressed by the reaction of curculin with the divalent cations (Ca^{2+} and Mg^{2+}) in the saliva. On exposure to water, the sweet taste is induced again as water washes the cations away. A study using the point mutation technique and x-ray analysis at a resolution of 1.50 Å (Kurimoto et al. 2007) has suggested that several amino acids located on the surface of this protein, such as Asp13, Asn36, Asn44, Arg47, Arg48, His67, Lys83, Lys90, Ser103, Pro106, and Arg111, are responsibe for its sweet taste and taste-modifying activity.

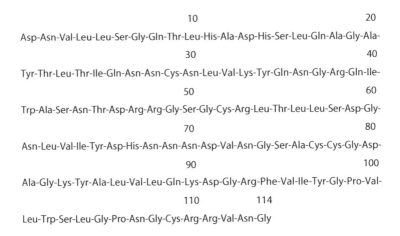

Figure 11.8 Amino acid sequence of curculin.

11.2.2 Peptides with Umami

The term "umami" was proposed by Ikeda (1909) to express the savory taste of kelp. Although umami was not widely recognized as a basic taste until recently, the Japanese word "umami" has now been established as a worldwide technical term for a basic taste as a result of various studies (Kawamura and Kare 1987). It has been established that: (1) the taste cells have receptors that bind with the umami substances and respond electrophysiologically to the umami substances; (2) there are taste nerves that transmit the umami stimuli; and (3) there are sites in the brain that respond to the umami stimuli in the same manner as to the other four basic tastes.

Representative umami substances include monosodium glutamate (MSG), 5′-inosine monophosphate (IMP), and 5′-guanosine monophosphate (GMP). MSG was isolated as the umami component of kelp soupstock (Ikeda 1909). IMP was isolated from dried-bonito soup stock (Kodama 1913). GMP was isolated from the hydrolysate of RNA (Kuninaka 1960) and was confirmed to be an umami-contributing substance in the shiitake mushroom.

Currently, two candidates for umami receptors have been proposed: mGluR4 (Roper 2003) and T1R1/T1R3 (Nelson et al. 2002). mGluR4 is a kind of metabotropic glutamate receptor, and it has been reported that a subtype with a lower affinity to glutamate contributes to umami perception. As mentioned in Section 11.2.1, three members of a novel subfamily of GPCRs, T1R1, T1R2, and T1R3, have been proposed to function as taste receptors based on their expression in the taste cells. Although each receptor solely expressed by the cultured cells does not react with the umami substances, the cells that coexpress T1R1 and T1R3 show reactivity to the umami substances. These results indicate that a heterodimer of T1R1 and T1R3 acts as an umami receptor (Brand 2000).

It was reported that α-glutamyl peptides, particularly peptides with hydrophilic amino acids, such as Glu-Asp, Glu-Thr, Glu-Ser, and Glu-Glu, elicited umami at a neutral pH (Arai, Yamashita, and Fujimaki 1972). These peptides were isolated from the umami constituents in the enzymatically hydrolyzed products of soy bean proteins. The same authors also reported that tripeptides, such as Glu-Gly-Ser, also elicited umami. The threshold value (0.15%) of these peptides is greater than that of MSG, indicating that the umami taste of these peptides is weaker than that of MSG. The flavor of a meat extract can be reproduced using these peptides with MSG and IMP.

11.2.3 Peptides with Bitterness

It is known that many kinds of peptides have hydrophobic amino acids, such as leucine, isoleucine, valine, phenylalanine, tyrosine, and tryptophan, which elicit bitterness. In this section, the studies of peptides with bitterness that have been reported to contribute to the taste of food are discussed. In particular, many studies have been performed on peptides with bitterness in various kinds of cheese. Murray and Baker (1952) reported that the bitter taste accumulated after casein was hydrolyzed by various proteases and that the principal components of this bitterness were peptides. Afterward, several groups successfully isolated the bitter peptides from a hydrolysate of casein. For example, the formation of the peptide with bitterness QNKIHPFAQTQSLVYPFPGPIP was identified during the maturation of Cheddar cheese (Richardson and Creamer 1973). On the other hand, the bitterness of several cyclopeptides in cacao has been reported. A diketopiperazine, cyclo (Trp-Leu) (Shiba and Nunami 1974), exhibits a bitter taste. Furthermore, cyclo (Val-Phe and Pro-Phe) amino acids have been reported to be bitter substances that are formed from cacao beans during roasting (Pickenhagen et al. 1975).

11.2.4 Proteins and Peptides with Kokumi

As shown in Figure 11.1, the kokumi taste is an important factor in determining food preferences. The term "kokumi" is used when a flavor cannot be represented by any of the five basic taste qualities. It has been reported that kokumi can be classified into four profiles, namely, thickness, continuity, mouthfulness, and harmony of taste (Kuroda, Yamanaka, and Miyamura 2004). Previous studies have revealed that several γ-glutamyl peptides, such as glutathione (γ-Glu-Cys-Gly), γ-Glu-Val, and γ-Glu-Leu, exhibit kokumi. In addition, a recent study indicated that the kokumi taste is perceived through a CaSR in humans (Ohsu et al. 2010). In this section, examples of the constituents that provide kokumi are described through illustrations of their structures.

11.2.4.1 γ-Glutamyl Peptides with Kokumi

Various foods are known to have kokumi-imparting properties, and several studies have been done to identify the kokumi-imparting compounds. Ueda et al. (1990) studied the kokumi-imparting properties of garlic and attempted to isolate the kokumi-imparting compound. As a result, sulfur-containing amino acids and peptides were characterized as kokumi-inducing constituents of garlic. As peptides, γ-glutamyl peptides such as glutathione (γ-glutamyl-cysteinylglycine) and γ-glutamyl-S-allyl-L-cysteine were identified. Ueda, Tsubuku, and Miyajima (1994) further investigated the kokumi-imparting constituents of onion using the same methodology and identified glutathione (γ-glutamyl-cysteinylglycine) and γ-glutamyl-S-(1-propenyl)-L-cysteine sulfoxide. These compounds only exhibit a slight flavor in water; however, when they are added to an umami solution or various kinds of foods, they substantially enhance the thickness, continuity, and mouthfulness of the taste of the food to which they have been added (Ueda et al. 1990). Using a neurophysiological approach in which the response of the mouse chorda tympani was observed, it was found that glutathione, which is a kokumi-inducing constituent, enhanced the umami flavor response, particularly for IMP (Yamamoto et al. 2009).

Recently, the addition of a nearly tasteless aqueous extract isolated from edible beans (*Phaseolus vulgaris* L.) to a model chicken broth enhanced the savory taste sensation (Dunkel, Koester, and Hofmann 2007). The key kokumi-imparting molecules were identified as γ-L-glutamyl-L-valine (γ-Glu-Val), γ-L-glutamyl-L-leucine (γ-Glu-Leu), and γ-L-glutamyl-L-cysteinyl-β-alanine (γ-Glu-Cys-β-Ala, homoglutathione).

In addition, kokumi peptides that impart long-lasting mouthfulness have been isolated from mature Gouda cheese (Toelstede, Dunkel, and Hofmann 2009). Several γ-glutamyl peptides, such as γ-glutamyl-glutamic acid (γ-Glu-Glu), γ-glutamyl-glycine (γ-Glu-Gly), γ-glutamyl-glutamine (γ-Glu-Gln), γ-glutamyl-methionine (γ-Glu-Met), γ-glutamyl-leucine (γ-Glu-Leu), and γ-glutamyl-histidine (γ-Glu-His), were identified as kokumi-imparting peptides, and it was concluded that these peptides contribute to the long-lasting mouthfulness of mature Gouda cheese.

Very recently, it was reported that the kokumi taste of several γ-glutamyl peptides is perceived through a CaSR in humans (Ohsu et al. 2010). These researchers confirmed that glutathione can activate a human CaSR, and they found that several γ-glutamyl peptides, such as γ-glutamyl-alanine (γ-Glu-Ala), γ-glutamyl-valine (γ-Glu-Val), γ-glutamyl-cysteine (γ-Glu-Cys), and γ-glutamyl-aminobutylyl-glycine (γ-Glu-Abu-Gly, ophthalmic acid), can also activate the CaSR and impart kokumi. In addition, it has been revealed that the CaSR activity of these γ-glutamyl peptides positively correlates with the intensity of the kokumi taste obtained by a sensory evaluation, suggesting that the kokumi taste is perceived through a CaSR in humans.

Among these kokumi-imparting γ-glutamyl peptides, the glutathione-enriched yeast extract has been commercialized and used for the enhancement of kokumi in the food industry. Kokumi-imparting γ-glutamyl peptides are summarized in Table 11.2.

Table 11.2   Structure and Origin of Kokumi-Imparting γ-Glutamyl Peptides

Peptides	Origin	References
γ-Glu-Cys-Gly (GSH)	Garlic, onion, yeast extract, scallop	Ueda et al. (1990)
γ-Glu-S-Allyl-Cys	Garlic	Ueda et al. (1990)
γ-Glu-S-(1-propenyl)-cysteine sulfoxide	Onion	Ueda, Tsubuku, and Miyajima (1994)
γ-Glu-Val	Edible beans (*Phaseolus vulgaris*)	Dunkel, Koester, and Hofmann (2007)
γ-Glu-Leu	Edible beans (*P. vulgaris*)	Dunkel, Koester, and Hofmann (2007)
γ-Glu-Cys-β-Ala	Edible beans (*P. vulgaris*)	Dunkel, Koester, and Hofmann (2007)
γ-Glu-Glu	Cheese	Toelstede, Dunkel, and Hofmann (2009)
γ-Glu-Gly	Cheese	Toelstede, Dunkel, and Hofmann (2009)
γ-Glu-Gln	Cheese	Toelstede, Dunkel, and Hofmann (2009)
γ-Glu-Met	Cheese	Toelstede, Dunkel, and Hofmann (2009)
γ-Glu-His	Cheese	Toelstede, Dunkel, and Hofmann (2009)
γ-Glu-Ala	Cheese	Ohsu et al. (2010)
γ-Glu-Cys	Yeast	Ohsu et al. (2010)
γ-Glu-Abu-Gly (ophthalmic acid)	Bovine, yeast	Ohsu et al. (2010)

It is expected that further progress in understanding the neurophysiology and molecular biology of the perception of kokumi may clarify this phenomenon at the molecular level.

11.2.4.2 Proteins with Kokumi

It has been indicated that proteinaceous substances contribute to kokumi in several foods. Kuroda and Harada (2004) investigated the kokumi-inducing effect of a beef meat extract, which is known to impart thickness, continuity, and mouthfulness, and attempted to isolate the kokumi-inducing compounds. They confirmed that the macromolecular fraction of the beef meat extract has kokumi, and they concluded that the proteinaceous material has kokumi based on the result that kokumi significantly decreased after protease treatment of the macromolecular extract. Although the chemical structure of the kokumi-imparting material has not been clarified, the investigation of the partial structure of this material suggests that collagen and tropomyosin are precursors of this material. It was suggested that the kokumi-imparting material is formed from collagen and tropomyosin during the heating process of the beef meat extract.

Recently, the kokumi-imparting substances in dried herring (*Migaki-nishin*) were investigated by Shah et al. (2009). It was reported that kokumi increased during the drying process. In addition, these researchers revealed that drying the dialyzed fraction of unmatured dried herring, which mainly consists of protein with docosahexisanoic acid, increased kokumi, which suggests that the kokumi-imparting substances in dried herring are formed from the protein and fatty acids during drying.

Since the concrete structure of these proteins with kokumi has not been clarified, further studies on the chemical structure and mechanism of perception of these proteins are expected.

11.3 Conclusion

This chapter described various proteins and peptides that are associated with sweetness, umami, bitterness, and kokumi. Since recent technological advances have enabled the analysis of the three-dimensional structure of the taste receptors, the molecular mechanisms of binding for several taste-active proteins and peptides to the taste receptors have been clarified. However, in many other cases, the mechanisms of binding remain a mystery. It is expected that further studies will clarify these mechanisms of taste for various proteins and peptides. In addition, it is also expected that the development of technology will enable the creation of novel taste-active proteins and peptides.

References

Arai, S., Yamashita, M., and Fujimaki, M. 1972. Glutamyl oligopeptides as factors responsible for tastes of a proteinase-modified soybean protein. *Agricultural and Biological Chemistry* 36: 1253–56.

Assadi-Porter, F. M., Aceti, D. J., Cheng, H., and Markley, J. L. 2000a. Efficient production of recombinant brazzein, a small, heat-stable, sweet-tasting protein. *Archives of Biochemistry and Biophysics* 376: 252–58.

Assadi-Porter, F. M., Aceti, D. J., and Markley, J. L. 2000b. Sweetness determinant sites of brazzein, a small, heat-stable, sweet-tasting protein. *Archives of Biochemistry and Biophysics* 376: 259–65.

Bohak, Z. and Li, S.-L. 1976. The structure of monellin and its relation to the sweetness of protein. *Biochimica et Biophysica Acta* 427: 153–70.

Brand, J. G. 2000. Receptor and transduction processes for umami taste. *Journal of Nutrition* 130: 942S–45S.

Caldwell, J. E., Abildgaard, F., Dzakula, Z., Ming, D., Hellekant, G., and Markley, J. L. 1998. Solution structure of the thermostable sweet-tasting protein brazzein. *Nature Structural and Molecular Biology* 5: 427–31.

Dunkel, A., Koester, J., and Hofmann, T. 2007. Molecular and sensory characterization of gamma-glutamyl peptides as key contributors to the kokumi taste of edible beans (Phaseolus vulgaris L.). *Journal of Agricultural and Food Chemistry* 55: 6712–19.

Edens, L., Heslinga, L., Klok, R., Ledeboer, A. M., Maat, J., Toonen, M. Y., Visser, C., and Verrips, C. T. 1982. Cloning of cDNA encoding the sweet-tasting plant protein thaumatin and the expression in *Escherichia coli. Gene* 18: 1–12.

Frank, G. and Zuber, H. 1976. The complete amino acid sequences of both subunits of the sweet protein monellin. *Hoppe-Seyler's Zeitschrift für Physiologische Chemie* 357: 585–92.

Harada, S., Ohtani, H., Maeda, S., Kai, Y., Kasai, N., and Kurihara, Y. 1994. Crystallization and preliminary X-ray diffraction studies of curculin. A new type of sweet protein having taste-modifying action. *Journal of Molecular Biology* 238: 286–87.

Heck, G. L., Mierson, S., and DeSimone, J. A. 1984. Salt taste transduction occurs through an amiloride-sensitive sodium transport pathway. *Science* 223: 403–5.

Huang, A. L., Chen, X., Hoon, M. A., Chandrashekar, J., Guo, W., Traenkner, D., et al. 2006. The cells and logic for mammalian. *Nature* 442: 934–38.

Hudson, G. and Biemann, K. 1976. Mass spectrometric sequencing of proteins. The structures of subunit I of monellin. *Biochemical and Biophysical Research Communications* 71: 212–20.

Igeta, H., Tamura, Y., Nakaya, K., Nakamura, Y., and Kurihara, Y. 1991. Determination of disulfide array and subunit structure of taste-modifying protein, miraculin. *Biochimica et Biophysica Acta* 1079: 303–7.

Ikeda, K. 1909. On the novel seasoning material. *Journal of the Tokyo Chemical Society* 30: 820–36.

Ishikawa, K., Ota, M., Ariyoshi, Y., Sasaki, H., Tanokura, M., Ming, D., et al. 1996. Crystallization and preliminary X-ray analysis of brazzein, a new sweet protein. *Acta Crystallographica: Section D, Biological Crystallography* 52: 577–78.

Ishimaru, Y., Inada, H., Kubota, M., Zhuang, H., Tominaga, M., and Matsunami, H. 2006. Transient receptor potential family members PKD1L3 and PKD2L1 form, a candidate sour taste receptor. *Proceedings of the National Academy of Sciences of the United States of America* 103: 12569–74.

Ito, K., Asakura, T., Morita, Y., Nakajima, K., Koizumi, A., Shimizu-Ibuka, A., et al. 2007. Microbial production of sensory-active miraculin. *Biochemical and Biophysical Research Communications* 360: 407–11.

Iyengar, R. B., Smits, P., van der Ouderaa, F., van der Wel, H., van Brouwershaven, J., Ravestein, P., et al. 1979. The complete amino acid sequence of the sweet protein thaumatin I. *European Journal of Biochemistry* 96: 193–204.

Izawa, H., Ota, M., Kohmura, M., and Ariyoshi, Y. 1996. Synthesis and characterization of the sweet protein brazzein. *Biopolymers* 39: 95–101.

Jiang, P., Ji, Q., Liu, Z., Snyder, L. A., Benard, L. M., Margolskee, R. F., and Max, M. 2004. The cysteine-rich region of T1R3 determines responses to intensely sweet proteins. *Journal of Biological Chemistry* 279: 45068–75.

Kawamura, Y. and Kare, M. R., eds. 1987. *Umami: A Basic Taste*. New York: Marcel Dekker.

Kodama, S. 1913. On the separation of inosinic acid. *Journal of the Tokyo Chemical Society* 34: 751–57.

Kohmura, M., Nio, N., and Ariyoshi, Y. 1990a. Complete amino acid sequence of sweet protein monellin. *Agricultural and Biological Chemistry* 54: 2219–24.

Kohmura, M., Nio, N., and Ariyoshi, Y. 1990b. Solid-phase synthesis and crystallization of monellin, an intensely sweet protein. *Agricultural and Biological Chemistry* 54: 1521–30.

Kohmura, M., Nio, N., and Ariyoshi, Y. 1990c. Solid-phase synthesis and crystallization of [Asn22, Gln25, Asn26]-A chain-[Asn49, Gln50]-B-chain-monellin, an analogue of sweet protein monellin. *Agricultural and Biological Chemistry* 54: 3157–62.

Kohmura, M., Nio, N., and Ariyoshi, Y. 1991. Solid-phase synthesis of crystalline [Ser41] B-chain monellin, an analogue of sweet protein monellin. *Agricultural and Biological Chemistry* 55: 1831–38.

Kohmura, M., Ota, M., Izawa, H., Ming, D., Hellekant, G., and Ariyoshi, Y. 1996. Assignment of the disulfide bonds in the sweet protein brazzein. *Biopolymers* 38: 553–56.

Kuninaka, A. 1960. Studies on taste of ribonucleic acid derivatives. *Nippon Nougei Kagaku Kaishi* 34: 489–92.

Kurimoto, E., Suzuki, M., Ameyama, E., Yamaguchi, Y., Nirasawa, S., Shimba, N., Xu, N., et al. 2007. Curculin exhibits sweet-tasting and taste-modulating activities through its distinct molecular surfaces. *Journal of Biological Chemistry* 282: 33252–56.

Kuroda, M. and Harada, T. 2004. Fractionation and characterization of the macromolecular meaty flavor enhancer from beef meat extract. *Journal of Food Science* 69: 542–48.

Kuroda, M., Yamanaka, T., and Miyamura, N. 2004. Change in taste and flavor of food during the aging with heating process. Generation of "KOKUMI" flavor during the heating of beef soup and beef extract. *Japanese Journal of Taste and Smell Research* (Nippon Aji to Nioi Gakkaishi) 11: 175–80.

Li, D.-H., Jiang, P., Zhu, D.-Y., Hu, Y., Max, M., and Wang, D.-C. 2008. Crystal structure of mabinlin II: A novel structural type of sweet proteins and the main structural basis for its sweetness. *Journal of Structural Biology* 162: 50–62.

Maehashi, K. and Udaka, S. 1998. Sweetness of lysozymes. *Bioscience Biotechnology & Biochemistry* 62: 605–6.

Masuda, T., Ide, N., and Kitabatake, N. 2005. Effects of chemical modification of lysine residues on the sweetness of lysozyme. *Chemical Senses* 30: 667–81.

Masuda, T., Ueno, Y., and Kitabatake, N. 2001. Sweetness and enzymatic activity of lysozyme. *Journal of Agricultural and Food Chemistry* 49: 4937–41.

Ming, D. and Hellekant, G. 1994. Brazzein, a new high-potency thermostable sweet protein from *Pentadiplandra brasseana* B. *FEBS Letters* 355: 106–8.

Morini, G., Bassoli, A., and Temussi, P. A. 2005. From small sweeteners to sweet proteins: Anatomy of the binding sites of the human T1R2–T1R3 receptor. *Journal of Medicinal Chemistry* 48: 5520–29.

Morris, J. A. and Cagan, R. H. 1972. Purification of monellin, the sweet principle of *Dioscoreophyllum cumminsii*. *Biochimica et Biophysica Acta* 261: 114–22.

Mueller, K. L., Hoon, M. A., Erlenbach, I., Chandrashekar, J., Zuker, C. S., and Ryba, N. J. P. 2005. The receptors and coding logic for bitter taste. *Nature* 434: 225–29.

Murray, T. K. and Baker, B. E. 1952. Studies on protein hydrolysis. I. Preliminary observations on the taste of enzymic hydrolysates. *Journal of the Science of Food and Agriculture* 3: 470–75.

Nelson, G., Chandrashekar, J., Hoon, M. A., Feng, L., Zhao, G., Ryba, N. J. P., and Zuker, C. S. 2002. An amino-acid taste receptor. *Nature* 416: 199–202.

Nelson, G., Hoon, M. A., Chandrashekar, J., Zhang, Y., Ryba, N. J., and Zuker, C. S. 2001. Mammalian sweet taste receptors. *Cell* 106: 381–90.

Nirasawa, S., Liu, X., Nishino, T., and Kurihara, Y. 1993. Disulfide bridge structure of the heat-stable sweet protein mabinlin II. *Biochimica et Biophysica Acta* 1202: 277–80.

Nirasawa, S., Nishino, T., Katahira, M., Uesugi, S., Hu, Z., and Kurihara, Y. 1994. Structure of heat-stable and unstable homologues of the sweet protein mabinlin. The difference in the heat stability is due to the replacement of a single amino acid residue. *European Journal of Biochemistry* 223: 989–95.

Ogata, C. M., Cordon, P. F., de Vos, A. M., and Kim, S.-H. 1992. Crystal structure of a sweet protein thaumatin I, at 1.65 A resolution. *Journal of Molecular Biology* 228: 893–908.

Ogata, C., Hatada, M., Tomlinson, G., Shin, W. C., and Kim, S.-H. 1987. Crystal structure of the intensely sweet protein monellin. *Nature* 328: 739–42.

Ohsu, T., Amino, Y., Nagasaki, H., Yamanaka, T., Takeshita, S., Hatanaka, T., et al. 2010. Involvement of the calcium-sensing receptor in human taste perception. *Journal of Biological Chemistry* 285: 1016–22.

Ohta, K., Masuda, T., Ide, N., and Kitabatake, N. 2008. Critical molecular regions for elicitation of the sweetness of the sweet-tasting protein, thaumatin I. *FEBS Journal* 275: 3644–52.

Pickenhagen, W., Dietrich, P., Keil, B., Polonsky, J., Nouaille, F., and Lederer, E. 1975. Identification of bitter principle of cocoa. *Helvetica Chimica Acta* 58: 1078–86.

Richardson, B. C. and Creamer, L. K. 1973. Casein proteolysis and bitter peptides in cheddar cheese. *New Zealand Journal of Dairy Science and Technology* 8: 46–51.

Roper, S. D. 2003. Monosodium glutamate (MSG) and taste-mGluR4, a candidate for an umami taste receptor. *Forum of Nutrition* 56: 87–89.

Shah, A. K. M. A., Tokunaga, M., Ogasawara, M., Kurihara, H., and Takahashi, K. 2009. Changes in chemical and sensory properties of migaki-nishin (dried herring fillet) during drying. *Journal of Food Science* 74: S309–14.

Shiba, T. and Nunami, K. 1974. Structure of bitter peptides in casein hydrolysate by bacterial proteinase. *Tetrahedron Letters* 6: 509–12.

Somoza, J. R., Jiang, F., Tong, L., Rang, C.-H., Cho, J. M., and Kim, S.-H. 1993. Two crystal structure of potently sweet protein. Natural monellin at 2.75 A resolution and single-chain monellin at 1.7 A resolution. *Journal of Molecular Biology* 234: 390–404.

Theerasilp, S., Hitotsuya, H., Nakajo, S., Nakaya, K., Nakamura, Y., and Kurihara, Y. 1989. Complete amino acid sequence and structure characterization of the taste-modifying protein, miraculin. *Journal of Biological Chemistry* 264: 6655–59.

Theerasilp, S. and Kurihara, Y. 1988. Complete purification and characterization of the taste-modifying protein, miraculin, from miracle fruit. *Journal of Biological Chemistry* 263: 11536–39.

Toelstede, S., Dunkel, A., and Hofmann, T. 2009. A series of kokumi peptides impart the long-lasting mouthfulness of matured Gouda cheese. *Journal of Agricultural and Food Chemistry* 57: 1440–48.

Tonosaki, K., Miwa, K., and Kanemura, F. 1997. Gustatory receptor cell responses to the sweeteners, monellin and thaumatin. *Brain Research* 748: 234–36.

Ueda, Y., Sakaguchi, M., Hirayama, K., Miyajimai, R., and Kimizuka, A. 1990. Characteristic flavor constituents in water extract of garlic. *Agricultural and Biological Chemistry* 54: 163–69.

Ueda, Y., Tsubuku, T., and Miyajima, R. 1994. Composition of sulfur-containing components in onion and their flavor characters. *Bioscience Biotechnology & Biochemistry* 58: 108–10.

de Vos, A. M., Hatada, M., van der Wel, H., Krabbendam, H., Peerdeman, A. F., and Kim, S.-H. 1985. Three-dimensional structure of thaumatin I, an intensely sweet protein. *Proceedings of the National Academy of Sciences of the United States of America* 82: 1406–9.

van der Wel, H. 1980. Physiological action and structure characteristics of the sweet-tasting proteins thaumatin and monellin. *Trends in Biochemical Sciences* 5: 122–23.

Yamamoto, T., Watanabe, U., Fujimoto, M., and Sako, N. 2009. Taste preference and nerve response to 5′-inosine monophosphate are enhanced by glutathione in mice. *Chemical Senses* 34: 809–18.

Yamashita, H., Akabane, T., and Kurihara, Y. 1995. Activity and stability of a new sweet protein with taste-modifying action, curculin. *Chemical Senses* 20: 239–43.

Yamashita, H., Theerasilp, S., Aiuchi, T., Nakaya, K., Nakamura, Y., and Kurihara, Y. 1990. Purification and complete amino acid sequence of a new type of sweet protein taste-modifying activity, curculin. *Journal of Biological Chemistry* 265: 15770–75.

Chapter 12

Bioavailability and Safety of Food Peptides

Elvira Gonzalez de Mejia, Cristina
Martinez-Villaluenga, Dina Fernandez,
Daisuke Urado, and Kenji Sato

Contents

12.1 Introduction

Bioactive food peptides derived from animal and plant proteins have been found to potentially impact human health (Mine and Shahidi 2005). Therefore, food peptides have already gained considerable commercial interest for their application as functional foods or pharmaceutical preparations (Meisel 2007). To exert their physiological effect, bioactive peptides need to remain in an active form after gastrointestinal digestion and be absorbed from the gastrointestinal tract. Therefore, it is important to determine the bioavailability of the food bioactive peptides before they are made widely accessible to the consumers (Hartmann et al. 2007). Dipeptides and tripeptides can be absorbed intact from the gastrointestinal tract (Daniel and Kottra 2004). In addition, studies in animals have shown that larger peptides (10–51 amino acids), generated by food protein digestion, can also be absorbed intact through the intestine and produce biological effects (Roberts et al. 1999). However, the bioavailability of peptides and their mechanism of absorption are still a topic under investigation.

This chapter provides an update on the bioavailability of food bioactive peptides, including their absorption and occurrence in human blood, paying special attention to the proposed mechanisms of intestinal absorption. Since these peptides are aimed for human consumption, the scientific evidence on their safety is also reviewed.

12.2 Bioavailability and Absorption of Food-Derived Peptides

Proteins are the most important structural and functional elements in living organisms and are formed by chains of α-amino acids connected by peptide bonds (polymers) (Roskoski and Greenberg 2008). Since proteins are the main building blocks of the cells, tissues, organs, and systems of the body, scientists of the human and animal sciences have expressed their interest in defining the mechanisms that maximize the bioavailability of proteins and their relationship with health benefits. Peptides are chains of less than 50 amino acids and their compositions and lengths depend on their nature. Hormones, such as oxytocin, vasopressin, and glucagon, as well as some hormones secreted by the hypothalamus, are peptides, and the most abundant peptide in the mammalian tissue is glutathione (Manning 2008).

In the last few decades, several bioactive peptides have been derived from food proteins by heat processing with alkali or acids or by fermentation (Shimizu and Son 2007). The discovery of the health benefits of bioactive peptides has changed the theories about protein nutrition and metabolism (Moughan et al. 2007).

According to Moughan et al. (2007), the bioactive peptides released from foods have several physiological effects, such as mineral binding, immunomodulation, antithrombotic activity, antibacterial properties, modulation of gastrointestinal motility, analgesic, stimulation of secretory processes, and other neuroactive effects. However, the ability of the bioactive peptides to reach their target sites intact is key to having a physiological effect. The bioactive peptides that are not transported from the gastrointestinal tract into the bloodstream via the intestinal epithelium may operate by modulating the nutrient absorption and influencing the gastrointestinal function (Miguel et al. 2008). However, for some peptides, like antihypertensive peptides, the absorption through the intestinal epithelium is absolutely necessary (Vermeirssen et al. 2002).

Western diets have an intake of 70–100 g/day of proteins derived from foods, from which 50–60 g will be secreted along the gastrointestinal tract. Before these proteins become short peptides and amino acids, they will be hydrolyzed by proteases and peptidases from the stomach and the pancreas. A person can consume 320–480 g of protein within 8 h without any assimilation problem (Daniel 2004).

The oral bioavailability of proteins and peptides is commonly very low because they are sensitive to chemical and enzymatic hydrolysis and present a poor cellular uptake. Additionally, the mucosal peptide hydrolases of the brush-border surface and the cytosol reduce the absorption of intact peptides, converting them into amino acids (Delie and Blanco-Preto 2005).

Several studies have demonstrated the advantage of different strategies to increase the percentage of the absorption of peptides; for example, the use of chemical modification, formulation vehicles, enzyme inhibitors, absorption enhancers, and mucoadhesive polymers (Shaji and Patole 2008).

Several studies support that small (dipeptides and tripeptides) and large peptides generated from the diet can be absorbed intact through the intestine and produce biologic effects at the tissue level (Roberts et al. 1999). According to a study conducted by Massey University's Institute of Food Nutrition and Human Health in New Zealand, there is enough evidence that lactoproteins contain several bioactive peptides, which have an impact on the opioid, immunostimulatory, and antihypertensive activities (Rutherfurd and Gill 2000). Other studies show a relationship between the platelet aggregation, the displayed antithrombotic activity and the consumption of peptides derived from κ-casein and lactoferrin (Rutherfurd and Gill 2000). In our laboratory, we have shown that lunasin, a peptide from plant foods, can be absorbed in the human plasma (Dia et al. 2009). A summary of the studies on the absorption and bioavailability of peptides and proteins is presented in Table 12.1.

12.2.1 Mechanisms of Absorption

The understanding of the mechanisms of digestion and absorption of proteins and peptides by the human body is still limited; scientists have found more solid bases to understand the mechanism of transport of peptides and the importance of enzymes in the hydrolysis of proteins. Many of the advances in the digestion and absorption of proteins and peptides are attributed to D. M. Matthews, Westminster Medical

Table 12.1 In Vivo and In Vitro Studies of Bioavailability of Food Peptides and Proteins

Description	Brief Results	In Vitro/In Vivo Model	References
SKWQHQQDSCRKQL QGVNLTPCEKHIMEKI QGRGDDDDDDDD	Lunasin is bioavailable in humans	Male 18–25 years old. *Dose:* 50 g soy protein/day during 5 days	Dia et al. (2009)
	In vivo assay: 35% of the oral dose of lunasin is absorbed and ends up in the various tissues of mice and rats 6 h after administration	In vivo (SENCAR mice). *Dose:* 250 µg/week (topic)	De Lumen (2008)
	In vitro assay: lunasin was found intact and bioactive in the blood and liver	In vitro foci formation assay. Concentrations from 10 nM to 10 µM	De Lumen (2008)
CSNLSTCVLGKLSQ ELHKLQTYPRTNT GSGTP-NH₂	0.8 mg calcitonin + 50 mL of water results in greater bioavailability than 0.8 mg calcitonin + 200 mL of water. The treatment resulted in greater bioavailability after 60 min premeal than 10 min premeal	Healthy postmenopausal women. *Doses:* SMC021 (0.8 mg SCT + 200 mg 5-CNAC). SMC021 placebo. 200 IU Miacalcic®. NS nasal spray. Water volume (50 or 200 mL water). Time between dosing and meal (10, 30, or 60 min)	Karsdal et al. (2008)
	Oral delivery of SCT is feasible with reproducible absorption and systemic biological efficacy. Such an oral formulation could facilitate the use of SCT in the treatment of osteoporosis	Healthy humans. *Doses:* 400, 800, and 1200 mg SCT orally, a placebo, and a 10 mg (50 IU) SCT intravenous infusion	Buclin et al. (2002)

RGIKIWFQNRRMKWKK	L-Penetrin markedly increased the permeability of insulin across the nasal membrane	Male Sprague Dawley rats. *Dose 1*: 10 IU/kg for insulin and 0.5 mM for each CPP. *Dose 2*: 0.2, 0.5, 1, and 2 mM of L- and D-penetrin.	Khafagy et al. (2008)
LKPNM	Tripeptide Ile-Pro-Pro selectively escaped from intestinal degradation and reached the circulation undegraded	Male subjects. *Dose 1*: 250 mL beverage. With 57 mg of LTP after a 10 h overnight fast and 30 min prior to intake of a standardized breakfast. *Dose 2*: 250 mL beverage. With 57 mg of LTP, 30 min after the start of a standardized breakfast. *Dose 3*: 250 mL beverage without added LTP after a 10 h overnight fast and 30 min prior to intake of a standardized breakfast.	Foltz et al. (2007)
Artificially synthesized polymer similar to DNA or RNA	The bioavailability using PNA modifications is limited as relatively high dosages (10–50 mg/kg) are required to obtain modest effects.	Sprague Dawley rats. *Dose*: 400 µCi (14.8 MBq) of PNA	Hamzavi et al. (2003)
Artificially synthesized polymer similar to DNA or RNA	In vivo bioavailability of PNA in mice is improved using lysine conjugation	In vivo, EGFP-654 transgenic mouse model. *Dose*: 50 mg/kg i.p. of each oligopolymers in PBS (2'-O-methoxyethyl (2'-O-MOE)-phosphorothioate and PNA-4K oligomers	Sazani et al. (2002)
Tripeptide/decapeptide EHWSYGLRPG	The potency of the entirely administered peptides decreased as the chain length increased	Male Sprague Dawley rats. *Dose*: Intravenous (50 mg) and enteral (125 and 500 mg) thyrotropin-releasing hormone (TRH). Intravenous (100 mg) and enteral (100 and 500 mg) luteinizing hormone-releasing hormone (LHRH)	Roberts et al. (1999)

School, London (Daniel 2004), who believed that the contribution of peptides to total amino acid absorption had been ignored for far too long (Gilbert et al. 2008).

According to Gilbert et al. (2008), it is absolutely necessary to hydrolyze the proteins into amino acids prior to absorption. Although the majority of peptides are hydrolyzed into the intestinal tract and epithelium, some of the peptides maintain their biological functions due to their resistance to proteases and peptidases (Shimizu and Son 2007). The end products of stomach digestion reach the small intestine, where trypsin, chymotrypsin, and elastase make them easy to absorb (Gilbert et al. 2008).

The epithelial cells have the important role of metabolizing nutrients and food substances and act as barriers to harmful substances, such as pathogens, allergens, and chemicals (Shimizu and Son 2007). The epithelial cells are located in the villi of the small intestine. Figure 12.1a provides a graphic understanding of the site where most of the nutrients are absorbed.

After the action of the enzymes in the small intestine, the short peptides are absorbed into the epithelial cells with the help of very specific transporters. The basic mechanism of absorption of the peptides is illustrated in Figure 12.1b. The large peptides are separated into amino acids and small peptides in the lumen, and the peptide transporter carries the small peptides (dipeptides and tripeptides) into the epithelial cells. Then, the dipeptide and tripeptide basolateral carriers move the absorbed peptides into the bloodstream.

12.2.2 Transporters

According to Gilbert et al. (2008), the three basic steps of a transport system are recognition, binding, and relocating of the substrates across the cell membranes. Transporters have been found in the endothelial cells and in the apical and basolateral membranes of the epithelial cells throughout the body. According to Sadee and Anderle (2006), PEPT1, encoded by the human gene SLC15A1, is one of the most known transporters and belongs to the solute carrier protein family SLC15. The SLC series comprises three kinds of transporters: the passive transporters, the ion-coupled secondary active transporters, and the exchangers. Each kind of transporter works with a specific SLC family, which has at least 20%–25% amino acid sequence similarity to other members of that family. The SLC family of interest to peptide absorption is the SLC15 family, which belongs to the proton-dependent oligopeptide transporter family (POT). The members of the SLC family are SLC15A1 (PEPT1), SLC15A2 (PEPT2), SLC15A3 (PHT2, PTR3), and SLC15A4 (PHT1, PTR4). The POT proteins work as symporters with the protons, which provide the driving force for transport across the plasma membrane (Sadee and Anderle 2006). Table 12.2 summarizes the members of the POT family and their respective aliases, substrates, and tissue distribution.

PEPT1 is expressed in the apical membrane of the enterocytes in the duodenum, jejunum, and ileum, with little or no expression in the normal colon

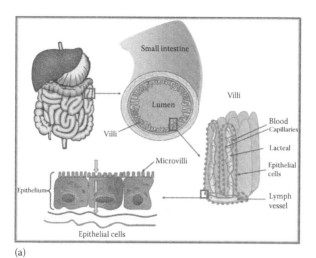

(a)

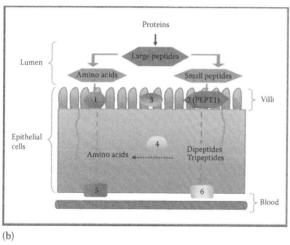

(b)

Figure 12.1 (a) Structures of the small intestine. The surface of the small intestine contains tiny projections called villi and microvilli, which constitute the most important area for absorption of the peptides. Blood capillaries and lacteals move the absorbed nutrients from the epithelium into circulation. (b) Protein and peptide absorption. (1) Brush-border membrane amino acid transporters; (2) Brush-border membrane dipeptide and tripeptide transporters; (3) Brush-border membrane peptidases; (4) Intracellular peptidases; (5) Basolateral-membrane amino acid carriers; (6) Basolateral-membrane dipeptide and tripeptide carriers. (c) Potential routes of peptide absorption. (1) Cell-penetrating peptides (CPPs) are able to carry cargo and peptides to the inside of the cells and could be independent or dependent of endocytosis and independent of energy and a specific receptor; (2) Increased permeability of tight junctions permits the uptake of peptides via the paracellular route.

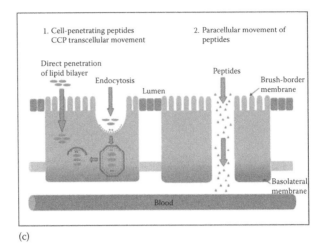

(c)

Figure 12.1   (Continued)

(Groneberg et al. 2001; Hu et al. 2008). The resistance of peptides to hydrolysis depends on the composition and amino acid sequence and also on the posttranslational modifications, such as glycosylation (Hannelore 2004). Glycosylation is a linking process of saccharides by an enzymatic process that produces glycans, which may be free or attached to proteins and lipids. During the enzymatic process of glycosylation, the cotransport and posttransport mechanisms take place, which are very important for the structure and function of proteins (Jakubowski 2008).

The uptake of dipeptides and tripeptides by the proton-coupled transporters is the most accepted route of dietary nitrogen absorption from the small intestine (PEPT1) and nitrogen reabsorption (PEPT1 and PEPT2) (Meredith and Boyd 2000). PEPT1 has been described as carrying not only dipeptides and tripeptides, but also peptide-like β-lactam antibiotics, orally active angiotensin-converting enzyme (ACE) inhibitors, and a variety of peptidomimetic drugs (Brodin et al. 2002; Shu et al. 2001). PEPT1 does not recognize tetra and large oligopeptides (Fei et al. 1994; Shimizu and Son 2007). PEPT2 is less studied than PEPT1, but it is known that it accepts a large number of substrates compared with most transporters. PEPT2 is most abundant in the kidney (Lu and Klaassen 2006) and has a higher affinity and a narrower substrate range than PEPT1 (Biegel et al. 2005; Meredith 2008). Both PEPT1 and PEPT2 are thought to be capable of transporting the 400 dipeptides and 8000 tripeptides that are derived from the 20 natural L-amino acids, the main source for food-derived proteins (Daniel and Kottra 2004). In addition to being transported through PEPT1, peptides may also be absorbed through alternative routes, including paracellular movement and cell-penetrating peptides (CPPs). Figure 12.1c describes these alternative mechanisms of absorption of peptides.

Table 12.2 Mammalian Proton Oligopeptide Cotransporter Family (POT)

Human Gene Name	Protein	Aliases	Substrates	Tissue Distribution/Cellular Expression
SLC15A1	PEPT1	Oligopeptide transporter 1 and H$^+$/peptide transporter 1	Dipeptide and tripeptide protons	Intestine and kidney apical membrane and lysosomal membrane
SLC15A2	PEPT2	Oligopeptide transporter 2 and H$^+$/peptide transporter 2	Dipeptide and tripeptide protons	Kidney, lung, brain, mammary gland, and bronchial epithelium
SLC15A3	hPTR3	Peptide/histidine transporter 2, human peptide transporter 3, and PHT2	Histidine, dipeptide, and tripeptide protons	Lung, spleen, thymus, brain, liver, adrenal gland, and heart
SLC15A4	PTR4	Peptide/histidine transporter 1, human peptide transporter 4, and PHT1	Histidine, dipeptide, and tripeptide protons	Brain, retina, and placenta

Source: Adapted from Daniel, H. and Kottra, G., *Pflugers Archiv: Europena Journal of Physiology* 447, 610–18, 2004.

12.2.2.1 Cell-Penetrating Peptides

The CPPs, also called protein transduction domains (PTDs), are capable of mediating the cellular uptake of hydrophilic macromolecules. According to Heitz, Morris, and Divita (2009), CPPs are peptides consisting of less than 30 amino acids and are capable of moving a charge across the cell membrane into the cytoplasm of the cells. Investigations by Heitz, Morris, and Divita (2009) and Chugh, Amundsen, and Eudes (2009) confirm that the uptake of CPPs does not need the endocytic pathways. Other scientists, such as Veldhoen, Laufer, and Restle (2008), contradict the theory that CPPs consist of less than 30 amino acids and that they are able to mediate the delivery of a cargo. They confirmed that a peptide is a CPP if it shows the ability to cross a biological membrane. In addition to their contradictory discoveries, they have gathered enough evidence to show that endocytosis is involved in the internalization process of various CPPs. Veldhoen, Laufer, and Restle (2008) emphasize that CPPs are grouped into three origin categories: (1) CPPs from naturally occurring proteins, (2) chimeric CPPs with different protein domains, and (3) model CPPs developed by a structure–function relationship instead of homology to natural sequences. All of these CPPs present a net positive charge at a physiological pH. The most representative CPPs based on natural amino acids are summarized in Table 12.3.

According to Heitz, Morris, and Divita (2009), two categories of CPPs are defined; the first category requires a covalent chemical linkage, and the second category forms stable noncovalent complexes. Tat, penetrin, polyarginine peptide Arg 8 sequence, transportan, VP22 protein, and calcitonin-derived peptide are examples of covalent CPPs. The noncovalent group includes GALA, GAL, JTS1, PPPTG1, MPG, and histidine-rich peptides.

12.2.2.2 Paracellular Pathway

The paracellular pathway is a nondegradative transport route for oligopeptides, keeping them intact, which means that this pathway plays a very important role in the absorption of functional peptides (Shimizu and Son 2007). The paracellular pathway is also referred to as the tight junction (TJ) permeability. The size, charge, and hydrophylicity of the peptides are the characteristics desired for paracellular absorption. Since the paracellular pathway is an aqueous extracellular route across the epithelium, a high hydrophylicity is the most important prerequisite for a peptide to be absorbed through this pathway (Gonzalez-Mariscal, Hernandez, and Vega 2008).

TJs are very dynamic structures. They can open and close to facilitate or restrict the paracellular passage of the molecules and their permeability varies significantly within different epithelia. For example, the lower parts of the small intestine, the jejunum and the ileum, are considered the major site of drug–peptide absorption with this pathway. The epithelial TJ divides the apical and basolateral plasma membrane domains and is a selectively permeable barrier to regulate paracellular diffusion (Shen, Weber, and Turner 2008; Gopalakrishnan et al. 2009).

Table 12.3   Examples of Cell-Penetrating Peptides (CPPs)

Origin	Sequences	Study	References
HIV-Tat protein	PAGRKKRRQRRPPQ	C57BL/6 mice. *Dose:* Intraperitoneally 1.7 nmol TAT-FITC peptide or control-free FITC	Schwarze et al. (1999)
Homodomain	RQIKIWFQNRRMKWKK	Male Sprague Dawley rats and mice NMRI-nude. *Dose:* Intravenous 2.5 mg/kg D-penetratin and SynB1	Rousselle et al. (2000)
Galanin–mastoparan	GWTLNSAGYKI NLKALAALAKKIL	In vitro — (COS-7 cells). *Dose:* 0.5 mM GFP-transportan for 10 min; 0.2 mM GFP-transportan for 1 h; 0.2 mM GFP-transportan for 1 h	Pooga et al. (2001)
HIV Gp41-SV40 NLS	GALFLGFLGAAGST MGAWSQPKKKRKV	In vitro — (COS-1, Vero, BHK-21 cells, and Hela cells). *Dose:* Microinjection into nuclei of cell: VP22 (AGV30) and WT VP22	Elliott and O'Hare (1997)
Trp-rich motif-SV40 NLS	KETWWETWWT EWSQPKKKRKV	Male ICR mice. *Dose:* Topical 50 μg control SOD and PEP-1-SOD. Male Mongolian gerbils. *Dose:* 100–500 μg PEP-1-SOD. In vitro (astrocyte cells). *Dose:* 2 μM native PEP-1-SOD	Eum et al. (2004)
Chimeric	KALAKALAKALA	Chinese hamster ovary (CHO) and A431 human epidermoid carcinoma cells. *Dose:* 5 μg/mL of FITC alone, FITC-MAP, or FITC-YG(R) 9 for 2 h at 37°C	Zaro et al. (2009)

(continued)

Table 12.3 (Continued)   Examples of Cell-Penetrating Peptides (CPPs)

Origin	Sequences	Study	References
Chimeric	GLFRALLRLLRSLWRLLLRA	Hela cells. *Dose:* ppTG1, KALA, or JTS1-K13 at charge ratios between ±0.2 and 10. B6SJL mice. *Dose:* Intravenous injection — 60 or 50 μg of pTG11236 complexed with ppTG1, ppTG20, or ppTG32	Rittner et al. (2002)
Chimeric	Agr8 or Agr9	Jurkat cells (human), murine B cells (CH27), or human PBL cells. *Dose:* 1–100 μM of arginine and oligomers of guanidine-substituted peptoids.	Wender et al. (2000)
Human calcitonin	LGTYTQDFNKTFPQTAIGVGAP	In vitro — (bovine nasal mucosa). *Dose:* 10–200 μM of human calcitonin (hCT), salmon calcitonin (sCT), and the somatostatin analog octreotide (SMS)	Lang et al. (1998)

In recent decades, progress has been made in understanding the mechanism of absorption of peptides and proteins, but despite being a topic of high interest to the scientific community, it remains unclear. Constant research efforts through in vivo and in vitro studies try to explain their potential benefits to animal and human health and to improve their bioavailability.

Several models of combined procedures have been proposed to explain the peptide absorption mechanisms. Researchers in this field have a special interest in characterizing the uptake of bioactive peptides because of their potential health benefits, such as antihypertensive, inmunostimulatory, antithrombotic, antibacterial, neuroactive, analgesic, and anti-inflammatory. Several in vivo and in vitro studies have tested the bioavailability and absorption of bioactive peptides (Dia et al. 2009; Karsdal et al. 2008; Khafagy et al. 2008; De Lumen 2008; Foltz et al. 2007; Sazani et al. 2002); however, overall, the mechanisms of peptide transport are not fully understood. Scientists continue working to define the best pathways that facilitate the absorption of the intact bioactive peptides, to enhance the human health benefits of these important components.

12.3 Comprehensive Identification of Food-Derived Peptides in Human Blood

As previously described, the bioavailability of potentially active peptides has been examined by using in vitro and in vivo models. If the target peptides consist of more than five amino acid residues, immunological techniques could be used. Based on this approach, lunasin, a 43 amino acid anti-inflammatory and anticancer peptide from soybean, was detected at 10–20 nM levels in the plasma of men after soy protein consumption (Dia et al. 2009). The occurrence of lunasin in the plasma was confirmed by matrix-assisted laser desorption ionization time-of-flight mass spectrometry (MALDI-TOF-MS). However, these techniques are insufficient for the detection of smaller peptides. Some potential ACE inhibitors, tripeptides (Ile-Pro-Pro) and dipeptides (Val-Tyr), have been detected at low nanomolar levels in the human blood by column switching (Matsui et al. 2002) or by the MS techniques (Foltz et al. 2007). Those studies examined the bioavailability of the potentially active peptide that had been identified by in vitro studies. However, the proteins and peptides in the diet can be degraded into smaller peptides during the digestion and absorption processes. Therefore, there is a possibility that some food-derived peptides, which had not been detected by the in vitro assay, may be absorbed into the circulating system. In general, structural information is not available for these peptides. Therefore, a comprehensive de novo identification of the food-derived peptides is necessary. Very recent advances in the methods for de novo identification of food-derived peptides in the human blood are introduced in the following sections.

12.3.1 Capturing Food-Derived Peptides from Biological Samples

Blood and organ extracts are extensively complex matrixes consisting of a variety of cells, endogenous proteins, and many other constituents. The capture of the food-derived oligopeptides from the complex matrix is a critical step for their isolation. In most cases, the food-derived oligopeptides are separated from the endogenous proteins in the plasma by ultrafiltration (Matsui et al. 2002) or by the selective precipitation technique using 75% ethanol (Iwai et al. 2005; Ohara et al. 2007). Other protein denaturants, such as trichloroacetic acid (TCA), perchloric acid, or salicylic acid, may be used for this purpose. However, the recovery of food-derived peptides by these techniques has not been optimized. The protein denaturant, which does not interfere with the following purification steps, should be selected. The deproteinized fraction is still a complex matrix. To capture the oligopeptides from the deproteinized fractions, size exclusion chromatography (Iwai et al. 2005; Ohara et al. 2007) and ion exchange chromatography using mini-spin columns (Aito-Inoue et al. 2007) and magnetic beads (Dia et al. 2009) have been used. The spin column packed with strong cation exchanger (AG50W × 8, Bio-Rad, Hercules, CA) can capture and clarify oligopeptides in a few milliliters of plasma with the exception of pyroglutamyl peptide due to the lack of an amino group and a phosphate-binding peptide (Aito-Inoue et al. 2007).

12.3.2 Isolation and Identification of Food-Derived Peptides

In most cases, food-derived peptides are isolated by reversed-phase high-performance liquid chromatography (HPLC). The elution of oligopeptides is frequently monitored by the absorption of ultraviolet light in the range of 210–230 nm, which nonspecifically detects the biological samples. Therefore, as shown in Figure 12.2, large peaks

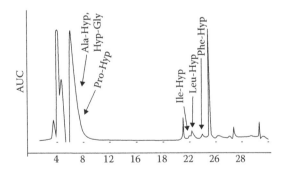

Figure 12.2 Reversed-phase HPLC patterns of oligopeptide fraction from human plasma after ingestion of gelatin hydrolysate. Peptides were resolved by a Cosmosil MS C18 column by linear gradient of acetonitrile in the presence of 0.1% trifluoroacetic acid. Absorption at 214 nm was monitored. See the experimental detail in the paper by Iwai, K., et al., *Journal of Agricultural and Food Chemistry* 53, 6531–36, 2005.

appear near the column void even after the subfractionation by size exclusion chromatography (Iwai et al. 2005). In addition, the resolution of some hydrophilic peptides is not satisfactory.

To improve the detection and resolution of peptides, precolumn derivatization techniques have been reported. To detect the hydrophilic peptides in the water-soluble fraction of an enzyme-modified cheese, the peptides were derivatized with 9-fluorenylmethoxycarbonyl chloride (FMOC-Cl) (Roturier, Le Bars, and Gripon 1995). Another derivatizing reagent is naphthalene-2,3-dialdehyde (NDA), which has been used to detect and resolve trace levels of food-derived antihypertensive peptides (Val-Tyr and Ile-Val-Tyr) in the animal plasma and human plasma (Matsui et al. 2002). These derivatives can be resolved by reversed-phase HPLC and detected with a high sensitivity. If an authentic peptide can be available as a standard, these techniques are powerful tools to detect and quantify the food-derived peptides in complex biological samples. However, these derivatives have not been applied to a de novo sequence analysis of food-derived peptides. A classic reagent for peptide modulation, phenyl isothiocyanate (PITC), can also be used for improving the resolution and detection of peptides by reversed-phase HPLC. The resultant derivative,

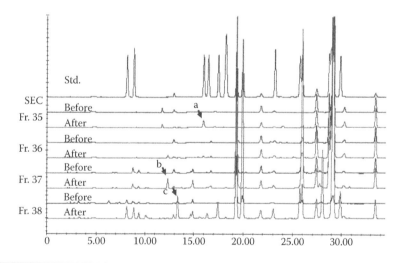

Figure 12.3   Reversed-phase HPLC patterns of PTC derivatives. Human plasma was deproteinized by 75% ethanol. Peptides and amino acids were captured by a mini-spin column packed with a strong cation exchanger (AG50 × 8). The peptide fraction was fractionated by size exclusion chromatography (SEC) using a Superdex Peptide 10/30 at 0.5 mL/min. The SEC fractions were derivatized with PITC and resolved by reversed-phase HPLC using Inertosil ODS3. See the experimental detail in the paper by Aito-Inoue, M., et al., *Journal of Agricultural and Food Chemistry* 54, 5261–66, 2007. Peaks marked with arrows were subjected to the Erdman sequencer. Peaks a, b, and c were identified as Pro-Hyp, Hyp-Gly, and free Hyp, respectively.

namely, phenyl thiocarbamyl (PTC) peptides, can be directly analyzed by the peptide sequencer based on the Erdman degradation. The sequencer program is changed to start from the cleavage reaction (Aito-Inoue et al. 2007), which is an advantage of PITC over other derivatization reagents. As shown in Figure 12.3, PTC hydrophilic peptides, hydroxylprolyl-glycine (Hyp-Gly) and prolyl-hydroxyproline (Pro-Hyp), with PTC-Hyp can be detected and identified in the human plasma after the ingestion of a collagen peptide, whereas these peptides without derivatization were not resolved from a large peak eluted in the column void (Figure 12.2).

For the identification of smaller amounts of food-derived peptides, the precolumn derivatization technique coupled with the electrospray ionization (ESI)-MS/MS technique can be used. Our preliminary study reveals that only a weak ion of the PTC peptide is observed by the ESI-MS. Subsequently, the peptide derivatives, which can easily be resolved from reagent peaks and ionized in the ESI-MS, have been studied. Consequently, we found that 6-aminoquinolyl-N-hydroxysuccinimidyl carbamate (AccQ) can be used for this purpose. As shown in Figure 12.4, authentic AccQ-Val-Thr-Leu yield a strong positive ion peak (m/z 502.17) and their daughter ions by ESI-MS and MS/MS analyses, which enables the sequencing of small peptides at low picomol levels (Figure 12.5).

Coupled with the capturing techniques for peptides using ion exchange, subfractionation by size exclusion and reversed-phase HPLC, precolumn derivatization of the peptides with PITC or AccQ, followed by reversed-phase HPLC separation, allowed us to identify some of the food-derived peptides in the human peripheral blood after the ingestion of the enzymatic hydrolysates of the porcine

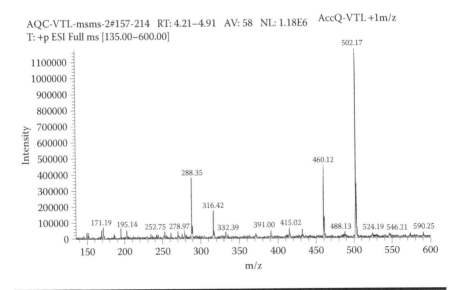

Figure 12.4 ESI-MS patterns of AccQ-Val-Thr-Leu. The authentic Val-Thr-Leu was derivatized with AccQ reagent and dissolved in 0.01% TFA solution.

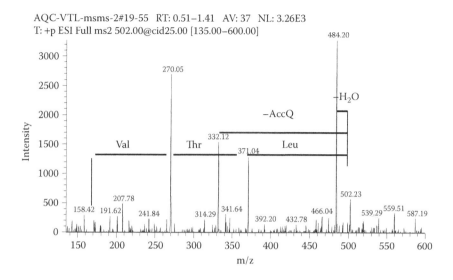

Figure 12.5 ESI-MS/MS patterns of AccQ-Val-Thr-Leu. The sequence of Val-Thr-Leu can be obtained from the daughter ions.

hemoglobulin and fish elastin. The content of these peptides in the blood is approximately 1000 times higher than the reported values of the ACE inhibitory peptides in the blood (Matsui et al. 2002; Foltz et al. 2007). Once the food-derived peptides are identified, the LC-MS or LC-MS/MS technique can be directly used for the accurate determination of a specific food-derived peptide. The information on the structure and bioavailability of food-derived peptides would reveal the real bioactive peptides by ingestion. In vitro study using the identified peptide in target organ or blood would enhance elucidation of mechanism of action by ingestion.

12.4 Safety Evaluation of Food Bioactive Peptides

The food bioactive peptides are released from the dietary proteins by enzymatic proteolysis during gastrointestinal digestion. Therefore, humans are continuously exposed to food-derived peptides without experiencing any adverse effects. Fermentation and enzymatic hydrolysis are the main processing methods to obtain products containing bioactive peptides. The addition of novel peptides to foods requires a safety assessment, which would be addressed through the regulations governing novel foods, as these would be considered ingredients with no history of safe use. By the contrast, the protein hydrolysates and traditionally fermented products that have been on the market for many years are considered as "Generally Recognized As Safe" products rather than novel foods.

A determination of safety for the proposed use(s) of food ingredients is required before the manufacturers and distributors can market them in the United States.

The safety evaluation of bioactive peptides, as functional ingredients, include a structure–toxicity analysis, the evaluation of the historical and intended exposure, animal studies, clinical–epidemiological studies, and the evaluation of special considerations, such as the potential for adverse food or drug interactions (Kruger and Mann 2003).

The application of biologically active food peptides in consumer products is a phenomenon of recent years and knowledge of their potential toxicity is limited. To date, toxicological studies have been conducted mostly in the ACE inhibitory peptides derived from dietary proteins because they are already on the market in Japan, China, Europe, and the United States (Gilani, Xiao, and Lee 2008). The safety of these bioactive peptides is supported by the results of in vitro and in vivo toxicity studies and human efficacy studies with commercially available products. These products include trypsin bovine αs1-casein hydrolysate (CH) (Lactium™) containing α-casozepine (αs1-casein f91–100), casein hydrolyzed by proteases from *Aspergillus oryzae* (referred as CH) containing Val-Pro-Pro (VPP) plus Ile-Pro-Pro (IPP), *Lactobacillus helveticus* fermented milk powder (referred as FM) also containing VPP plus IPP, cow milk protein hydrolysate (Tensguard™) containing IPP, chicken breast extract (CBEX™) containing anserin and carnosin, and sardine protein hydrolysate incorporated into a vegetable drink containing Val-Tyr (VY).

12.4.1 In Vitro Toxicity Studies

For the confirmation of the safety-in-use of food bioactive peptides, commercial products have been examined in several genetic toxicological studies (Table 12.4). The in vitro genotoxicity studies show that Lactium™ and Tensguard™ containing α-casozepine (αs1-casein f91–100) and IPP, respectively, do not induce mutagenicity as assessed in the L5178Y mouse lymphoma assay, tested up to 5000 μg product/mL (Boudier 2004; Doorten, vd Wiel, and Jonker 2009). In addition, synthesized VPP, FM, and CH containing IPP plus VPP did not exert a genotoxic potential in vitro as observed by the chromosomal aberration test in the Chinese hamster lung fibroblast and/or bacterial reverse mutation test up to 5000 μg product/mL (Bernard 2005; Maeno et al. 2005b; Doorten, vd Wiel, and Jonker 2009). Finally, CBEX™ containing the dipeptides carnosin and anserin did not show any mutagenic effect in vitro as assessed by a bacterial reverse mutation test up to 5000 μg product/mL (Sato et al. 2008).

12.4.2 In Vivo Toxicity Studies

Acute and subchronic toxicities following oral exposure of several bioactive peptides have been investigated in animals (Table 12.5). CH and FM did not induce mutagenicity in vivo as analyzed in the micronucleus test in rats and mice, tested up to 2000 μg product/kg body weight (BW), respectively (Matsuura et al. 2005). In the acute oral toxicity test in animals, no treatment-related systemic or local toxicity was detected,

Table 12.4 In Vitro Genotoxic Studies of Bioactive Food Peptides

Material	Bioactivity	Model	Dose	Observations	References
Bovine α_{s1} casein hydrolysate (Lactium™) containing 1.8% decapeptide α_{s1} casein (f91–100)	ACE inhibition	Mouse lymphoma L5178Y cells	0–5000 μg/mL media	Lactium™ exerted no mutagenic activity in the presence or absence of metabolic activation	Boudier (2004)
Tripeptides (VPP, IPP) from *Aspergillos oryzae* casein hydrolysate (CH) and *Lb. helveticus* fermented milk (FM) powder	ACE inhibition	Cultured Chinese hamster lung (CHL) cells	0–5000 μg/plate in the presence or absence of metabolic activation	Neither short-term nor continuous exposure to either CH or FM caused the induction of significant changes in cell growth, incidences of chromosomal aberrations or polyploids	Maeno et al. (2005a)
Synthesized tripeptide (VPP)	ACE inhibition	*Salmonella typhimurium* strains TA98, TA100, TA1535, TA1537, and *E. coli* strain WP2uvrA	0–5000 μg/plate in the presence or absence of metabolic activation	There was no evidence that VPP induced increases in the incidences of revertant colonies in any bacterial strain with and without metabolic activation	Bernard (2005)

(continued)

Table 12.4 (Continued)   In Vitro Genotoxic Studies of Bioactive Food Peptides

Material	Bioactivity	Model	Dose	Observations	References
Chicken breast extract (CBEX™) containing carnosin and anserine as primary constituents	Antiaging, decrease muscle fatigue	*S. typhimurium* strains TA98, TA100, TA1535, TA1537, and *E. coli* strain WP2uvrA	0–5000 µg/plate in the presence or absence of metabolic activation	CBEX™ exerted no mutagenic activity in the presence or absence of metabolic activation	Sato et al. (2008)
Protein hydrolysate from the glycomacropeptide fraction of cow's milk (Tensguad™) containing a tripeptide (IPP)	ACE inhibition	Mouse lymphoma L5178Y cells	0–5000 µg/plate in the presence or absence of metabolic activation	Tensguad™ did not increase the mutation frequency in the absence or presence of metabolic activation. No effects on the growth and proliferation of lymphoma L5178Y cells were observed	Doorten, vd Wiel, and Jonker (2009)
		S. typhimurium strains TA98, TA100, TA1535, TA1537, and *E. coli* strain WP2uvrA	0–5000 µg/plate in the presence or absence of metabolic activation	Tensguad™ exerted no mutagenic activity in the presence or absence of metabolic activation	

Table 12.5 Mutagenic, Acute and Subchronic Toxicity and Teratogenic Studies of Food Bioactive Peptides in Animal Models

Food Peptides	Bioactivity	Model	Dosage and Study Design	Duration	Observations	References
Mutagenicity						
Tripeptides (VPP, IPP) from *A. oryzae* casein hydrolysate (CH) and *Lb. helveticus* fermented milk (FM) powder	ACE inhibition	Male Sprague Dawley rats ($n = 5$) and mice ($n = 6$)	Single dose: 0–2000 mg CH (0–12 mg VPP plus IPP)/kg BW/day in rats; or 0–2000 mg FM (0–3.3 mg VPF plus IPP)/kg BW/day in mice	2 days	No signs of systemic toxicity. Similarly, neither CH nor FM caused statistically significant variations in the incidences of either polychromatic erythrocytes (PCEs) or micronucleated PCEs	Matsuura et al. (2005)
Acute Toxicity						
Bovine α_{s1} casein hydrolysate (Lactium™) containing 1.8% decapeptide α_{s1} casein (f91–100)	ACE inhibition	Sprague Dawley rats ($n = 6$)	Oral single dose of 2 g of Lactium™/kg BW	15 days	No changes in BW or behavior. No negative findings on necropsy. LD$_{50}$ >2000 mg/kg	Boudier (2004)

(continued)

Table 12.5 (Continued) Mutagenic, Acute and Subchronic Toxicity and Teratogenic Studies of Food Bioactive Peptides in Animal Models

Food Peptides	Bioactivity	Model	Dosage and Study Design	Duration	Observations	References
Tripeptides (VPP, IPP) from *A. oryzae* casein hydrolysate (CH) and *Lb. helveticus* fermented milk (FM) powder	ACE inhibition	Sprague Dawley rats (n = 5)	Single dose of CH (2000 mg/kg), powdered FM (2000 or 4000 mg/kg), or VPP (40, 200, or 400 mg/kg)	14 days	No antemortem or postmortem (necropsy) evidence of either systemic or local toxicity. LOEL of CH, FM, and VPP >4000, 2000, and 400 mg/kg, respectively	Maeno et al. (2005b)
Chicken breast extract (CBEX™) containing carnosin and anserine	Anti aging, decrease muscle fatigue	Specific pathogen-free bred Wistar rats (n = 5)	Single dose of CBEX at 6000 mg/kg BW	14 days	No adverse effects on BW, food consumption, behavioral effects, hematological and clinical parameters, organ weights, and microscopy LD_{50} >6000 mg/kg BW	Sato et al. (2008)
Subchronic Toxicity						
Bovine α_{s1} casein hydrolysate (Lactium™) containing 1.8% decapeptide α_{s1} casein (f91–100)	ACE inhibition	Sprague Dawley rats (n = 6)	Repeated dose of 40–1000 mg Lactium™/kg BW/day	28 days	No significantly relevant changes in hematology. No treatment-related gross pathology, histopathology, or organ weight changes. NOAEL >1000 mg/kg/day	Boudier (2004)

Pure tripeptide (VPP)	ACE inhibition	Charles River rats (n=20) and dogs (n=5)	Repeated dose study of 0, 2, 8, or 16 mg/kg BW/day	52 days	No evidence of either local or systemic toxicity was observed. No evidence of neurotoxicity that might have been detected by the appearance of physical or behavioral changes. No target organs of VPP toxicity were identified. NOAEL > 16 mg/kg/day	Nakamura et al. (2005)
Tripeptides (VPP, IPP) from *A. oryzae* casein hydrolysate (CH)	ACE inhibition	Male and female Charles River rats (n=12)	Repeated dose of casein hydrolysate containing from 0 to 6 mg VPP plus IPP/kg BW	91 days	No evidence of target organ toxicity. NOEL > 6 mg/kg BW/day	Mizuno et al. (2005)
Chicken breast extract (CBEX™) containing carnosin and anserine	Anti aging, decrease muscle fatigue	Specific pathogen-free bred Wistar rats (n=5)	Repeated dose of CBEX from 0 to 2000 mg/kg BW/day	90 days	No adverse effects on BW, food consumption, behavioral effects, hematological and clinical parameters, organ weights, and microscopy NOAEL > 2000 mg/kg BW/day	Sato et al. (2008)

(continued)

Table 12.5 (Continued) Mutagenic, Acute and Subchronic Toxicity and Teratogenic Studies of Food Bioactive Peptides in Animal Models

Food Peptides	Bioactivity	Model	Dosage and Study Design	Duration	Observations	References
Cow milk protein hydrolysate (Tensguad™) containing especially tripeptide (IPP)	ACE inhibition	Wistar rats (Crl:WI (WU), outbred) (n = 40)	Repeated dose of Tensguad™ 0–2000 mg (0–40 mg IPP)/ kg BW/day	90 days	No adverse effects on BW, food consumption, behavioral effects, hematological and clinical parameters, organ weights, and microscopy NOAEL > 40 mg IPP/kg BW/day	Doorten, vd Wiel, and Jonker (2009)
Reprotoxicity, Embriotoxicity and Developmental Toxicity						
Tripeptides (VPP, IPP) from *Lb. helveticus* fermented milk powder (FM)	ACE inhibition	Male and female Sprague Dawley rats (n = 5) and CD-1 mice (n = 6)	Repeated dose of FM containing VPP and IPP (0, 0.8, 1.6, or 3.3 mg/kg)		No effect on estrus cycle, mating behavior, fertility index, or reproductive competence. No effects on postimplantation survival-loss, sex ratio, or birth weights of fetuses. No effect on the behavioral and sexual maturation and reproductive capability of F1 generation. NOAEL > 3.3 mg/kg BW/day	Kurosaki et al. (2005)

Tripeptides (VPP, IPP) from *A. oryzae* casein hydrolysate (CH) and *Lb. helveticus* fermented milk (FM) powder	ACE inhibition	Pregnant Wistar rats (n = 8)	Developmental toxicological test—150 mg Lactium™/kg BW/day during gestation or during week 1, 2, or 3 of gestation	No effects on the gestation length, maternal behavior, care of offspring, or physical, leucomotor, behavioral, or cognitive development in the offspring. NOAEL >150 mg/kg	Maeno et al. (2005b)
		Female Himalayan rabbit (n = 80)	Teratogenicity—150 mg Lactium™/kg BW/day during gestation	No effects on the general external, internal visceral, and skeletal malformations in the F2 offspring. No histological changes were noted. NOAEL >150 mg/kg	
Tripeptides (VPP plus IPP) from *A. oryzae* casein hydrolysate (CH)	ACE inhibition	Male and female Wistar rats (n = 5)	Repeated dose of IPP plus VPP from 0 to 33.6 mg/kg BW/day	No adverse effects were seen in a subchronic repeat-dose toxicity study. No adverse effects of treatment on the progress and outcome of pregnancy. NOAEL >8.4 mg IPP/kg BW/day	Dent et al. (2007)

resulting in a median lethal dose (LD_{50}) of more than 2000 mg CH/kg BW, 2000 mg Lactium™/kg BW, 4000 mg FM/kg BW, and 6000 mg CBEX™/kg BW (Boudier 2004; Maeno et al. 2005b; Sato et al. 2008). Moreover, subchronic repeated-dose studies did not show any treatment-related adverse effects. A 28-day subchronic toxicity test in rats showed a no-observed adverse-effect level (NOAEL) of more than 1000 mg Lactium™ (more than 18 mg α-casozepine)/kg BW/day and 2000 mg FM (more than 3.3 mg VPP + IPP)/kg BW/day (Boudier 2004; Maeno et al. 2005a). A 52-day subchronic toxicity test in rats and dogs resulted in an NOAEL of more than 16 mg synthesized VPP/kg BW/day. A 90-day subchronic toxicity test in rodents showed an NOAEL of more than 1000 CH (more than 6 mg VPP + IPP)/kg BW/day and 2000 mg Tensguard™ (more than 40 mg IPP)/kg BW/day (Mizuno et al. 2005; Doorten, vd Wiel, and Jonker 2009). No information is available on the toxicity after chronic oral exposure.

A specific safety concern with the use of ACE-inhibiting drugs during pregnancy is the possibility of fetopathy (Guron and Friberg 2000). This effect is characterized by fetal hypotension, disruption in the development of the fetal kidneys, and a reduction in the production of amniotic fluid, which can lead to further malformations or fetal death (Guron and Friberg 2000). In addition, children affected during pregnancy may suffer oliguria or anuria (Pryde et al. 1993). Therefore, the teratogenic and reprotoxic effects of the available commercial products containing ACE inhibitor peptides have also been investigated (Table 12.5). These studies showed no changes indicative of treatment-related fetopathy. Moreover, the fertility, developmental, and F1 and F2 generation studies showed an NOAEL of more than 150 mg Lactium™ (more than 2.4 mg α-casozepine)/kg BW/day (Boudier 2004). The fertility, reproductive performance, embryotoxicity, and F1 generation studies in male and female rats showed an NOAEL of more than 2000 mg FM (more than 3.3 mg VPP + IPP)/kg BW/day (Kurosaki et al. 2005). Finally, the embryo–fetal development study in rabbits and the prenatal and postnatal studies in rats resulted in an NOAEL of more than 1000 mg (8.4 VPP + IPP) and 4000 mg (33.6 mg VPP + IPP) CH/kg BW/day, respectively (Dent et al. 2007).

12.4.3 Clinical Human Studies

Human clinical studies have been conducted to evaluate the possible consequences of the biological activity of ACE inhibitor peptides (Table 12.6). ACE inhibition involves the modulation of the renin–angiotensin system, which regulates the vascular tone, sodium and water balance, and blood pressure (Trufo-McReddie et al. 1995; Hilgers, Norwood, and Gomez 1997). However, ACE inhibitory drugs may cause adverse side effects, such as hypotension, dry cough, skin rash, hyperkalemia, acute renal failure, proteinuria, glycosuria, neutropenia, angioedema, or hepatotoxicity (Jackson 2005). Indicators of such effects were not observed in the human studies summarized in Table 12.6. In these studies, subjects with a blood pressure status ranging from healthy normotensive to hypertensive consumed doses of

Table 12.6   Clinical Studies on the Adverse Effects of Food Bioactive Peptides in Humans

Food Peptides	Bioactivity	Subjects	Study Design and Dosage	Duration (Days)	Observations	References
Dipeptide (VY) from sardine protein hydrolysate	ACE inhibition	Male and female mild high-normal hypertension (n = 21), and normotensive (n = 21)	Randomized double-blind placebo-controlled study. *Dose:* 585 g of vegetable drink containing 1.2 g of VP (three times the recommended amount of intake) once a day	14 days	Blood pressure of the normotensive subjects was not affected; biochemical analysis of the blood and urine showed no abnormalities	Kawasaki et al. (2002)
Bovine α$_{s1}$ casein hydrolysate (Lactium™) containing the 1.8% decapeptide α$_{s1}$ casein (f91–100)	ACE inhibition	Male normotensive (n = 42)	Randomized double-blind placebo-controlled study. *Dose:* 200 mg (3.9 mg decapeptide) each morning and evening	2 days	Lactium™ exhibited anxiolytic-like effects based on the systolic and diastolic blood pressures and heart rate. Moreover, it significantly reduced the plasma cortisol content during stress test	Boudier (2004)

(continued)

Table 12.6 (Continued)   Clinical Studies on Adverse Effects of Food Bioactive Peptides in Humans

Food Peptides	Bioactivity	Subjects	Study Design and Dosage	Duration (Days)	Observations	References
		Male normotensive (n = 52)	Randomized double-blind placebo-controlled study. *Dose:* Ingestion of 150 mg (2.7 mg decapeptide) in the evening from days 1 to 31, followed by a 12-day washout period	43 days	At days 11–31, blood pressure reactivity was lower in the treated group. No significant effect on the ambulatory cardiovascular parameters or urinary cortisol levels or anxiety. No adverse effects on the recovery periods	
Powdered fermented milk with *L. helveticus* CM4 (FM) containing IPP plus VPP	ACE inhibition	Mild- (n = 40) and high-normal hypertension (n = 40) subjects	Randomized double-blind placebo-controlled study. *Dose:* Ingestion of 12 g FM (4.7 mg IPP and 8.3 mg VPP) daily	28 days	Ingestion of the tablets containing FM in the subjects with high-normal blood pressure or mild hypertension reduces elevated blood pressure without any adverse effects	Aihara et al. (2005)
L. helveticus LBK-16H fermented milk (FM) containing IPP plus VPP	ACE inhibition	High-normal hypertension subjects (n = 97)	Randomized double-blind placebo-controlled study. *Dose:* Ingestion of 150 mL of FM (IPP 7.5 mg/100 g and VPP 10 mg/100 g) twice daily	70 days	*L. helveticus* LBK-16H fermented milk containing bioactive peptides, in daily use does have a BP-lowering effect in hypertensive subjects without any adverse effects	Jauhiainen et al. (2005)

ACE inhibitory peptides up to 3.9 mg of α-casozepine for up to 43 days of treatment and 7.5 mg of IPP and 10 mg of VPP for up to 70 days of administration. A recently conducted study reported that high-normal or mild hypertensive subjects receiving 750 mg Tensguard™ (containing 15 mg IPP) daily for 28 days showed no adverse effects (Doorten, vd Wiel, and Jonker 2009). In addition, normal ingestion of 195 g of a vegetable drink containing 0.5 g of VP from hydrolyzed sardine protein resulted in a decrease in the blood pressure in hypertensive subjects, but not in normotensive individuals. Additionally, excessive ingestion of VP (595 g of a vegetable drink containing 1.2 g of VP) had no side effects either in hypertensive or normotensive subjects (Kawasaki et al. 2002).

In summary, the exposure to food bioactive peptides reported in this chapter shows no cytotoxicity, genotoxicity, or clastogenicity based on in vitro experiments. In addition, consumption of food products containing bioactive peptides at a dietary level exerted no adverse effects, as demonstrated in acute, subchronic repeated dose, reproductive, embryo developmental, and prenatal and postnatal toxicity studies performed in experimental animal models. Finally, the food peptides exhibiting ACE inhibitory activity (VY, VPP, IPP, decapeptide αs1-casein [f91–100]) are not associated with any adverse effects based on human clinical studies.

12.5 Conclusions

Although the mechanisms of absorption and bioavailability of peptides are still under investigation, there is enough evidence to conclude that food bioactive peptides are bioavailable and can be absorbed into the body. Various mechanisms, such as the transcellular movement of CPPs, the paracellular pathway, or the use of specifics transporters like PEPT1 and PEPT2, allow these compounds into the bloodstream via the epithelial cells. Scientific studies on the safety of food bioactive peptides, such as VY, anserine and carnosine, IPP, VPP, α-casozepine, and σs1 casein (f91–100) peptide, have demonstrated that their consumption resulted in no adverse effects. This conclusion was based on in vitro and in vivo studies either in animal models or in human clinical trials. It is therefore concluded that food peptides are safe under the conditions of intended use. The market for functional foods containing bioactive peptides for health promotion is expected to grow. At the same time, as novel production technologies advance, the assessment of the bioactive potency and function and the understanding of the interaction of peptides with other components of the diet are needed. Therefore, more data from experimental, epidemiological, and clinical studies are necessary for further investigation of the safety of the new bioactive peptides that are intended to be marketed as food ingredients. For this purpose, the industry and other stakeholders can follow the guidelines established by the FDA to determine which toxicological studies are recommended in order to evaluate the safety of food ingredients (FDA Redbook 2000).

References

Aihara, K., Kajimoto, O., Hirata, H., Takahashi, R., and Nakamura, Y. 2005. Effect of powdered fermented milk with *Lactobacillus helveticus* on subjects with high normal blood pressure or mild hypertension. *Journal of the American College of Nutrition* 24: 257–65.

Aito-Inoue, M., Ohtsuki, K., Nakamura, Y., et al. 2007. Improvement in isolation and identification of food-derived peptides in human plasma based on precolumn derivatization of peptides with phenyl isothiocyanate. *Journal of Agricultural and Food Chemistry* 54: 5261–66.

Bernard, K. B. 2005. Studies of the toxicological potential of tripeptides (L-Valyl-L-prolyl-L-proline and Isoleucyl-L-prolyl-L-proline): I. Executive summary. *International Journal of Toxicology* 24: 1–3.

Biegel, A., Gebauer, S., Hartrodt, B., Brandsch, M., Neubert, K., and Thondorf, I. 2005. Three-dimensional quantitative structure–activity relationship analyses of beta-lactam antibiotics and tripeptides as substrates of the mammalian HC/peptide cotransporter PEPT1. *Journal of Medicinal Chemistry* 48: 4410–19.

Boudier, J. 2004. New Dietary Ingredient Notification for Lactium™, 1–21. http://www.fda.gov/OHRMS/DOCKETS/DOCKETS/95s0316/95s-0316-rpt0242-08-vol176.pdf.

Brodin, B., Nielsen, C., Steffansen, B., and Frøkjaer, S. 2002. Transport of peptidomimetic drugs by the intestinal Di/tri-peptide transporter, PepT1. *Pharmacology and Toxicology* 90: 285–96.

Buclin, T., Cosma Rochat, M., Burckhardt, P., Azria, M., and Attinger, M. 2002. Bioavailability and biological efficacy of a new oral formulation of salmon calcitonin in healthy volunteers. *Journal of Bone and Mineral Research* 17: 1478–85.

Chugh, A., Amundsen, E., and Eudes, F. 2009. Translocation of cell-penetrating peptides and delivery of their cargoes in triticale microspores. *Plant Cell Reports* 28: 801–10.

Daniel, H. 2004. Molecular and integrative physiology of intestinal peptide transport. *Annual Review of Physiology* 66: 361–84.

Daniel, H. and Kottra, G. 2004. The proton oligopeptide cotransporter family SLC15 in physiology and pharmacology. *Pflugers Archiv: Europena Journal of Physiology* 447: 610–18.

De Lumen, B. 2008. Lunasin; A cancer-preventive soy peptide. *Journal of AOAC International* 91: 932–35.

Delie, F. and Blanco-Prieto, M. 2005. Polymeric particulates to improve oral bioavailability of peptide drugs. *Molecules* 10: 65–80.

Dent, M. P., O'Hagan, S., Braun, W. H., Schaetti, P., Marburger, A., and Vogel, O. 2007. A 90 day subchronic toxicity study and reproductive toxicity studies on ACE-inhibiting lactotripeptide. *Food and Chemical Toxicology* 45: 1468–77.

Dia, V., Torres, S., De Lumen, B., Erdman, J. Jr, and Gonzalez de Mejia, E. 2009. Presence of lunasin in plasma after soy protein consumption. *Journal of Agricultural and Food Chemistry* 57: 1260–66.

Doorten, A. Y. P. S., vd Wiel, J. A. G., and Jonker, D. 2009. Safety and evaluation of an IPP tripeptide-containing milk protein hydrolysate. *Food and Chemical Toxicology* 47: 55–61.

Elliott, G. and O'Hare, P. 1997. Intercellular trafficking and protein delivery by a herpes virus structural protein. *Cell Press* 88: 223–33.

Eum, W., Kim, D., Hwang, I., et al. 2004. In vivo protein transduction: Biologically active intact pep-1-superoxide dismutase fusion protein efficiently protects against ischemic insult. *Free Radical Biology and Medicine* 37: 1656–69.

FDA. 2007. Toxicological principles for the safety assessment of food ingredients Redbood 2000. http://www.cfsan.fda.gov/guidance.html. Accessed April 1, 2009.

Fei, Y. J., Kanai, Y., Nussberger, S., et al. 1994. Expression cloning of a mammalian proton-coupled oligopeptide transporter. *Nature* 368: 563–66.

Foltz, M., Meynen, E., Bianco, V., Van Platerink, C., Koning, T., and Kloek, J. 2007. Angiotensin converting enzyme inhibitory peptides from a lactotripeptide-enriched milk beverage are absorbed intact into the circulation. *Journal of Nutrition* 137: 953–58.

Gilani, G. S., Xiao, C., and Lee, N. 2008. Need for accurate and standardized determination of amino acids and bioactive peptides for evaluating protein quality and potential health effects of foods and dietary supplements. *Journal of AOAC International* 91: 894–900.

Gilbert, E., Li, H., Emmerson, D., Webb, K. Jr, and Wong, E. 2008. Dietary protein quality and feed restriction influence abundance of nutrient transporter messenger RNA in the small intestine of broiler chicks. *Journal of Nutrition* 138: 1–10.

Gonzalez-Mariscal, L., Hernandez, S., and Vega, J. 2008. Inventions designed to enhance drug delivery across epithelial and endothelial cells through the paracellular pathway. *Recent Patents on Drug Delivery and Formulation* 2: 145–76.

Gopalakrishnan, S., Pandey, N., Tamiz, A., et al. 2009. Mechanism of action of ZOT-derived peptide AT-1002, a tight junction regulator and absorption enhancer. *International Journal of Pharmaceuticals* 365: 121–30.

Groneberg, D., Doring, F., Eynott, P., Fischer, F., and Daniel, H. 2001. Intestinal peptide transport: Ex vivo uptake studies and localization of peptide carrier PEPT1. *American Journal of Physiology – Gastrointestinal and Liver Physiology* 281: G697–704.

Guron, G. and Friberg, P. 2000. An intact renin-angiotensin system is a prerequisite for normal renal development. *Journal of Hypertension* 18: 123–37.

Hamzavi, R., Dolle, F., Tavitian, B., Dahl, O., and Nielsen, P. 2003. Modulation of the pharmacokinetic properties of PNA: Preparation of galactosyl, mannosyl, fucosyl, N-acetylgalactosaminyl, and N-acetylglucosaminyl derivatives of aminoethylglycine peptide nucleic acid monomers and their incorporation into PNA oligomers. *Bioconjugate Chemistry* 14: 941–54.

Hannelore, D. 2004. Molecular and integrative physiology of intestinal transport. *Annual Review of Physiology* 66: 361–84.

Hartmann, R., Wal, J-M., Bernard, H., and Pentzien, A-K. 2007. Cytotoxicity and allergenic potential of bioactive proteins and peptides. *Current Pharmaceutical Design* 13: 897–920.

Heitz, F., Morris, M., and Divita, G. 2009. Twenty years of cell-penetrating peptides: From molecular mechanism to therapeutics. *British Journal of Pharmacology* 157: 195–206.

Hilgers, K. F., Norwood, V. F., and Gomez, R. A. 1997. Angiotensin's role in renal development. *Seminars in Nephrology* 17: 492–501.

Hu, Y., Smith, D., Ma, K., Jappar, D., Thomas, W., and Hillgren, K. 2008. Targeted disruption of peptide transporter Pept1 gene in mice significantly reduces dipeptide absorption in intestine. *Molecular Pharmaceutics* 5: 1122–30.

Iwai, K., Hasegawa, T., Taguchi, Y., et al. 2005. Identification of food-derived collagen peptides in human blood after oral ingestion of gelatin hydrolysates. *Journal of Agricultural and Food Chemistry* 53: 6531–36.

Jackson, E. K. 2005. Renin and angiotensin: Drugs affecting renal and cardiovascular function. In *Goodman & Gilman's The Pharmacological Basis of Therapeutics*, eds. L. L. Brunton, J. S. Lazo, and K. Parker, 789–808. New York: McGraw-Hill.

Jakubowski, H. 2008. Carbohydrates and glycosylation. The Virtual Library of Biochemistry. http://www.biochemweb.org/carbohydrates.shtml. Accessed April 12, 2009.

Jauhiainen, T., Vapaatalo, H., Poussa, T., Kyrönpalo, S., Rasmussen, M., and Korpela, R. 2005. *Lactobacillus helveticus* fermented milk lowers blood pressure in hypertensive subjects in 24 h ambulatory blood pressure measurement. *American Journal of Hypertension* 18: 1600–5.

Karsdal, M., Byrjalsen, I., Riis, B., and Christiansen, C. 2008. Optimizing bioavailability of oral administration of small peptides through pharmacokinetic and pharmacodynamic parameters: The effect of water and timing of meal intake on oral delivery of Salmon calcitonin. *BMC Clinical Pharmacology* 10: 1186–1472.

Kawasaki, T., Jun, C. J., Fukushima, Y., et al. 2002. Antihypertensive effect and safety evaluation of vegetable drink with peptides derived from sardine protein hydrolysates on mild hypertensive, high-normal and normal blood pressure subjects. *Fukuoka Acta Medica* 93: 208–18.

Khafagy, E., Morishita, M., Isowa, K., Imai, I., and Takayama, K. 2008. Effect of cell-penetrating peptides on the nasal absorption of insulin. *Journal of Controlled Release* 133: 103–8.

Kruger, S. W. and Mann, S. W. 2003. Safety evaluation of functional ingredients. *Food Chemistry and Toxicology* 41: 793–805.

Kurosaki, T., Maeno, M., Mennear, J. H., and Bernard, B. K. 2005. Studies of the toxicological potential of tripeptides (L-Valyl-L-prolyl-L-proline and L Isoleucyl-L-prolyl-L-proline): VI. Effects of Lactobacillus helveticus-fermented milk powder on fertility and reproductive performance of rats. *International Journal of Toxicology* 41: 61–89.

Lang, S., Rothen-Rutishauser, B., Perriard, J., Schmidt, M., and Merkle, H. 1998. Permeation and pathways of human calcitonin (hCT) across excised bovine nasal mucosa. *Peptides* 19: 599–607.

Lu, H. and Klaassen, C. 2006. Tissue distribution and thyroid hormone regulation of Pept1 and Pept2 mRNA in rodents. *Peptides* 27: 850–57.

Maeno, M., Mizuno, S., Mennear, J. H., and Bernard, B. K. 2005a. Studies of the toxicological potential of tripeptides (L-valyl-L-prolyl-L-proline and L-isoleucyl-L-prolyl-L-proline): VIII. Assessment of cytotoxicity and clastogenicity of tripeptides-containing casein hydrolysate and *Lactobacillus helveticus*-fermented milk powders in Chinese hamster lung cells. *International Journal of Toxicology* 24: 97–105.

Maeno, M., Nakamura, Y., Mennear, J. H., and Bernard, B. K. 2005b. Studies of the toxicological potential of tripeptides (L-valyl-L-prolyl-L-proline and L-isoleucyl-L-prolyl-L-proline): III. Single- and/or repeated-dose toxicity of tripeptides containing *Lactobacillus helveticus* fermented milk powder and casein hydrolysate in rats. *International Journal of Toxicology* 24: 13–23.

Manning, J. 2008. Peptide. Access Science, McGraw-Hill Encyclopedia of Science and Technology online. http://www.accessscience.com. Accessed March 20, 2008.

Matsui, T., Tamaya, K., Seki, E., Osajima, K., Matsumoto, K., and Kawasaki, T. 2002. Absorption of Val-Tyr with in vitro angiotensin I-converting enzyme inhibitory activity into the circulating blood system of mild hypertensive subjects. *Biological and Pharmaceutical Bulletin* 25: 1228–30.

Matsuura, K., Mennear, J. H., Maeno, M., and Bernard, B. K. 2005. Studies of the toxicological potential of tripeptides (L-valyl-L-prolyl-L-proline and L-isoleucyl-L-prolyl-L-proline): VII. Micronucleous test of tripeptides-containing casein hydrolysate and *Lactobacillus helveticus*-fermented milk powders in rats and mice. *International Journal of Toxicology* 24: 91–96.

Meisel, H. 2007. Editorial: Food-derived bioactive proteins and peptides as potential components of nutraceuticals. *Current Pharmaceutical Design* 13: 771–72.

Meredith, D. 2008. The mammalian proton-coupled peptide cotransporter PepT1: Sitting on the transporter–channel fence? *Philosophical Transactions of Royal Society B.* 364: 203–7.

Meredith, D. and Boyd, C. 2000. Structure and function of eukaryotic peptide transporters. *Cellular and Molecular Life Sciences* 57: 754–78.

Miguel, M., Valos, A., Manso, M., de la Peña, G., Lasuncion, M., and López-Fandiño, R. 2008. Transepithelial transport across Caco-2 cell monolayers of antihypertensive egg-derived peptides. PepT1-mediated flux of Tyr-Pro-Ile. *Molecular Nutrition and Food Research* 52: 1507–13.

Mine, Y. and Shahidi, F. 2005. *Nutraceutical Proteins and Peptides in Health and Disease*, vol. 4, *Nutraceutical Science and Technology*. Boca Raton, FL: Taylor & Francis.

Mizuno, S., Mennear, J. H., Matsuura, K., and Bernard, B. K. 2005. Studies of the toxicological potential of tripeptides (L-valyl-L-prolyl-L-proline and L-isoleucyl-L-prolyl-L-proline): V. A 13-week toxicity study of tripeptides-containing casein hydrolysate in male and female rats. *International Journal of Toxicology* 24: 41–59.

Moughan, P., Fuller, M., Sik Han, K., Kies, A., and Miner-Williams, W. 2007. Food-derived bioactive peptides influence gut funtion. *International Journal of Sport Nutrition and Exercise Metabolism* 17: 5–22.

Nakamura, Y., Bando, I., Menear, J. H., and Bernard, B. K. 2005. Studies of the toxicological potential of tripeptides (L-Valyl-Lprolyl-L-proline and L-Isoleucyl-L-prolyl-L-proline): IV. Assessment of the repeated-dose toxicological potential of synthesized L-Valyl-L-prolyl-L-proline in male and female rats and dogs. *International Journal of Toxicology* 24: 25–39.

Ohara, H., Matsumoto, H., Ito, K., Iwai, K., and Sato, K. 2007. Comparison of quantity and structures of hydroxyproline-containing peptides in human blood after oral ingestion of gelatin hydrolysates from different sources. *Journal of Agricultural and Food Chemistry* 55: 1532–35.

Pooga, M., Kut, C., Kihlmark, M., et al. 2001. Cellular translocation of proteins by transportan. *FASEB Journal* 8: 1451–53.

Pryde, P. G., Nugent, C. E., Sedman, A. B., and Barr, M. 1993. Angiotensin converting enzyme inhibitor fetopathy. *Journal of American Nephrology* 3: 1575–82.

Rittner, K., Benavente, A., Bompard-Sorlet, A., et al. 2002. New basic membrane-destabilizing peptides for plasmid-based gene delivery in vitro and in vivo. *Molecular Therapy* 5: 104–14.

Roberts, P., Burney, J., Black, K., and Zaloga, G. 1999. Effect of chain length on absorption of biologically active peptides from the gastrointestinal tract. *Digestion* 60: 332–37.

Roskoski, R. Jr. and Greenberg, D. M. 2008. Protein metabolism. In *McGraw-Hill Encyclopedia of Science and Technology*, 10th ed., Chapter 14, pp. 514–18. New York: McGraw-Hill.

Roturier, J. M., Le Bars, D., and Gripon, J. C. 1995. Separation and identification of hydrophilic peptides in dairy products using FMOC derivatization. *Journal of Chromatography A* 696: 209–17.

Rousselle, C., Clair, P., Lefauconier, J. M., Kaczorek, M., Scherrmann, J. M., and Temsamani, J. 2000. New advances in the transport of doxorubicin through the blood-brain barrier by a peptide vector-mediated strategy. *Molecular Pharmacology* 57: 679–86.

Rutherfurd, K. and Gill, H. 2000. Peptides affecting coagulation. *British Journal of Nutrition* 84: 99–102.

Sadee, W. and Anderle, P. 2006. PEPT1. UCSD-Nature Molecule. doi:10.1038/mp.a002590.011038-mp.

Sato, M., Karasawa, M., Shimizu, M., Morimatsu, F., and Yamada, R. 2008. Safety evaluation of chicken breast extract containing carnosine and anserine. *Food Chemistry and Toxicology* 46: 480–89.

Sazani, P., Gemignani, F., Kang, S., et al. 2002. Systemically delivered antisense oligomers upregulate gene expression in mouse tissues. *Nature Biotechnology* 20: 1228–33.

Schwarze, S., Ho, A., Vocero-Akbani, A., and Dowdy, S. 1999. In vivo protein transduction: Delivery of a biologically active protein into the mouse. *Science* 285: 1569–72.

Shaji, J. and Patole, V. 2008, Protein and peptide drug delivery: Oral approaches. *Indian Journal of Pharmacological Science* 70: 269–77.

Shen, L., Weber, C., and Turner, J. 2008. The tight junction protein complex undergoes rapid and continuous molecular remodeling at steady state. *Journal of Cell Biology* 181: 683–95.

Shimizu, M. and Son, O. 2007. Food-derived peptides and intestinal functions. *Current Pharmaceutical Design* 13: 885–95.

Shu, C., Shen, H., Hopfer, U., and Smith, D. E. 2001. Mechanism of intestinal absorption and renal reabsorption of an orally active ace inhibitor: Uptake and transport of fosinopril in cell cultures. *Drug Metabolism and Disposition* 29: 1307–15.

Trufo-McReddie, A., Romano, L. M., Harris, J. M., Ferder, L., and Gomez, R. A. 1995. Angiotensin II regulates nephrogenesis and real vascular development. *American Journal of Physiology* 269: F110–15.

Veldhoen, S., Laufer, S., and Restle, T. 2008. Recent developments in peptide-based nucleic acid delivery. *International Journal of Molecular Science* 9: 1276–1320.

Vermeirssen, V., Deplancke, B., Tappenden, K. A., and Van Camp, J. 2002. Intestinal transport of the lactokinin Ala-Leu-Pro-Met-His-Ile-Arg through a Caco-2 monolayer. *Journal of Peptide Science* 8: 95–100.

Wender, P., Mitchell, D., Pattabiraman, K., Pelkey, E., Steinman, L., and Rothbard, J. 2000. The design, synthesis and evaluation of molecules that enable or enhance cellular uptake: Peptoid molecular transporters. *Proceedings of the National Academy of Sciences of the United States of America* 97: 13003–8.

Zaro, J., Vekich, J., Tran, T., and Shen, W. 2009. Nuclear localization of cell-penetrating peptides is dependent on endocytosis rather than cytosolic delivery in CHO cells. *Molecular Pharmacology* 6: 337–44.

Chapter 13

Database of Biologically Active Proteins and Peptides

Bartłomiej Dziuba, Piotr Minkiewicz,
and Małgorzata Darewicz

Contents

13.1 Introduction

Due to their nutritional value and functional properties, proteins are the major structural components of food. They are used as a source of energy, but primarily the amino acids in proteins are essential for the synthesis of body proteins. In addition, many proteins show specific biological activities that can have an influence on the functional or health-promoting attributes of foodstuffs. These proteins and the products of their hydrolysis—peptides—may affect both the food properties and the bodily functions. Every protein may perform the role of a precursor of biologically active and functional peptides. Recently, increasing attention has been paid to the use of bioactive peptides as physiologically active food ingredients, which are important elements for the prevention and treatment of various lifestyle diseases. Worldwide, there is a growing demand for this kind of food and for research studies on protein hydrolysates and peptides. The peptides derived from proteins may lower blood pressure; inhibit the activity of proline endopeptidases; stimulate the immune system; exhibit opioid activity; act as opioid antagonists; contract the smooth muscles; inhibit platelet aggregation; inhibit HIV proteinase and oxidation processes; show antibacterial and fungicidal activity and surface activity; bond ions and participate in the transport of minerals, determine the sensory properties of food products and improve their nutritional value; and help control body weight. According to Karelin, Blishchenko, and Ivanov (1998), in addition to its primary function, each protein may be a reserve source of peptides, controlling the life processes of organisms. For this reason, the determination of a new, additional criterion for evaluating proteins as a potential source of biologically active peptides contributes to a more comprehensive and objective definition of their biological value (Dziuba and Iwaniak 2006). Potentially, biologically active protein fragments, whose structural motifs resemble those of bioactive peptides, remain inactive in precursor protein sequences. These fragments, released from the precursor by proteolytic enzymes, may interact with selected receptors and affect the physiological functions of the body as well as the functional properties of food products (Figure 13.1).

In addition to the analytical methods, computer-aided techniques are also employed to evaluate food components, including proteins. The process of modeling the physicochemical properties of proteins (Lackner et al. 1999), predicting their secondary structure (Bairoch and Apweiler 2000), or searching for a homology between proteins to identify their functions (Kriventseva et al. 2001; Bray et al. 2000) requires analyses supported by databases of protein sequences or sequence motifs (Bennett et al. 2004; Colinge and Masselot 2004). A complementary part of such research is the strategy of examining food proteins as precursors of biologically active peptides, with the use of the BIOPEP database of proteins and bioactive peptides (http://www.uwm.edu.pl/biochemia). The database contains information on 44 types of bioactive peptides, their EC_{50} values, and source of origin. Proteins are evaluated as bioactive peptide precursors based on new, additional criteria: the profile of potential biological activity, the frequency of the bioactive fragments occurrence, and the potential biological protein activity (Dziuba, Iwaniak, and Minkiewicz 2003). This original approach, not described by other authors, has already been successfully applied. BIOPEP can be interfaced with global databases investigating the protein structure and functions, such as ProDom, PROSITE, TrEMBL, SWISS-PROT, EROP, and PepBank.

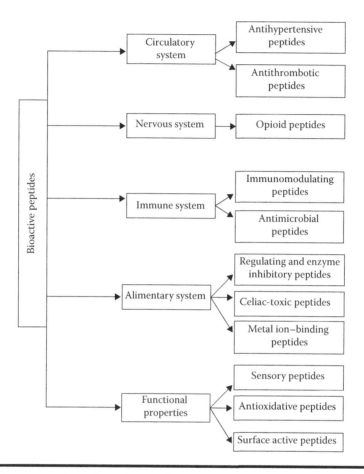

Figure 13.1   Function of peptides derived from food proteins.

13.2 Motifs with Potential Physiological Activity in Food Proteins

The physiologically active peptides form a complex and highly diversified group of compounds, with regard to their terminology, structure, and functions. Many physiologically active peptides are multifunctional, as they perform regulatory functions and directly affect various developmental and metabolic processes. It should be stressed that specific physiological effects are not always manifested following the oral administration of a peptide or as a result of peptide release by digestive enzymes; for instance, opioid and immunostimulating peptides derived from bovine milk proteins most probably undergo further hydrolysis in the body (Gobetti et al. 2002).

The richest potential sources of bioactive peptides are mostly milk proteins, wheat α- and β-gliadins, and collagen. Tables 13.1 and 13.2 show the profiles of

Table 13.1   Profile of Potential Biological Activity of Bovine Lactoferrin (*Bos taurus*) [BIOPEP ID 1212]

Lactoferrin sequence:
MKTLLILTILAMAITIGTANIQVDPSGQVQWLQQQLVPQLQQPLSQQPQQTFPQ PQQTFPHQPQQQVPQPQQPQQPFLQPQQPFPQQPQQPFPQTQQPQQPFPQ QPQQPFPQTQQPQQPFPQQPQQPFPQTQQPQQPFPQLQQPQQPFPQPQQ QLPQPQQPQQSFPQQQRPFIQPSLQQQLNPCKNILLQQCKPASLVSSLWSIIW PQSDCQVMRQQCCQQLAQIPQQLQCAAIHSVVHSIIMQQQQQQQQQQG MHIFLPLSQQQQVGQGSLVQGQGIIQPQQPAQLEAIRSLVLQTLPSMCNVYVPP ECSIMRAPFASIVAGIGGQ

Activity	Sequence	Location
Activating ubiquitin-mediated proteolysis	LA	[247–248], [411–412], [434–435], [533–534], [589–590], [648–649], [39–40], [47–48], [236–237], [603–604]
	RA	[39–40], [47–48], [236–237], [603–604]
	WA	[560–561]
Antihypertensive	AA	[78–79], [604–605]
	AF	[40–41], [195–196], [207–208], [241–242], [482–483], [529–530], [541–542], [685–686]
	AG	[67–68], [123–124], [203–204], [402–403], [465–466], [506–507]
	AH	[605–606]
	AP	[1–2], [31–32], [237–238], [492–493], [592–593]
	AY	[165–166]
	DA	[55–56], [240–241], [390–391]
	DG	[60–61], [201–202], [261–262], [395–396], [575–576]
	EA	[66–67], [388–389], [583–584], [682–683]
	EI	[80–81]

Table 13.1 (Continued)   Profile of Potential Biological Activity of Bovine Lactoferrin (*Bos taurus*) [BIOPEP ID 1212]

Activity	Sequence	Location
	EK	[51–52], [220–221], [276–277], [521–522]
	EV	[337–338]
	EW	[15–16]
	EY	[659–660], [664–665]
	FAL	[41–43]
	FG	[190–191], [278–279], [289–290], [618–619]
	FGK	[278–280], [618–620]
	FR	[530–531], [569–570]
	GA	[30–31], [147–148], [194–195], [202–203], [494–495], [528–529]
	GF	[306–307]
	GGY	[396–398]
	GI	[130–131]
	GK	[262–263], [279–280], [403–404], [619–620], [622–623]
	GL	[118–119], [406–407], [445–446], [472–473], [511–512]
	GM	[62–63]
	GP	[351–352]
	GQ	[294–295], [366–367]
	GR	[68–69], [111–112], [120–121], [653–654]
	GRP	[653–655]
	GS	[101–102], [290–291], [321–322], [479–480]

(*continued*)

Table 13.1 (Continued)   Profile of Potential Biological Activity of Bovine Lactoferrin (*Bos taurus*) [BIOPEP ID 1212]

Activity	Sequence	Location
	GT	[83–84], [576–577], [662–663]
	GW	[124–125], [466–467]
	GY	[191–192], [397–398], [432–433], [525–526]
	HL	[246–247], [588–589]
	HY	[91–92]
	IA	[49–50], [669–670]
	IP	[127–128], [310–311], [469–470]
	IR	[46–47]
	IRA	[46–48]
	IW	[267–268]
	IY	[81–82], [399–400]
	KA	[53–54], [221–222], [273–274], [339–340], [441–442]
	KE	[85–86], [210–211], [243–244], [263–264], [520–521]
	KF	[151–152], [277–278], [628–629]
	KG	[100–101], [174–175], [386–387]
	KL	[28–29], [73–74], [269–270], [650–651]
	KP	[579–580]
	KY	[522–523]
	KYY	[522–524]
	LA	[247–248], [411–412], [434–435], [533–534], [589–590], [648–649]

Table 13.1 (Continued)   Profile of Potential Biological Activity of Bovine Lactoferrin (*Bos taurus*) [BIOPEP ID 1212]

Activity	Sequence	Location
	LF	[288–289], [299–300], [617–618], [631–632], [640–641]
	LG	[29–30], [119–120], [305–306], [320–321], [651–652], [661–662]
	LN	[232–233], [392–393], [564–565]
	LQ	[109–110], [145–146], [199–200]
	LRP	[74–76], [132–134], [427–429]
	LVL	[383–385]
	LY	[318–319]
	MG	[129–130], [471–472]
	NF	[103–104]
	NG	[553–554], [621–622]
	PG	[293–294], [493–494]
	PL	[144–145], [679–680]
	PR	[2–3]
	PP	[292–293]
	PT	[429–430], [655–656]
	PQ	[88–89]
	QK	[355–356]
	RA	[39–40], [47–48], [236–237], [603–604]
	RL	[500–501], [570–571]
	RP	[75–76], [133–134], [428–429], [654–655]
	RR	[20–21], [38–39]

(*continued*)

Table 13.1 (Continued)   Profile of Potential Biological Activity of Bovine Lactoferrin (*Bos taurus*) [BIOPEP ID 1212]

Activity	Sequence	Location
	RW	[7–8], [21–22]
	RY	[323–324], [341–342]
	SF	[285–286]
	TE	[139–140], [430–431], [582–583], [645–646], [663–664]
	VAA	[77–79]
	VAP	[591–593]
	VAV	[95–97], [436–438]
	VG	[350–351], [537–538]
	VK	[98–99], [209–210], [338–339], [439–440], [543–544], [607–608]
	VP	[158–159], [250–251], [408–409], [516–517]
	VR	[6–7], [37–38]
	VW	[346–347], [548–549]
	YA	[93–94]
	YG	[82–83], [524–525]
	YK	[72–73]
	YL	[135–136], [319–320], [324–325], [433–434], [660–661]
	YP	[166–167]
Antibacterial	APRKNVRW	[1–8]
	APRKNVRWCTI	[1–11]
	APRKNVRWCTISQPEW	[1–16]
	CIRA	[45–48]
	CRRWQWRMKKLGAPSITCV	[19–37]

Table 13.1 (Continued)   Profile of Potential Biological Activity of Bovine Lactoferrin (*Bos taurus*) [BIOPEP ID 1212]

Activity	Sequence	Location
	FKCRRWQWRMKKLG	[17–30]
	FKCRRWQWRMKKLG APSITCVRRAF	[17–41]
	FKCRRWQWRMKKL GAPSITCVRRAFA	[17–42]
	FKCRRWQWRMKKL GAPSITCVRRAFAL	[17–43]
	FKCRRWQWRMKKLG APSITCVRRAFALECIR	[17–47]
Anticancer	FKCRRWQWRMK KLGAPSITCVRRAF	[17–41]
	RRWQWR	[20–25]
Antioxidative	HL	[246–247], [588–589]
	LH	[612–613]
Antithrombotic	GP	[351–352]
	PG	[293–294], [493–494]
Antiviral	ADRDQYELL	[222–230]
	EDLIWK	[264–269]
Bacterial permease ligand	KK	[27–28], [52–53], [99–100], [356–357], [440–441], [454–455], [673–674]
Dipeptidyl-aminopeptidase IV inhibitor	AP	[1–2], [31–32], [237–238], [492–493], [592–593]
	FA	[41–42]
	GP	[351–352]
	GQ	[294–295], [366–367]
	HA	[253–254], [595–596]

(continued)

Table 13.1 (Continued)   Profile of Potential Biological Activity of Bovine Lactoferrin (*Bos taurus*) [BIOPEP ID 1212]

Activity	Sequence	Location
	KA	[53–54], [221–222], [273–274], [339–340], [441–442]
	LA	[247–248], [411–412], [434–435], [533–534], [589–590], [648–649]
	LL	[229–230], [270–271], [298–299], [571–572], [611–612], [639–640], [680–681]
	LP	[218–219]
	PP	[292–293]
	VA	[77–78], [95–96], [149–150], [206–207], [256–257], [436–437], [540–541], [591–592]
	VP	[158–159], [250–251], [408–409], [516–517]
	VV	[97–98], [255–256], [345–346], [438–439], [597–598]
Heparin-binding peptide	APRKNVRWCT	[1–10]
	FKCRRWQWRMKKLGA	[17–31]
	PSITCVRRAF	[32–41]
	WQWRMKKLGA	[22–31]
Immunomodulating	GFL	[306–308]
	RKP	[578–580]
	RKSSK	[415–419]
	TRKP	[577–580]
	YG	[82–83], [524–525]
Opioid	YG	[82–83], [524–525]
	YL	[135–136], [319–320], [324–325], [433–434], [660–661]

Table 13.1 (Continued)   Profile of Potential Biological Activity of Bovine Lactoferrin (*Bos taurus*) [BIOPEP ID 1212]

Activity	Sequence	Location
Regulating phosphoinositol mechanism at bovine αs1-casein	GFL	[306–308]
Regulating the stomach mucosal membrane activity	GP	[351–352]
	PG	[293–294], [493–494]

the potential biological activity of bovine lactoferrin and wheat γ-gliadin, respectively. Dipeptide and tripeptide motifs with potential antihypertensive and dipeptidyl aminopeptidase-IV inhibiting activities predominate in both the proteins. The polypeptide chain of lactoferrin contains 97 fragments and γ-gliadin contains 40 fragments with antihypertensive activity. The number of dipeptides that act as dipeptidyl aminopeptidase-IV inhibitors is 13 in lactoferrin and 11 in γ-gliadin. Lactoferrin may also be a source of 10 peptides with antibacterial activity, 5 peptides with immunomodulatory activity, and 4 peptides with heparin-binding activity. The above proteins also contain three fragments potentially activating ubiquitin-dependent proteolysis; two fragments with anticoagulant, opioid, anti-amnesic, antioxidant, and antimicrobial activity; and one fragment with bacterial permease ligand activity. In addition, γ-gliadin contains 11 potentially celiac-toxic fragments.

Motifs corresponding to the peptides with antihypertensive activity are the best-represented group in the sequences of food proteins (Dziuba and Darewicz 2007). The majority of known peptides showing antihypertensive activity are inhibitors of the angiotensin I-converting enzyme (ACE), peptidyl dipeptide hydrolase (EC 3.4.15.1). Peptide inhibitors of ACE are present in the amino acid sequences of many food proteins and represent their fragments on release (Table 13.3). The relationship between the structure and the functions of these peptides has not been fully elucidated; however, they share certain common features. The investigations into the effect of the structure of antihypertensive peptides on their inhibitory activity have shown that enzyme–inhibitor interactions are strongly dependent on the C-terminal tripeptide sequence (Meisel 2005). Antihypertensive peptides are competitive inhibitors of peptidyl dipeptide hydrolase and contain hydrophobic (aromatic or branch-chained) amino acid residues in at least one of the three C-terminal positions. Many peptides contain Pro as the C-terminal amino acid. The short-chain peptides, dipeptides, and tripeptides, having Tyr, Phe, Trp, or Pro as a C-terminal residue, are particularly effective. Trp exerts the strongest influence

Table 13.2   Profile of Protein Potential Biological Activity of γ-Gliadin from Wheat (*Triticum aestivum*) [BIOPEP ID 1397]

γ-Gliadin sequence:
MKTLLILTILAMAITIGTANIQVDPSGQVQWLQQQLVPQLQQPLSQQPQQTFPQ PQQTFPHQPQQQVPQPQQPQQPFLQPQQPFPQQPQQPFPQTQQPQQPFP QPQQPFPQTQQPQQPFPQQPQQPFPQTQQPQQPFPQLQQPQQPFPQPQQ QLPQPQQPQQSFPQQQRPFIQPSLQQQLNPCKNILLQQCKPASLVSSLWSIIW PQSDCQVMRQQCCQQLAQIPQQLQCAAIHSVVHSIIMQQQQQQQQQQG MHIFLPLSQQQQVGQGSLVQGQGIIQPQQPAQLEAIRSLVLQTLPSMCNVYV PPECSIMRAPFASIVAGIGGQ

Activity	Sequence	Location
Antihypertensive	AA	[232–233]
	AG	[322–323]
	AP	[315–316]
	EA	[288–289]
	FP	[52–53], [59–60], [84–85], [92–93], [102–103], [110–111], [120–121], [128–129], [138–139], [148–149], [165–166]
	GG	[325–326]
	GI	[277–278], [323–324]
	GM	[254–255]
	GQ	[27–28], [268–269], [275–276], [326–327]
	GT	[17–18]
	GS	[270–271]
	IF	[257–258]
	IG	IG [16–17], [324–325]
	IP	[225–226]
	IR	[290–291]
	IW	[205–206]
	KP	[193–194]
	LA	[10–11], [222–223]

Table 13.2 (Continued) Profile of Protein Potential Biological Activity of γ-Gliadin from Wheat (*Triticum aestivum*) [BIOPEP ID 1397]

Activity	Sequence	Location
	LAMA	[10–13]
	LN	[181–182]
	LNP	[181–183]
	LQ	[32–33], [40–41], [78–79], [141–142], [177–178], [189–190], [229–230], [295–296]
	LQQ	[32–34], [40–42], [141–143], [177–179], [189–191]
	LW	[201–202]
	LVL	[293–295]
	PH	[60–61]
	PL	[43–44], [260–261]
	PP	[307–308]
	PQ	[38–39], [48–49], [53–54], [55–56], [63–64], [68–69], [70–71], [73–74], [80–81], [85–86], [88–89], [93–94], [98–99], [103–104], [106–107], [111–112], [116–117], [121–122], [124–125], [129–130], [134–135], [139–140], [144–145], [149–150], [151–152], [156–157], [158–159], [161–162], [166–167], [207–208], [226–227], [281–282]
	QG	[253–254], [269–270], [274–275], [276–277]
	RA	[314–315]
	RP	[170–171]
	SF	[164–165]
	SG	[26–27]
	TQ	[95–96], [113–114], [131–132]

(continued)

Table 13.2 (Continued) **Profile of Protein Potential Biological Activity of γ-Gliadin from Wheat (*Triticum aestivum*) [BIOPEP ID 1397]**

Activity	Sequence	Location
	VG	[267–268]
	VP	[37–38], [67–68], [306–307]
	VPP	[306–308]
	VY	[304–305]
	YVP	[305–307]
Activating ubiquitin-mediated proteolysis	LA	[10–11], [222–223]
	RA	[314–315]
	VV	[237–238]
Celiac toxic	GQ	[27–28], [268–269], [275–276], [326–327]
	PFPQPQQQL	[147–155]
	PFPQTQQPQ	[91–99], [109–117], [127–135]
	PP	[307–308]
	PQPQQQLPQ	[149–157]
	PQQPQQSFPQQQRPF	[158–172]
	PQQQLPQPQ	[151–159]
	QPQQSFPQQQ	[160–169]
	QQPFPQQPQQPFPQ	[81–94], [99–112], [117–130]
	QTQQPQQPF	[94–102], [112–120], [130–138]
	VQGQGIIQPQQPAQL	[273–287]
Dipeptidyl-aminopeptidase IV inhibitor	AP	[315–316]
	FA	[317–318]

Table 13.2 (Continued)   Profile of Protein Potential Biological Activity of γ-Gliadin from Wheat (*Triticum aestivum*) [BIOPEP ID 1397]

Activity	Sequence	Location
	FP	[52–53], [59–60], [84–85], [92–93], [102–103], [110–111], [120–121], [128–129], [138–139], [148–149], [165–166]
	LA	[10–11], [222–223]
	LL	[4–5], [188–189]
	LP	[155–156], [259–260], [298–299]
	MA	[12–13]
	PA	[194–195], [284–285]
	VA	[321–322]
	VP	[37–38], [67–68], [306–307]
	VV	[237–238]

on the inhibitory activity of peptides. In peptides with longer chains, the efficiency of the ACE inhibitor interactions is also affected by their conformation.

Some tripeptides and peptides with longer chains have Arg or Lys as a C-terminal residue. As demonstrated by FitzGerald and Meisel (2000), the activities of the above-mentioned peptides are also determined by the positive charge of the guanidine or ε-amine residues of amino acids. The replacement of the Arg residue at the C-terminus reduces the activity of the analog. This suggests that the interaction between the inhibitor and the anion group situated beyond the active site of the enzyme may affect the mechanism of ACE inhibition. Some peptide inhibitors have glutamic acid at the C-terminal position.

Animal proteins provide a potentially richer source of antihypertensive peptides than vegetable proteins. The average frequency of occurrence of motifs with hypertensive activity in the protein sequences of chicken meat, β-lactoglobulin, β-casein, soybean globulins, and wheat gliadins is 0.086, 0.092, 0.194, 0.051, and 0.090, respectively (Table 13.3). However, the differences in the frequency of occurrence of potentially active sequences in the vegetable and animal proteins are usually small. There is no simple relationship between the frequency of occurrence of bioactive fragments in a given protein and its potential activity. β-Casein is the richest source of antihypertensive peptides, and the frequency of their occurrence in this protein is almost twofold higher than in rice prolamins. However, the potential antihypertensive activity of β-casein is approximately threefold lower, compared with rice prolamins. These result from the fact that the fragments showing antihypertensive

Table 13.3 Structure and Frequency of Motifs with Potential Biological Activity Occurrence in Animal and Plant Proteins (BIOPEP; http://www.uwm.edu.pl/biochemia)

Activity/Structure	Proteins	A Value
Antihypertensive motifs: KYPVQPFTESQSLTL; LPP; VY; LA; FP; RY; YG; LQQ; LF; PAP; LLYQQPVLGPVRGPFPIIV; PYP; AVPYPQR; VG; GINYWLAHK; AP; and AA	Bovine αs1-casein, genetic variant A	0.134
	Bovine αs1-casein, genetic variant B	0.146
	Bovine αs1-casein, genetic variant C	0.146
	Bovine αs1-casein, genetic variant D	0.146
	Bovine αs2-casein, genetic variant A	0.086
	Bovine β-casein, genetic variant A^1	0.196
	Bovine β-casein, genetic variant A^2	0.230
	Bovine β-casein, genetic variant A^3	0.196
	Bovine β-casein, genetic variant B	0.191
	Bovine β-casein, genetic variant C	0.196
	Bovine β-casein, genetic variant E	0.191
	Bovine β-casein, genetic variant F	0.220
	Bovine κ-casein, genetic variant A	0.059
	Human κ-casein	0.101
	Bovine β-lactoglobulin, genetic variant A	0.092
	Bovine α-lactalbumin, genetic variant A	0.032
	Human lactoferrin	0.061
	Human lysozyme C	0.081
	Bovine elastin	0.037
	Bovine retinol-binding protein	0.045
	Chicken connectin	0.086
	Rice prolamin, clone PPROL 7	0.170
	Rice 13 kDa prolamin	0.124

Table 13.3 (Continued) Structure and Frequency of Motifs with Potential Biological Activity Occurrence in Animal and Plant Proteins (BIOPEP; http://www.uwm.edu.pl/biochemia)

Activity/Structure	Proteins	A Value
	Rice prolamin, clone PPROL 14	0.147
	Wheat α/β-gliadin, clone PW 1215	0.090
	Wheat α/β-gliadin, class a-V	0.077
	Wheat γ-gliadin	0.042
	Barley γ-hordein	0.066
	Pumpkin 11S globulin	0.083
	Soybean 7S globulin	0.049
	Sunflower 11S globulin	0.051
	Pea legumin A	0.048
	Pea legumin B	0.047
	Pea 11S legumin	0.107
	Faba bean legumin B chain	0.071
	Oat 12S globulin	0.042
	Serendipity berry monellin A chain	0.111
	Serendipity berry monellin B chain	0.120
Antithrombotic motifs: MAIPPK; NQDK; GPRG; GP; PG; and DGEA	Bovine β-casein, genetic variant E	0.003
	Bovine κ-casein, genetic variant A	0.003
	Bovine hemoglobin α-chain	0.007
	Bovine collagen α1-chain	0.274
	Bovine tropoelastin	0.067
	Chicken connectin	0.043
	Chicken myosin	0.043
	Wheat α/β-gliadin (clone PW 8142)	0.022
	Serendipity berry monellin A chain	0.003

(continued)

Table 13.3 (Continued)   Structure and Frequency of Motifs with Potential Biological Activity Occurrence in Animal and Plant Proteins (BIOPEP; http://www.uwm.edu.pl/biochemia)

Activity/Structure	Proteins	A Value
Opioid motifs: RYLGYLE; YPSF; YPWTQRF; and GYYPT	Bovine αs1-casein, genetic variant A	0.032
	Bovine β-casein, genetic variant A^2	0.029
	Bovine β-casein, genetic variant F	0.019
	Bovine κ-casein, genetic variant A	0.018
	Bovine α-lactalbumin, genetic variant B	0.033
	Bovine β-lactoglobulin, genetic variant A	0.012
	Rabbit α-lactalbumin	0.082
	Bovine α-hemoglobin	0.014
	Horse α-hemoglobin	0.014
	Goat α-hemoglobin	0.071
	Bovine troponin C	0.004
	Chicken myosin, subunit 1	0.005
	Wheat glutenin, HMW subunit	0.121
	Wheat α/β-gliadin, clone PW 1215	0.003
	Wheat α/β-gliadin, clone PW 8142	0.007
	Wheat α/β-gliadin, class I	0.011
	Pea legumin A2	0.002
Opioid antagonistic motifs: YPSYGLN; YIPIQYVLSR; and YPYY	Bovine αs1-casein, genetic variant A	0.005
	Bovine αs1-casein, genetic variant B	0.005
	Bovine αs1-casein, genetic variant C	0.005
	Bovine κ-casein, genetic variant A	0.011
	Caprine κ-casein, genetic variant A	0.010
Immunomodulating motifs: LLY; HCQRPR; YKPR; and YGG	Bovine β-casein, genetic variant A^2	0.024
	Bovine β-casein, genetic variant A^3	0.019

Table 13.3 (Continued)   Structure and Frequency of Motifs with Potential Biological Activity Occurrence in Animal and Plant Proteins (BIOPEP; http://www.uwm.edu.pl/biochemia)

Activity/Structure	Proteins	A Value
	Bovine β-casein, genetic variant B	0.019
	Bovine β-casein, genetic variant E	0.024
	Bovine α-lactalbumin	0.025
	Human α-lactalbumin	0.024
	Caprine α-lactalbumin	0.024
	Horse myoglobin	0.022
	Chicken tropomyosin, β chain	0.011
	Rice prolamin, clone PPROL 14	0.020
	Rice prolamin, clone PPROL 7	0.020
	Rice prolamin, clone PPROL 4A	0.020
	Faba bean HMW legumin	0.015
	Serendipity berry monellin chain B	0.020
Peptidase inhibitory motifs: IHPFAQTQ; IHPFAQTQ; GP; FP; LLSPWNINA; VP; and LPPV	Bovine αs1-casein, genetic variant B	0.085
	Bovine αs2-casein, genetic variant A	0.048
	Bovine β-casein, genetic variant A[1]	0.148
	Ovine β-lactoglobulin	0.085
	Bovine tropoelastin	0.146
	Bovine α-1 (I) collagen	0.258
	Chicken α-1 (I) collagen	0.205
	Wheat α/β-gliadin, clone PTO	0.054
	Wheat α/β-gliadin, clone PW8142	0.089
	Wheat γ-gliadin	0.066
	Barley γ-hordein	0.105

(continued)

Table 13.3 (Continued) Structure and Frequency of Motifs with Potential Biological Activity Occurrence in Animal and Plant Proteins (BIOPEP; http://www.uwm.edu.pl/biochemia)

Activity/Structure	Proteins	A Value
	Rice prolamin, clone PPROL 17	0.040
	Rice prolamin, 10 kDa fraction	0.097
	Oat 12S globulin	0.060
	Soybean albumin 1	0.077
	Pea legumin A2	0.040
	Serendipity berry monellin chain A	0.111
Celiac-toxic motifs: QPYP; QQPY; PSQQ; QQQP; PFPPQQPYPQPQPF; and QPFRPQQPYPQPQP	Wheat α/β-gliadin, class A-I	0.042
	Wheat α/β-gliadin, class A-II	0.027
	Wheat α/β-gliadin, class A-III	0.025
	Wheat α/β-gliadin, class A-IV	0.020
	Wheat α/β-gliadin, class A-V	0.019
	Wheat α/β-gliadin, MM1	0.029
	Wheat α/β-gliadin, clone PW 1215	0.020
	Wheat α/β-gliadin, clone PW 8142	0.030
	Wheat α-gliadin fragment	0.040
	Wheat α-gliadin	0.040
	Wheat γ-gliadin, class B-1	0.020
	Wheat ω-gliadin	0.040
	Wheat ω-gliadin B	0.034
	Wheat glutenin, type II, 3 group subunit	0.020
	Oat avenin	0.038
	Oat avenin N9	0.038
	Barley γ3-hordein	0.031
	Barley B1 hordein	0.017
	Barley B3 hordein	0.015

Table 13.3 (Continued)   Structure and Frequency of Motifs with Potential Biological Activity Occurrence in Animal and Plant Proteins (BIOPEP; http://www.uwm.edu.pl/biochemia)

Activity/Structure	Proteins	A Value
Antibacterial and antiviral motifs: RPKHPIKHQGLPEQVLNENLLRF; LKKISQRYQKFALPQY; CKDDQNPHISCDKF; and AASDISLLDAQSAPLR	Bovine αs2-casein, genetic variant A	0.019
	Bovine α-lactalbumin	0.025
	Ovine α-lactalbumin	0.032
	Arabian camel α-lactalbumin	0.008
	Bovine lactoferrin	0.017
	Human lactoferrin	0.004
	Caprine lactoferrin	0.006
	Bovine hemoglobin α-chain	0.007
Antioxidative motifs: LH; HHPLL; and HL	Horse hemoglobin β-chain	0.034
	Bovine κ-casein, genetic variant A	0.030
Chemotactic motifs: VGAPG; PGAIPG; and LGTIPG	Bovine tropoelastin	0.024
	Bovine α-1 (I) collagen	0.001
Embryotoxic motifs: RGD and YIGSR	Chicken α-1 (I) collagen	0.002
	Bovine α-1 (I) collagen	0.001
	Bovine α-1 (III) collagen	0.001
	White lupine PR-10 protein	0.006
	Gingko 11S globulin	0.002
Ion flow–regulating motifs: DY and TSLYR	Vetch narbonin	0.010
	Narbon bean narbonin	0.010
	Serendipity berry monellin chain A	0.022
Gastric mucosa function–regulating motifs: GP; PG; GPGG; and PGP	Chicken α-1 (I) collagen	0.210
	Bovine α-1 (I) collagen	0.267
	Bovine α-1 (III) collagen	0.272

(continued)

Table 13.3 (Continued)   Structure and Frequency of Motifs with Potential Biological Activity Occurrence in Animal and Plant Proteins (BIOPEP; http://www.uwm.edu.pl/biochemia)

Activity/Structure	Proteins	A Value
Ubiquitin-mediated proteolysis-activating motifs: RA; LA; and WA	Human lysozyme C	0.038
	Chicken connectin (titin)	0.046
	Sorghum kafirin PSKR2	0.041
	Sorghum kafirin PSK8	0.037
Phosphoinositol action–regulating motifs: GFW; GFL; LGY; and GLY	Human α-lactalbumin	0.008
	Bovine α-lactalbumin	0.008
	Caprine α-lactalbumin	0.008
	Ovine α-lactalbumin	0.008
	Arabian camel α-lactalbumin	0.008
Bacterial permease ligand motifs: KK and KKKA	Bovine troponin C	0.052
	Horse myoglobin	0.037
Anorectic motifs: PGP and APGPR	Bovine β-casein, all genetic variants	0.005
	Chicken α-1 (I) collagen	0.036
	Bovine α-1 (I) collagen	0.042
	Bovine α-1 (III) collagen	0.045
	Bovine elastin	0.001
	Gingko 11S globulin	0.002

activity found in rice prolamins are more potent inhibitors of ACE than the corresponding fragments of β-casein.

ACE peptide inhibitors are present in many food raw materials and products, including fish (Fujita, Yokoyama, and Yoshikawa 2000), sake (Saito et al. 2005), casein (Silva and Malcata 2005), whey (Abukabar et al. 1998), and cereal proteins (Fujita, Yokoyama, and Yoshikawa 2000). Maruyama and Suzuki (1982) isolated four peptides from a tryptic hydrolysate of casein. These peptides, with the structure of dodecapeptide EFVAPFPEVFGK, pentapeptide FFVAP, hexapeptide TTMPLW, and heptapeptide AVPYPQR, strongly inhibited the ACE activity in vitro. Their sequences corresponded to the following fragments of αs1-casein: 23–34, 23–27, 194–199, and to fragment 177–183 of β-casein. The oral administration of a tryptic hydrolysate of casein decreased the blood pressure in spontaneously hypertensive rats, but had no effect on the electrocardiogram, heart histopathology, and serum lipid concentrations. A similar response was observed in rats given sour milk (Xu 1998). The clinical examinations involving hypertensive human subjects also showed that a daily administration of 95 mL of sour milk significantly reduced the systolic blood pressure (Yamamoto 1997). In a group of prehypertensive human subjects, 4 weeks of repeated daily intake of 3.8 g of C12-peptide (containing a bovine casein hydrolysate) resulted in a substantial decrease in the systolic and diastolic pressures (in comparison with the placebo group) as well as in the plasma angiotensin II levels (Cadée et al. 2007). In another study, three strong ACE inhibitors with the LRP, LSP, and LQP sequences were isolated from α-zein hydrolyzed with thermolysin. Six hours after the oral administration of these peptides (30 mg kg⁻¹ body weight), the systolic blood pressure was found to decrease up to a maximum of 15 mmHg (Kim et al. 2001; Meyer et al. 2001).

Examples of peptides with anticoagulant activity, derived from food proteins, are presented in Table 13.3. The sequences of these peptides are dominated by glycine and proline. This is natural because both glycine and proline occur in the N-terminus, the tripeptide sequence of fibrin (α-chain), which is important for the process of fibrin polymerization. The richest source of peptides with anticoagulant activity is collagen. The collagen sequence consists of a number of repeating motifs containing glycine and proline residues. The peptide profile of chicken collagen contains 303 fragments with potential anticoagulant activity, dominated by dipeptides (126 PG fragments and 123 GP fragments) and tripeptides (52 PGP fragments). Two DGEA fragments and one KDGEA fragment are also present. The secondary structure of the motifs containing anticoagulant peptides in collagen and in other proteins is unordered in more than 60%, with α-helices accounting for around 3%. The strongly hydrophilic character of the above-mentioned motifs (mean hydrophobicity of approximately −1.2 for proteins in the BIOPEP database) indicates that they are available to the action of proteolytic enzymes and can easily liberate anticoagulant peptides.

Nonaka, Tanaka, and Maruyama (1995) and Nonaka et al. (1997) examined the effect of peptides isolated from enzymatic hydrolysates of collagen and of the corresponding synthetic analogs on fibrin polymerization and platelet aggregation. The GPR peptide was isolated from a bacterial collagenase hydrolysate of porcine skin collagen by ultrafiltration, reverse osmosis, and reversed-phase liquid chromatography. These authors demonstrated that the GPR tripeptide and its synthetic analogs corresponding to the motifs in the collagen sequence and to the peptides in the thermolysin hydrolysates of collagen inhibited adenosine disphosphate (ADP)-induced platelet aggregation. Such peptides as GPRG, GPRGP, GPRPP, and GPRPPP at a concentration of 0.3 mM suppressed human platelet aggregation by more than 50%. This suppression was not directly related to thrombin inhibition (Nonaka et al. 1997). The synthetic peptides, the GPR analogs extended at their C-terminus, and the thermolysin hydrolysates of collagen did not inhibit thrombin.

The peptide analogs of GPR extended at their N-terminus, such as AGPR and GPAGPR, as well as those with single amino acid substitution, including SarPR (Sar = sarcosine), GPK, GAR, and AGPR, at a concentration from 0.1 to 0.8 mM, had no inhibitory effect on platelet aggregation.

Many peptides derived from κ-casein possess anticoagulant activity. According to an interesting hypothesis, κ-casein and fibrin γ-chain have evolved from a common ancestor 450 million years ago (Jolles, Loucheux-Lefebvre, and Henschen 1978). Indeed, there is a homology between the mechanisms of chymosin-induced milk coagulation and thrombin-induced blood coagulation (Fiat, Migliore, and Jolles 1993). At the first stage of the enzymatic coagulation of milk, chymosin specifically hydrolyzes κ-casein. As a result of the hydrolysis of the Phe105–Met106 bond, the N-terminal fragment, para κ-casein, combines with other casein fractions to form an insoluble curd, and caseinmacropeptide (CMP), a C-terminal soluble fragment of κ-casein, is released into whey. Its tryptic hydrolysate contains anticoagulant peptides (Fosset and Tome 2000). The main isolated anticoagulant peptide corresponds to a motif of κ-casein (fr. 106–116) with the MAIPPKKNQDK sequence. This peptide and its KNQDK (112–116) and NQDK (113–116) fragments, known as casoplatelins, are both structurally and functionally similar to the C-terminal dodecapeptide of the γ-chain of human fibrin with the HHLGGAKQAGDV sequence. The amino acid residues, Ile108, Lys112, and Asp115, in κ-casein occupy homologous positions, compared with the γ-chain of human fibrin. The above-mentioned amino acid residues present in the anticoagulant peptides derived from κ-casein are responsible for the competitive inhibition of the binding of fibrin γ-chains with the receptor sites on the surface of the blood platelets (Schlimme and Meisel 1995). There is also a homology between the fibrin α-chain tetrapeptide RGDX sequence and the KRDS sequence corresponding to lactoferrin fragment 39–42. The KRDS peptide displays anticoagulant activity (Rutherford and Gill 2000).

The precursor proteins of the opioid peptides, similar to the opioid receptors, are synthesized primarily in the central and peripheral nervous systems as well as in the immune system and in the endocrine system. A characteristic feature of the structure of typical opioid peptides is an identical N-terminal amino acid sequence, YGGF, responsible for the interactions with the opioid receptors (Chaturvedi et al. 2000).

The endogenous opioidergic system is supplemented with the opioid peptides originating from the food proteins. The opioid peptides, whose precursors are food proteins, show an affinity for opioid receptors and a similar type of activity as endogenous opiates. They are formed in the digestive tract as a result of the digestion of precursor proteins and exert a direct effect on specific receptors in the alimentary tract. Moreover, following their absorption into the bloodstream, they may interact with the endogenous opioid receptors. The opioid peptides found in the hydrolysates of food proteins or produced during digestion have been termed exorphines, due to their exogenous origin and morphine-like activity. The structure of the exorphines resembles that of atypical endogenous opioid peptides, with the N-terminal tyrosine residue, important for the ligand–receptor interactions. In some cases, a tyrosine residue at the N-terminal end is replaced with some other amino acid residue (e.g., R, G, V, and L).

The main sources of the opioid peptides derived from food proteins are gluten (or gliadin), αs-casein, β-casein, and hemoglobin (Table 13.3). The opioid peptides present in foods are resistant to further hydrolysis by the small intestinal enzymes and exert a direct influence on specific receptors in the digestive tract. When absorbed into the circulatory system, these peptides supplement the endogenous opioidergic system. The peptides with opioid activity were identified and described for the first time in β-casein hydrolysates. Those that have been studied most intensely are β-casomorphins (1–11), that is, fragments of the β-casein sequence 60–70 (YPFPGPIPNSL). Fragments of this motif also show opioid activity (Brantl et al. 1981). β-Casomorphins are important, biologically active compounds present in the diet. They are resistant to the action of the digestive enzymes and may affect the physiological functions of the small intestine (Hayes et al. 2007). Apart from their opioid activity, β-casomorphins play a vital role in the gastrointestinal tract. Due to the enhancement of water and electrolyte absorption in the small intestine, β-casomorphins prolong the gastrointestinal transit time, which is a component of their antidiarrheal action. They can also affect nutrient absorption and insulin secretion. Moreover, they show a natural affinity for the μ-receptor and, as demonstrated in experiments on rats, exert a strong analgesic effect, induce apnea and irregular breathing, and affect sleep patterns (Meisel and FitzGerald 2000).

The casomorphins and lactorphins are present in many fermented dairy products obtained by using lactic acid bacteria. This results from the specific activity of the proteolytic enzymes of these microorganisms. In dairy products manufactured using *Lactococcus lactis* strains, β-casein is hydrolyzed by extracellular PI-type proteinase into many different oligopeptides, including β-casomorphin (1–9) (Hayes et al. 2007). Particularly high concentrations of β-casomorphin (1–11) have been reported in fermented products containing milk contaminated with proteolytic bacteria, such as *Pseudomonas aeruginosa* and *Bacillus cereus* (Hamel, Kielwein, and Teschemacher 1985). Opioid peptides, derivatives of αs-casein and β-casein, are released into fermented UHT milk by the action of the proteolytic enzymes of *Lactobacillus rhamnosus* GG. A variety of peptides are formed during the ripening of Edam cheese and Australian Cheddar cheese, including β-casomorphins (Dionysius et al. 2000; Sabikhi and Matur 2001). In fermented whey produced using yeasts, *Kluyveromyces marxianus* var. *marxianus*, lactorphins are formed as a result of the enzymatic hydrolysis of whey proteins (Belem, Gibas, and Lee 1999).

Some peptides derived from αs1-casein and κ-casein show antagonist activity against enkephalins and casomorphins. These are casoxins A, B, C, and D, whose structure corresponds to fragments of κ-casein (35–41, 58–61, 25–34) and αs1-casein. Clare and Swaisgood (2000) have isolated casoxin D from the αs1-casein hydrolysate. The metoxyl derivatives of casoxins A, B, and C show substantially higher activity than unmodified casoxins.

Wheat gluten is the richest source of opioid peptides. Zioudrou, Streaty, and Klee (1979) were the first to describe the opioid activity of gluten peptides. The peptides isolated from the peptic and thermolysin hydrolysates have been termed gluten exorphines A_4, A_5, B_4, and B_5. Opioid peptides may also be formed as a result of gluten digestion with pepsin, trypsin, and chymotrypsin (Kitts and Weiler 2003).

The cases of food peptides strengthening or weakening the gastrointestinal immune system have been well documented. Rich sources of such peptides are rice prolamins, casein (primarily β-casein), and α-lactalbumin present in mammalian milk. The frequency of occurrence of immunomodulatory motifs in the sequences of the above-mentioned precursor proteins exceeds 0.02 (Table 13.3). The polypeptide chains of those proteins contain 3, 5, and 3 potentially active fragments, respectively. Motifs with potential immunomodulatory activity can be found in the hydrophilic fragments of protein molecules, as proven by their negative hydrophobicity (mean hydrophobicity of immunomodulatory motifs in the BIOPEP database is around −0.4). The structure of these motifs is characterized by the presence of an α-helix (approximately 20%), a β-turn (27%), a β-structure (25%), and an unordered structure (28%). The peptides isolated from the tryptic hydrolysates of rice and soybean protein activate the superoxide anions, which stimulate the nonspecific immune response (Kitts and Weiler 2003). The peptide isolated from a hydrolysate of the albumin fraction of rice proteins, oryzatensin (GYPMYPLPR), is multifunctional. It affects the cells of the immune system and the smooth muscles via a single receptor. Oryzatensin and oryzatensin-related peptides contain arginine residues at the C-terminus, leucine residues at position 3, and hydrophobic residues (leucine or tyrosine) at position 5 from the C-end. Another immunomodulatory peptide is HCQRPR, derived from a tryptic hydrolysate of soybean protein, which stimulates phagocytosis and tumor necrosis factor (TNF) production.

The sulfated glycopeptides, formed as a result of the hydrolysis of egg albumin with trypsin, activate the macrophages from male mice in vitro. They also enhance the proliferation of the macrophages, the production of interleukin-1 (IL-1) and superoxide anions in leukocytes. The glycoside residues present in the peptides include N-acetylgalactosamine, galactose, and N-acetylneuraminic acid. The glycoside and sulfonic residues play a key role in the interactions with the macrophage components (Tanizaki et al. 1997).

The immunopeptides originating from αs1-casein, β-casein, κ-casein, and α-lactalbumin may both suppress and enhance the immune response. Jolles et al. (1981) were the first to demonstrate that tryptic hydrolysates of human β-casein display immunostimulating activity. For instance, the hexapeptide VEPIPY, corresponding to fragment 54–59 of human β-casein, induces the activity of macrophages in mice and humans and enhances the natural resistance of young mice to infections caused by *Klebsiella pneumoniae.* The peptides formed as a result of the hydrolysis of κ-casein with trypsin inhibit the immune response of mouse spleen lymphocytes and rabbit Peyer's cells (Kitts and Weiler 2003). The *L. helveticus*-fermented milk products show an immunostimulating effect on the lymphocyte proliferation in vitro and the ability to stimulate the phagocytic activity of the lung macrophages (Laffineur, Genett, and Leonil 1996). These products contain numerous oligopeptides that are formed as a result of the hydrolysis of milk proteins, mainly β-casein, by the extracellular proteolytic enzymes of *L. helveticus.* The above-mentioned strain is characterized by high proteolytic activity. The results of the experiments on mice have revealed the immunostimulating and anticarcinogenic properties of the peptide fractions isolated from fermented milk in the presence of the strain *L. helvetius* R389. Many fermented dairy products, including yogurt, enhance the immune response when milk is inoculated with *L. casei* and *L. acidophilus.* In addition, the yogurt filtrate, free from microbes, stimulates interferon-γ (IFN-γ) production and the activity of the human natural killer (NK) cells (Hayes et al. 2007).

Interesting results have been reported concerning the properties of casein-derived phosphopeptides, which are known to bind calcium. These phosphopeptides enhance IgG secretion, acting in vitro on the human lymphocytes. The immunostimulating activity of casein phosphopeptides has been ascribed to the presence of phosphoserine residues in their sequence, and phosphorylation sites have been found to be allergenic epitopes of casein (Meisel 2005).

Some of the physiological properties of peptides (antihypertensive, immuno-stimulating, and anticoagulant peptides and peptides inhibiting HIV-1 proteinase) result from their ability to block the proteolytic enzymes. The processes of protein hydrolysis and the inhibition of those processes are of key physiological significance for digestion, blood clotting and fibrinolysis, blood pressure regulation, hormonal neuromodulation, and phagocytosis, but they also play an important role in the pathological states (such as pulmonary emphysema, carcinoma, and hypertension) and in various infections (HIV infections and parasitic invasions). A major group of food protein-derived peptides are prolyl oligopeptidase (POP) inhibitors (EC 3.4.21.26), including a peptide with the HLPPPV sequence, the product of zein hydrolysis with subtilisin (Maruyama and Suzuki 1982), and a peptide with the LLSPWNINA sequence, isolated from the by-products of sake production (Saito et al. 1997). The typical structural motifs in most POP inhibitors and other proline proteases are proline residues.

The richest potential source of peptide inhibitors of proline proteases are β-casein and collagen α-chains (Table 13.3). The frequency of occurrence of motifs with POP inhibitory activity in the sequences of the discussed precursor proteins ranges from 0.15 to 0.25. Motifs with the potential activity of proline protease inhibitors occur in the hydrophobic fragments of protein molecules (their mean hydrophobicity in proteins listed in the BIOPEP database is approximately 0.18). The structure of these motifs is characterized by the presence of a β-turn (around

23%), a β-structure (53%), and an unordered structure (23%) and by the complete absence of an α-helix.

N-acetylpepstatin is a potent inhibitor of aspartyl proteases. To a limited extent, this peptide also inhibits HIV-1 proteinase, thereby blocking the process of HIV-1 replication in the infected cell (Rival et al. 2001). The peptides isolated from cheese and a tryptic hydrolysate of zein (Gobetti et al. 2002) inhibit the activity of thermostable proteinases and endopeptidases synthesized by the psychrotrophic bacteria, *P. fluorescens*. This prevents the development of a bitter taste and gel formation in UHT milk during storage and extends the shelf-life of the dairy products. Control over the activity of proteolytic enzymes, including those originating from psychrotrophic bacteria, is important for improving the keeping quality of foods.

Wheat, rye, and barley grain proteins contain fragments of amino acid sequences that initiate pathophysiological processes and, in consequence, cause intestinal epithelial damage. Protein maldigestion may lead to coma, diarrhea (the most common symptom), anemia, and osteoporosis (Dewar, Pereira, and Ciclitira 2004). Celiac disease is a multigenic disorder, but the mode of its inheritance remains unknown. Studies have shown that the toxicity of cereal proteins to celiac patients is related to the peptides released during the digestion of these proteins. The wheat gliadin peptides have been most extensively investigated to date. According to Cornell (1996), the motifs with the sequences QQQP, QQPY, PSQQ, and QPYP, are responsible for celiac disease. These motifs are also present in many nontoxic proteins, which suggests that celiac disease may be caused by the expanded amino acid motifs found in wheat α-gliadins and containing the above-mentioned tetrapeptides (McLachlan, Gullis, and Cornell 2002). Such an assumption could provide a basis for an analysis of the potential toxicity of cereal proteins to celiac patients, involving the determination of similarities between the sequences of these proteins (Darewicz and Dziuba 2007). The amino acid sequences of the proteins of selected cereal and legume species, listed in the BIOPEP database, contain the tetrapeptides responsible for celiac disease (sequences: QQQP, QQPY, PSQQ, and QPYP). The tetrapeptides (from 6 to 11) responsible for celiac disease can be found in the sequences of wheat (*Triticum aestivum*) α-gliadin, and their number depends on the length of an amino acid sequence (Table 13.3). These tetrapeptides are present primarily in the N-terminal domain of the α-gliadin chain. The number of expanded amino acid motifs in this fragment ranges from 0 to 7, which is lower than the number of tetrapeptides because some of the expanded motifs contain two tetrapeptide sequences. For instance, the QPFPPQQPYPQPQP motif (α-gliadin, fr. 36–49) contains the fragments: QQPY (residues 41–44) and QPYP (residues 42–45). The expanded motifs responsible for celiac diseases are absent from or sporadically present in the polypeptide chains of the remaining gliadin fractions.

The sequences of many food proteins contain the motifs with low occurrence frequency, corresponding to the bioactive peptides, which are not discussed in detail in this chapter (Table 13.3).

13.3 Functionally Active Peptides Derived from Food Proteins

The functional properties of both proteins and the products of their hydrolysis have been investigated by numerous authors (Darewicz et al. 2000b; Foegeding et al. 2002; Kilara and Panyam 2003; Kristinsson and Rasco 2000; Phillips, Whitehead, and Kinsella 1994). The functional properties can be divided into surface properties, including the ability to form and stabilize emulsion (oil–water interface), the ability to form and stabilize foam (air–water interface) and solubility (water–protein and water–peptide interactions); and hydrodynamic properties (intermolecular interactions), including gelation properties, textural properties, and sensory properties (taste and aroma, flavor). These properties reflect the natural characteristics of the protein and peptide molecules, such as the size, shape, elasticity, susceptibility to denaturation, amino acid composition and sequence, charge distribution, hydrophobicity or hydrophilicity, type and number of microdomain structures, adaptivity of domains or the whole molecule to changing environmental conditions, and the specificity of the interactions between the protein–peptide and other food components. They are also related to the most important environmental features, such as pH, temperature, pressure, ionic composition, and the concentrations of particular ions. Major differences between proteins and peptides with respect to their functional properties stem from their different structural properties. This chapter discusses the peptides responsible for flavor attributes, antioxidant capacity, and surface activity.

Enzymatic hydrolysis of proteins results not only in a decrease in the relative molecular weight, but also in an increase in the number of functional groups capable of ionization and in the exposure of the hydrophobic surfaces, which affects the sensory properties of hydrolysates and peptides (Darewicz and Dziuba 2006). Table 13.4 presents a profile of the potential sensory activity of β-lactoglobulin. There are 18 potentially bitter dipeptides and tripeptides in the amino acid sequence of this protein, containing mostly leucine, isoleucine, valine, or phenylalanine. One of the problems associated with the enzymatic hydrolysis of proteins is the activation of a bitter taste due to peptide release (Kim, Kawamura, and Lee 2003; Lemieux and Simard 1992). The bitter off-flavor may develop when proteins are broken down into short-chain hydrophobic peptides. The bitterness and astringency of the peptides are dependent on the type of amino acids as well as on the sequence and conformation of the peptide chains. Bitterness is related to the hydrophobic character of the amino acid residues. Another factor that must be considered in the studies on bitter-tasting peptides is the length of the polypeptide chains (De Armas et al. 2004). The high values of the mean hydrophobicity reported for casein (1605 cal mol^{-1}), soybean proteins (1540 cal mol^{-1}), and maize zein proteins (1480 cal mol^{-1}) indicate that these proteins may be a natural source of the bitter-tasting peptides. Sigh et al. (2005) isolated a bitter peptide from the C-terminal end of β-casein (fragment 193–209). Aubes-Dufau et al.

Table 13.4   Profile of Potential Sensory Activity of β-Lactoglobulin, gen. var. A, Bovine (*Bos taurus*) [BIOPEP ID 1116]

β-Lactoglobulin sequence:

MKCLLLALALTCGAQALIVTQTMKGLDIQKVAGTWYSLAMAASDISLLDAQSAPLR
VYVEELKPTPEGDLEILLQKWENDECAQKKIIAEKTKIPAVFKIDALNENKVLVLDTDY
KKYLLFCMENSAEPEQSLACQCLVRTPEVDDEALEKFDKALKALPMHIRLSFNPTQL
EEQCHI

Activity	Sequence	Location
Bitter-tasting peptide	GL	[25–26]
	II	[87–88]
	IL	[72–73]
	IV	[18–19]
	KF	[151–152]
	KP	[63–64]
	LD	[26–27], [48–49], [111–112]
	LE	[70–71], [149–150], [172–173]
	LF	[120–121]
	LI	[17–18]
	LL	[4–5], [5–6], [47–48], [73–74], [119–120]
	LLL	[4–6]
	LV	[109–110], [138–139]
	VE	[59–60]
	VF	[97–98]
	VL	[108–109], [110–111]
	VY	[57–58]
	VD	[144–145]

(1995) isolated a bitter peptide from a peptic hydrolysate of hemoglobin (fragment 32–40 of the β-chain of bovine hemoglobin). Kukman, Zelenik-Blatnik, and Abram (1995) isolated 14 low-molecular-mass bitter peptides from soybean protein hydrolysates treated with Alcalase. These peptides contain three to six mostly hydrophobic amino acid residues, and leucine, valine, and tyrosine predominate in their C-terminal position.

The determination of the amino acid composition and the type of N- and C-terminal amino acids, followed by the calculation of the mean hydrophobicity (1100–1200 cal mol⁻¹), enabled the identification of fragments 26–45 and 169–193 in αs1-casein as astringent/sour peptides. Sörensen and Jepsen (1998) demonstrated that specific flavor profiles are a consequence of the presence of acidic amino acids in the structure of the peptides formed during cheese ripening. In general, the acidic amino acids present in the peptide structure are responsible for a sour taste and for the activation of a bouillon-like or, more specifically, umami taste. Umami is a key component of the flavor profile of various food products. Yamasaki and Maekawa (1978) were the first to describe the taste of a peptide isolated from beef gravy and to establish its sequence (KGDEESLA). Park et al. (2002) reported that the specific flavor of a Vietnamese fish sauce resulted from the presence of dipeptides, tripeptides, and tetrapeptides, which enhanced not only the umami taste, but also the sweetness, saltiness, and bitterness of the sauce. Schlichtherle-Cerny and Amadò (2002) analyzed the peptides activating the bouillon taste of the enzymatic hydrolysates of deamidated wheat gluten. The sequences of the four identified peptides are pEPS, pEP, pEPE, and pEPG.

Numerous studies have shown that some of the hydrolysates of proteins originating from various sources display antioxidant activity. These are the protein hydrolysates of whey, milk, wheat, soybean, egg, shrimp, Atlantic horse mackerel, mackerel, herring, and capelin, as well as common sole and Alaska pollock by-products (Klompong et al. 2007). Far fewer studies have examined the antioxidant capacity of precisely defined protein fragments. The essential components of the antioxidant peptides are the histidine or tyrosine residues, which are potent antioxidants. Similar properties are exhibited by methionine, lysine, arginine, phenylalanine, and tryptophan. The antioxidant capacity of a given peptide is also affected by its amino acid sequence and spatial configuration (Peña-Ramos and Xiong 2001). Antioxidant peptides are usually composed of 3–16 amino acid residues. The presence of the hydrophobic residues facilitates the interactions between peptides and, for example, linoleic acid. A relationship has been found between the mean hydrophobicity of the polypeptide chains and the antioxidant activity of the peptides (Saiga, Tanabe, and Nishimura 2003). Saito et al. (2003) constructed a library of tripeptides containing two histidine or tyrosine residues to analyze their antioxidant properties. The tripeptides possessing tryptophan or tyrosine residues at their C-terminus are characterized by exceptional free-radical scavenging activity. They have also been shown to interact synergistically with other antioxidants, such as phenolic compounds.

Many peptides with antioxidant properties have been isolated from dairy products. Such peptides always contain one or more histidine, proline, tyrosine, and tryptophan residues (Pihlanto 2006). Antioxidant peptides may be formed from milk proteins as a result of hydrolysis in the presence of digestive enzymes or as a result of milk fermentation in the presence of lactic acid bacteria. Casein possesses polar domains containing a characteristic sequence of three phosphorylated serine residues responsible for the ability to chelate cations and form complexes with calcium, iron, and zinc. Therefore, phosphorylated casein-derived peptides used for the formation and stabilization of oil–water emulsions may prevent lipid oxidation. Diaz et al. (2003) and Kim, Jang, and Kim (2007) demonstrated that the mechanism of their antioxidant activity encompasses not only the ability of the phosphoserine residues to chelate metal ions, but also their free-radical scavenging ability. Diaz and Decker (2004) observed that caseinophosphopeptides exert a pro-oxidative effect when applied at high concentrations. The antioxidant activity of β-casein, its peptides, and αs1-casein have been well documented and reviewed in the literature. Rival et al. (2001) obtained hydrolysates with an antioxidant capacity by hydrolyzing β-casein with trypsin and/or subtilisin. Fragment 169–176 with the KVLPVEK sequence has been found to possess the greatest ability to prevent linolenic acid oxidation. Suetsuna, Ukeda, and Ochi (2000) isolated a fragment with the YFYPEL sequence from αs1-casein and reported that the EL sequence had a significant influence on its antioxidant capacity. β-Lactoglobulin, similar to potentially all whey proteins, displays antioxidant properties (Farnfield, Smith, and Stockmann 2003). Hernández-Ledesma et al. (2005) isolated and identified a series of peptides from β-lactoglobulin hydrolyzed with COROLASE PP. One of those peptides, with the WYSLAMAASDI sequence, had more free-radical neutralizing capacity than butylated hydroxyanisole (BHA). The presence of methionine and tyrosine plays a particularly important role in neutralizing the free radicals.

Some of the peptides formed during the hydrolysis of β-conglycinin and glycinin in soybeans also show antioxidant activity (Chen et al. 1996). These peptides contain 5–16 amino acid residues, mostly valine or leucine residues at the N-terminal position, and proline, histidine, and tyrosine residues within the polypeptide chain. The tripeptide unit, PHH, and the pentapeptide unit, LLPHH, have been found to be highly antioxidative. These peptides show a greater antioxidant capacity than the native proteins, which may be explained by the structural changes that occur during hydrolysis and by the surface exposure of active arginine residues. The above-mentioned peptides and their synthetic analogues inhibit linoleic acid oxidation in vitro. According to Yang et al. (2000), the antioxidant activity of soybean peptides is determined by the presence of the PHH active site. Liu, Chen, and Lin (2005) described the antioxidant properties of soymilk-kefir. Yokomizo, Takenaka, and Takenaka (2002) isolated four antioxidant peptides, AY, ADF, SDF, and GYY, from the by-products generated during soymilk and tofu production. The antioxidant capacity of the last of these peptides is comparable to that of carnosine.

Zheng et al. (2006) subjected extruded corn gluten to enzymatic hydrolysis with Alcalase from *B. licheniformis*. The prepared hydrolysate contained peptides with a molecular weight of 660–3710 Da. The amino acid sequence of a fragment characterized by the highest antioxidant activity was FPLEMMPF. Jae-Young, Pyo-Jam, and Se-Kwon (2005) isolated a peptide with antioxidant properties (FPLEMMPF) from Alaska pollock by-products, using a mackerel's digestive enzyme. Dávalos et al. (2004) isolated four peptides included in the protein sequence of ovalbumin. The peptides were produced by enzymatic hydrolysis of the egg white with pepsin. One of the peptides, with the YAEERYPIL sequence, exhibits a high antioxidant capacity and a strong ACE inhibitory activity, which makes it useful for the control and prevention of cardiovascular diseases. Li et al. (2007) hydrolyzed porcine skin collagen with a mixture of protease from the bovine pancreas, protease from *Streptomyces*, and protease from *Bacillus* spp. Four peptides with the sequences QGAR, LQGM, LQGMHyp, and HylC were isolated from the hydrolysate. The first of these peptides displayed the highest antioxidant activity. In their previous study, Kim et al. (2001) showed that the GPHyp sequence is of primary significance for the antioxidant properties of the peptides isolated from the Alaska pollock skin collagen. Saiga, Tanabe, and Nishimura (2003) used papain and actinase to isolate the antioxidant peptides from porcine myofibrillar proteins. The obtained peptides had the sequences DAQEKLE, DSGVT, IEAEGE, EELDNALN, and VPSIDDQEELM.

Some peptides, the products of protein hydrolysis, can readily adsorb at interfaces, reduce surface tension, and form coherent layers around oil drops or air bubbles (Walstra and Roos 1993). Turgeon et al. (1992) reported that the peptide fractions of β-lactoglobulin, containing fragments 21–40 and 41–60, are characterized by better surface properties than the fractions containing fragments 61–69-S-S-149–162 and 61–70-S-S-149–162. The reasons for the worse functional properties of the last two fragments include the lack of differences in the charge distribution along the polypeptide chain, the similar hydrophobicity values of the amino acids, and the structure stiffening with disulfide bonds.

It has been found that the fractions containing amphiphilic peptides, obtained upon plasmin digestion of β-casein, have better emulsifying properties than β-casein (Darewicz, Dziuba, and Caessens 2000a). A deterioration in these properties has been observed in the fraction containing fragment 1–28 of β-casein. It has been suggested that this is a consequence of the synergistic effect of the fractions containing fragments 1–105/107 and 29–105/107, and hydrophobic fragments of β-casein. Popineau et al. (2002) used protease with a chymotryptic activity to prepare the hydrolysates and fractions of the hydrophobic and hydrophilic peptides. The hydrolysates displayed a foaming capacity, but the foams were not stable. The hydrophilic peptides generated only short-lived foams and exhibited no emulsifying properties, while the hydrophobic peptides formed foams with a good stability and were more efficient than the hydrolysates with respect to emulsion stabilization. Lee et al. (1987) examined the effect of the pH on the emulsifying properties of the hydrophilic (1–25) and hydrophobic (193–209) fragments of β-casein. In a neutral environment, both peptides showed low emulsifying activity. The hydrophobic fragment exhibited an increased emulsifying capacity in an alkaline and acidic environment, whereas the hydrophilic fragment exhibited an increased emulsifying capacity at a low pH. The authors suggested that an improvement in the studied properties was related to a higher surface activity noted at a pH of 3. Synergism was also observed between a hydrophobic fragment 193–209 of β-casein and a hydrophilic glycomacropeptide from κ-casein, as supported by the enhanced emulsifying activity (Lemieux and Simard 1992). The plasmin treatment of β-casein enables to obtain peptides with different molecular characteristics, for instance, peptides differing in molecular weight, hydrophobicity, and charge distribution along the polypeptide chains (Darewicz et al. 2000b). These peptides may be classified as hydrophilic: fragment 1–28 of β-casein; amphiphilic (with a clear-cut division between the hydrophobic and hydrophilic surfaces): fragments 1/29–105/107 of β-casein; and hydrophobic: fragments 106/108/114–209 of β-casein. The determination of the structural changes in these peptides, following interfacial adsorption, provides a basis for investigating the correlations between their molecular and surface properties. Studies on peptides derived from β-casein have revealed that despite a relatively high content of proline residues, their adsorption on the hydrophobic surfaces results in an increase in the proportion of the ordered α-helical structure, at the expense of the unordered structure.

13.4 Antimicrobial Peptides

Antibacterial peptides are present in the amino acid sequences of many food proteins, but the frequency of their occurrence is relatively low (Table 13.3). Studies conducted to date in this area have been limited to milk proteins and egg white proteins. Lactenine is the first antibacterial component isolated from milk treated with chymosin (Floris et al. 2003). Lactenine inhibits the growth of bacteria of the genus *Streptococcus*. Another group is formed by casocidins, the basic high-molecular-weight polypeptides released as a result of milk heating or casein digestion with chymosin. These peptides exert bactericidal activity against members of the genus *Lactobacillus* and many other pathogenic bacteria, including *Staphylococcus aureus*, *B. subtilis*, and *Diplococcus pneumoniae*. An antibacterial peptide released from αs1-casein upon the action of chymosin is isracidin, whose sequence corresponds to the N-terminal fragment of protein (1–23). Under in vitro conditions, isracidin at high concentrations (0.1–1 mg/mL) shows a bacteriostatic effect on *Lactobacillus* and Gram-positive bacteria. When applied at a dose of 10 μg, isracidin provides protection against infections caused by *S. aureus*, *Streptococcus pyogenes*, and *Listeria monocytogenes*, as confirmed during in vivo tests on mice. When administered to sheep and cows at doses comparable to antibiotic doses, this peptide effectively protected the animals against mastitis. αs2-Casein is an equally abundant source of antibacterial peptides. A cationic peptide that inhibits the growth of Gram-positive (*S. aureus*) and Gram-negative (*Escherichia coli*) bacteria has been isolated from heat-treated sour milk (Floris et al. 2003). This peptide, casocidin-I, has an amphipathic character and high pI values and its structure is homologous to fragment 150–188 of αs2-casein. The activity of casocidin-I disappears after the hydrolysis with trypsin, but not after the treatment with protease V8, which indicates that the full peptide sequence is not necessary to maintain the antibacterial properties. Recio and Visser (1999) identified two bioactive fractions from αs2-casein, fragment 164–179, which is part of the casocidin-I sequence, and fragment 183–207. A distinctive feature of both peptides is a high proportion of basic amino acids. In a neutral environment, fragment 164–179 exhibited weaker antibacterial activity than fragment 183–207, despite a higher positive charge. Both peptides were active against Gram-positive and Gram-negative bacteria. Liepke et al. (2001) identified, for the first time, a bioactive peptide from human κ-casein, located at position 63–117. This peptide was isolated from sour milk hydrolyzed with pepsin. Another example of a peptide originating from κ-casein is kappacin (Floris et al. 2003). Kappacin is a nonglycosylated phosphorylated fragment of the macropeptide (106–169), active against *Streptococcus mutans*, *Porphyromonas gingivalis*, and *E. coli*. The glycosylated forms of the macropeptide have no antibacterial properties, which suggests that its activity is dependent on the phosphate residues. Although kappacin does not show sequence homology with cationic antibacterial peptides, the mechanism of action of this peptide probably involves the ability to adopt an amphipathic helical structure in the presence of a bacterial membrane. It has been suggested that kappacin molecules may aggregate in their native environment and that anionic pores may be formed to increase the permeability of the bacterial cell membrane to cations. Fragment 43–97 of κ-casein, released upon the action of pepsin, also has antibacterial properties (Liepke et al. 2001). This cationic proline-rich peptide exhibits broad-spectrum activity, including the inhibition of the growth of bacteria and yeasts. The digestion of bovine κ-casein with trypsin leads to the liberation of the pentapeptide, named κ-casocidin, whose sequence is homologous to fragment 17–21 (Peña-Ramos and Xiong 2001). This peptide suppresses the growth of *S. aureus*, *E. coli*, and *S. typhimurium*. However, the range of its applications is limited because it may induce cytotoxic effects in some mammalian cell types, as observed with respect to human leukocyte cell cultures. Lòpez-Exposito et al. (2006) identified another six peptides with antibacterial activity in a pepsin hydrolysate of κ-casein. Fragments 18–24, 139–146, and 30–32 are most active against *L. innocua*, *S. carnosus*, and *E. coli*. Whey proteins are also a rich source of antibacterial peptides (Pellegrini 2003). Lactoferrin is an iron-binding protein. The antimicrobial activity of lactoferrin is due to its ability to chelate iron ions that are required for microbial growth. Numerous bioactive peptides are generated by pepsin digestion of human and bovine lactoferrin, including human lactoferricin H (fr. 1–47) and lactoferricin B (fr. 17–41) (Bellamy et al. 1992). The sequences of these peptides are located in different regions of lactoferrin from the sequences responsible for iron binding. In addition, this bioactive fragment may directly affect the body since lactoferrin is partly hydrolyzed in the digestive tract of humans and animals to fragments that contain lactoferricin sequences. Lactoferricin is active against many different bacteria, including bacterial pathogens like *L. monocytogenes*. This peptide has much stronger antibacterial properties than lactoferrin, and it is active against the majority of Gram-positive and Gram-negative bacteria. Apart from lactoferricin, a variety of other antibacterial peptides originating from lactoferrin have been identified (Floris et al. 2003), including fragments 1–11-S-S-17–47, 1–16-S-S-43–48, 1–16-S-S-45–48, and 1–42-S-S-43–48, which have been isolated from a pepsin hydrolysate of lactoferrin.

Two antimicrobial peptides, LDT1 (fr. 1–5) and LDT2 (fr. 17–31-S-S-109–114), are generated during the tryptic hydrolysis of α-lactalbumin, while the hydrolysis of the whey protein with chymotrypsin leads to the formation of a peptide named LDC (fr. 61–68-S-S-75–80). In contrast to the majority of known antibacterial peptides, those derived from α-lactalbumin have an anionic character. Their activity is directed mainly against the Gram-positive bacteria. Four antimicrobial peptides (fr. 15–20, 25–40, 78–83, 92–100) have been identified following the tryptic hydrolysis of β-lactoglobulin, and all are active only against Gram-positive bacteria (Pellegrini 2003).

The lysozyme and ovotransferrin present in hen egg white contain in their sequences antibacterial peptides that are released by digestive enzymes. The lysozyme maintains its antimicrobial properties despite denaturation and the natural loss of the enzymatic activity. This suggests that the antibacterial activity is not determined by the lysozyme molecule as a whole, but by the sequences of the bioactive peptides contained in the protein. Enzymatic hydrolysates of the lysozyme show bactericidal activity against many Gram-positive and Gram-negative bacteria. Pellegrini (2003) studied a variety of enzymatic hydrolysates of lysozyme. The most potent bactericidal hydrolysate was obtained with the use of clostripain. The antimicrobial activity was ascribed to fragment 98–112 of the polypeptide chain of the lysozyme. Both ovotransferrin and lysozyme are involved in the antimicrobial defense mechanism of egg white. Ovotransferrin is active against Gram-positive bacteria, Gram-negative bacteria, and fungi. Studies on ovotransferrin have revealed the fact that transferrin family proteins can inhibit bacterial growth regardless of their ability to remove iron from the bacteria. It has been found that a peptide composed of 92 amino acid residues, obtained by partial acidic hydrolysis, displays antimicrobial activity. The sequence of this cationic peptide corresponds to the sequence (fr. 109–200) located in the upper part of the active site of ovotransferrin. This peptide, known as OTAP-92, shows high homology to insect defensins, in particular to scorpion defensins—agiotoxin I and kaliotoxin 1—which are blockers of ion channels. Detailed data on defensins can be found in the Defensins Knowledge base (Seebah et al. 2007).

Antimicrobial peptides, both cationic and neutral, are also secreted by Gram-positive and Gram-negative bacteria. They are referred to as bacteriocins. Bacteriocins are ribosomally synthesized, extracellularly released low-molecular-mass peptides or proteins (usually 30–60 amino acids). In their review, Settanni and Corsetti (2008) described a wide variety of bacteriocins isolated from raw materials and foods. More information about bacteriocins can be found in the BACTIBASE database (Hammami et al. 2007).

13.5 Conclusion

This chapter contains exemplary results obtained using the BIOPEP database with comments. It serves for interpreting or predicting the experimental results. Some examples of the application of the BIOPEP database have recently been published. One of the possibilities is the identification of peptides using mass spectrometry and searching for their sequences in the BIOPEP database. This option was used by Bauchart et al. (2007) and Català-Clariana et al. (2010). The prediction of the ACE inhibitory activity of proteins followed by the experimental measurements of the activity was described as exemplified by the proteins from amaranthus (Vecchi and Añón 2009; Fritz et al. 2010) or peanut (Jimsheena and Gowda 2011). Examples of the application of the BIOPEP database encourage us to conclude that it has become a valuable tool for food scientists.

Acknowledgments

This work was supported by the Rector of the University of Warmia and Mazury in Olsztyn, research project no. 528-0712-0809. The authors are grateful to Professor Jerzy Dziuba for inspiring, helping, and providing background information for this chapter.

References

Abukabar, A., Saito, T., Kitazawa, H., Kawai, Y., and Itoh, T. 1998. Structural analysis of new antihypertensive peptides derived from cheese whey protein by proteinase K digestion. *Journal of Dairy Science* 81: 3131–38.

Aubes-Dufau, I., Capdevielle, J., Series, J. L., and Combes, D. 1995. Bitter peptide from hemoglobin hydrolysate: Isolation and characterization. *FEBS Letters* 364: 115–19.

Bairoch, A. and Apweiler, R. 2000. The SWISS-PROT protein sequence database and its supplement TrEMBL in 2000. *Nucleic Acids Research* 28(1): 45–48.

Bauchart, C., Morzel, M., Chambon, C., Mirand, P. P., Reynès, C., Buffière, C., et al. 2007. Peptides reproducibly released by *in vivo* digestion of beef meat and trout flesh in pigs. *British Journal of Nutrition* 98: 1187–95.

Belem, M. A. F., Gibas, B. F., and Lee, B. H. 1999. Proposing sequences for peptides derived from whey fermentation with potential bioactive sites. *Journal of Dairy Science* 82: 486–93.

Bellamy, W., Takase, M., Yamauchi, K., Wakabayashi, H., Kawase, K., and Tomita, M. 1992. Identification of the bactericidal domain of lactoferrin. *Biochimica et Biophysica Acta* 1121: 130–36.

Bennett, K. L., Brønd, J. C., Kristensen, D. B., Podtelejnikov, A. V., and Wiśniewski, J. R. 2004. Analysis of large-scale MS data sets: The dramas and the delights. *DDT: Targets* 3(2) (Suppl.): 43–49.

Brantl, V., Teschemacher, H., Henschen, A., Lottspeich, F., and Blaig, J. 1981. Opioid activities of β-casomorphins. *Life Sciences* 28: 1903–9.

Bray, J. E., Todd, A. E., Pearl, F. M., Thornton, J. M., and Orengo, C. A. 2000. The CATH dictionary of homologous superfamilies (DHS): A consensus approach for identifying distant structural homologues. *Protein Engineering* 13(3): 153–65.

Cadée, J. A., Chang, C. Y., Chen, C. W., Huang, C. N., Chen, S. L., and Wang, C. K. 2007. Bovine casein hydrolysate (C12 peptide) reduces blood pressure in prehypertensive subjects. *American Journal of Hypertension* 20(1): 1–5.

Català-Clariana, S., Benavente, F., Giménez, E., Barbosa, J., and Sanz-Nebot, V. 2010. Identification of bioactive peptides in hypoallergenic infant milk formulas by capillary electrophoresis-mass spectrometry. *Analytica Chimica Acta* 683: 119–25.

Chaturvedi, K., Christoffers, K. H., Sing, K., and Howells, R. D. 2000. Structure and regulation of opioid receptors. *Biopolymers* 55: 334–46.

Chen, H. M., Muramoto, K., Yamauchi, F., and Nokihara, K. 1996. Antioxidant activity of designed peptides based on the antioxidative peptide derived from digests of soybean peptide. *Journal of Agricultural and Food Chemistry* 44: 2619–23.

Clare, D. A. and Swaisgood, H. E. 2000. Bioactive milk peptides: A prospectus. *Journal of Dairy Science* 83: 1187–95.

Colinge, J. and Masselot, A. 2004. Mass spectrometry has married statistics: Uncle is functionality, children are selectivity and sensitivity. *DDT: Targets* 3(2) (Suppl.): 50–55.

Cornell, H. J. 1996. Coeliac disease: A review of the causative agents and their possible mechanisms of action. *Amino Acids* 10: 1–19.

Darewicz, M. and Dziuba, M. 2006. Effect of the structure of products of partial proteolysis of β-casein on their emulsifying properties. *Polish Journal of Natural Sciences* 20: 413–22.

Darewicz, M., Dziuba, J., and Caessens, P. W. J. R. 2000a. Effect of enzymatic hydrolysis on the emulsifying and foaming properties of milk proteins – A review. *Polish Journal of Food and Nutrition Science* 9: 3–8.

Darewicz, M., Dziuba, J., Caessens, P. W. J. R., and Gruppen, H. 2000b. Dephosphorylation-induced structural changes in β-casein and its amphiphilic fragment in relation to emulsion properties. *Biochimie* 82: 191–95.

Dávalos, A., Miguel, M., Bartolomè, B., and Lòpez-Fandiño, R. 2004. Antioxidant activity of peptides derived from egg white proteins by enzymatic hydrolysis. *Journal of Food Protection* 67: 1939–44.

De Armas, R. R., Diaz, H. G., Molina, R., González, M. P., and Ulriarte, E. 2004. Stochastic-based descriptions studying peptides biological properties: Modelling the bitter tasting threshold of dipeptides. *Bioorganic and Medicinal Chemistry* 12: 4815–22.

Dewar, D., Pereira, S. P., and Ciclitira, P. J. 2004. The pathogenesis of coeliac disease. *The International Journal Biochemistry and Cell Biology* 36: 17–24.

Diaz, M. and Decker, E. A. 2004. Antioxidant mechanisms of caseinophosphopeptides and casein hydrolysates and their application in ground beef. *Journal of Agricultural and Food Chemistry* 52: 8208–13.

Diaz, M., Dunn, C. M., McClements, D. J., and Decker, E. A. 2003. Use of caseinophosphopeptides as natural antioxidants in oil-in-water emulsions. *Journal of Agricultural and Food Chemistry* 51: 2365–70.

Dionysius, D. A., Marschke, R. J., Wood, A. J., and Milne, J. 2000. Identification of physiologically functional peptides in dairy products. *Australian Journal of Dairy Technology* 55: 103–9.

Dziuba, M. and Darewicz, M. 2007. Food proteins as precursors of bioactive peptides. Classification into families. *Food Science and Technology International* 13(6): 393–404.

Dziuba, J. and Iwaniak, A. 2006. Database of protein and bioactive peptide sequences. In *Nutraceutical Proteins and Peptides in Health and Disease*, eds. Y. Mine and F. Shahidi, 543–64. Boca Raton–London–New York: CRC Press, Taylor & Francis.

Dziuba, J., Iwaniak, A., and Minkiewicz, P. 2003. Computer-aided characteristics of proteins as potential precursors of bioactive peptides. *Polimery* 48: 50–53.

Farnfield, M. M., Smith, S., and Stockmann, R. 2003. Antioxidant activity of β-lactoglobulin and its modified derivatives. *Australian Journal of Dairy Technology* 58: 186–94.

Fiat, A. M., Migliore, D., and Jolles, P. 1993. Biologically active peptides from milk proteins with emphasis on two examples concerning antithrombotic and immunostimulating activities. *Journal of Dairy Science* 76: 301–10.

FitzGerald, R. J. and Meisel, H. 2000. Milk protein derived peptide inhibitors of angiotensin-I-converting enzyme. *British Journal of Nutrition* 84: 33–37.

Floris, R., Recio, I., Berkhout, B., and Visser, S. 2003. Antibacterial and antiviral effect of milk proteins and derivatives thereof. *Current Pharmaceutical Design* 9: 1257–75.

Foegeding, E. A., Davis, J. P., Doucet, D., and McGuffey, M. K. 2002. Advances in modifying and understanding whey protein functionality. *Trends in Food Science & Technology* 13: 151–59.

Fosset, S. and Tomè, D. 2000. Dietary protein-derived peptides with antithrombotic activity. *Bulletin of IDF* 353–58: 65–68.

Fritz, M., Vecchi, B., Rinaldi, G., and Añón, M. C. 2010. Amaranth seed protein hydrolysates have *in-vivo* and *in-vitro* antihypertensive activity. *Food Chemistry*. In press.

Fujita, H., Yokoyama, K., and Yoshikawa, M. 2000. Classification and antihypertensive activity of angiotensin-I-convering enzyme inhibitory activity derived from food proteins. *Journal of Food Science* 65: 564–69.

Gobetti, M., Stepaniak, L., De Angelis, M., Corsetti, A., and Di Cagno, R. 2002. Latent bioactive peptides in milk proteins: Proteolytic activation and significance in dairy. *Critical Reviews in Food Science and Nutrition* 42: 223–39.

Hamel, U., Kielwein, G., and Teschemacher, H. 1985. β-Casomorphin immunoreactive materials in cow's milk incubated with various bacterial species. *Journal of Dairy Research* 52: 139–41.

Hammami, R., Zouhir, A., Ben Hamida, J., and Fliss, I. 2007. BACTIBASE: A new web-accessible database for bacteriocin characterization. *BMC Microbiology* 7: article no. 89.

Hayes, M., Stanton, C., Fitzgerald, G. F., and Ross, R. P. 2007. Putting microbes to work: Dairy fermentation, cell factories and bioactive peptides. Part II: Bioactive peptides functions. *Biotechnological Journal* 2: 435–49.

Hernández-Ledesma, B., Dávalos, A., Bartolomè, B., and Amigo, L. 2005. Preparation of antioxidant enzymatic hydrolysates from alpha-lactalbumin and beta-lactoglobulin. Identification of active peptides by HPLC-MS/MS. *Journal of Agricultural and Food Chemistry* 53: 588–93.

Jae-Young, J., Pyo-Jam, P., and Se-Kwon, K. 2005. Antioxidant activity of a peptide isolated from Alaska pollock (*Theragra chalcogramma*) frame protein hydrolysate. *Food Research International* 38: 45–50.

Jimsheena, V. K. and Gowda, L. R. 2011. Angiotensin I-converting enzyme (ACE) inhibitory peptides derived from arachin by simulated gastric digestion. *Food Chemistry* 125: 561–69.

Jollès, P., Loucheux-Lefebvre, M., and Henschen, A. 1978. Structural relatedness of κ-casein and fibrinogen γ-chain. *Journal of Molecular Evolution* 11: 271–77.

Jolles, P., Parker, F., Floch, F., and Migliore, D. 1981. Immunostimulating substances from human casein. *Journal of Immunopharmacology* 3: 363–69.

Karelin, A. A., Blishchenko, E. Y., and Ivanov, V. T. 1998. A novel system of peptidergic regulation. *FEBS Letters* 428: 7–12.

Kilara, A. and Panyam, D. 2003. Peptides from milk proteins and their properties. *Critical Reviews in Food Science and Nutrition* 43: 607–33.

Kim, G.-N., Jang, H.-D., and Kim, C.-I. 2007. Antioxidant capacity of caseinophospho-peptides prepared from sodium caseinate using alcalase. *Food Chemistry* 104: 1359–65.

Kim, M.-R., Kawamura, Y., and Lee, C.-H. 2003. Isolation and identification of bitter peptides of tryptic hydrolysates of soybean 11S glycinin by reversed-phase high-performance liquid chromatography. *Journal of Food Science* 68: 2416–22.

Kim, S., Kim, Y., Byuan, H., Nam, K., Joo, D., and Shahidi, F. 2001. Isolation and char-acterization of antioxidative peptides from gelatin hydrolysate of Alaska pollock skin. *Journal of Agricultural and Food Chemistry* 49: 1984–89.

Kitts, D. D. and Weiler, K. 2003. Bioactive proteins and peptides from food sources. Applications of bioprocesses used in isolation and recovery. *Current Pharmaceutical Design* 9: 1309–23.

Klompong, V., Benjakul, S., Kantachote, D., and Shahidi, F. 2007. Antioxidative activity and functional properties of protein hydrolysate of yellow stripe trevally (*Selaroides leptolepis*) as influenced by the degree of hydrolysis and enzyme type. *Food Chemistry* 102: 1317–27.

Kristinsson, H. G. and Rasco, B. A. 2000. Fish protein hydrolysates: Production, bio-chemical, and functional properties. *Critical Reviews in Food Science and Nutrition* 40: 43–81.

Kriventseva, E. V., Fleischmann, W., Zdobnov, E. M., and Apweiler, R. 2001. CluStr: A database of clusters of SWISS-PROT+TrEMBL proteins. *Nucleic Acids Research* 29(1): 33–36.

Kukman, I. L., Zelenik-Blatnik, M., and Abram, V. 1995. Isolation of low-molecular-mass hydrophobic bitter peptides in soybean protein hydrolysates by reversed-phase high-performance liquid chromatography. *Journal of Chromatography A* 704: 113–20.

Lackner, P., Koppensteiner, W. A., Dominques, F. S., and Sippl, M. J. 1999. Numerical evaluation methods. Automated large scale evaluation of protein structure prediction. *Proteins: Structure, Function, and Genetics* 37 (Suppl. 3): 7–14.

Laffineur, E., Genett, N., and Leonil, J. 1996. Immunomodulatory activity of beta-casein permeate medium fermented by lactic acid bacteria. *Journal of Dairy Science* 79: 2112–20.

Lee, S. W., Shimizu, M., Kaminogawa, S., and Yamauchi, K. 1987. Emulsifying properties of a mixture of peptides derived from the enzymatic hydrolysates of bovine caseins. *Agricultural and Biological Chemistry* 51: 1535–40.

Lemieux, L. and Simard, R. E. 1992. Bitter flavour in dairy products 2. A review of bitter peptides from casein – Their formation, isolation and identification, structure masking and inhibition. *Lait* 72: 335–83.

Li, B., Chen, F., Wang, X., Ji, B., and Wu, Y. 2007. Isolation and identification of antioxida-tive peptides from porcine collagen hydrolysate by consecutive chromatography and electrospray ionization – Mass spectrometry. *Food Chemistry* 102: 1135–43.

Liepke, C., Zucht, H. D., Forsman, W. G., and Standker, L. 2001. Purification of novel peptide antibiotics from human milk. *Journal of Chromatography B* 752: 369–77.

Liu, J. R., Chen, M. J., and Lin, C. W. 2005. Antimutagenic and antioxidant properties of milk-kefir and soymilk-kefir. *Journal of Agricultural and Food Chemistry* 53: 489–98.

Lòpez-Exposito, I., Minervini, F., Amigo, L., and Recio, I. 2006. Identification of antibacterial peptides from bovine κ-casein. *Journal of Food Protection* 69: 2992–97.

Maruyama, S. and Suzuki, H. 1982. A peptide inhibitor of angiotensin I converting enzyme in the tryptic hydrolysate of casein. *Agricultural and Biological Chemistry* 46: 1393–94.

McLachlan, A., Gullis, P. G., and Cornell, H. J. 2002. The use of extended amino acid motifs for focusing on toxic peptides in coeliac disease. *Journal of Biochemistry, Molecular Biology and Biophysics* 6: 319–24.

Meisel, H. 2005. Biochemical properties of peptides encrypted in bovine milk proteins. *Current Medicinal Chemistry* 12: 1905–19.

Meisel, H. and FitzGerald, R. J. 2000. Opioid peptides encrypted in intact milk protein sequences. *British Journal of Nutrition* 84: 27–31.

Meyer, L., Sperber, K., Chan, L., Child, J., and Toy, L. 2001. Oral tolerance to protein antigens. *Allergy* 56: 12–15.

Nonaka, I., Katsuda, S., Ohmori, T., Nakagami, T., and Maruyama, S. 1997. *In vitro* and *in vivo* anti-platelet effects of enzymatic hydrolysates of collagen and collagen-related peptides. *Bioscience, Biotechnology, and Biochemistry* 61: 772–75.

Nonaka, I., Tanaka, H., and Maruyama, S. 1995. Production of fibrin polymerization inhibitor from collagen by some proteases. *Annals of the New York Academy of Science* 750: 412–14.

Park, J.-N., Ishida, K., Watanabe, T., Endoh, K.-I., Watanabe, K., Murakami, M., and Abe, H. 2002. Taste effects of oligopeptides in a Vietnamese fish sauce. *Fisheries Science* 68: 921–28.

Pellegrini, A. 2003. Antimicrobial peptides from food proteins. *Current Pharmaceutical Design* 9: 1225–38.

Peña-Ramos, E. A. and Xiong, Y. L. 2001. Antioxidative activity of whey protein hydrolysates in a liposomal system. *Journal of Dairy Science* 84: 2577–83.

Phillips, L. G., Whitehead, D. M., and Kinsella, J. 1994. *Structure–Function Properties of Food Proteins.* San Diego, CA: Academic Press.

Pihlanto, A. 2006. Antioxidative peptides derived from milk proteins. *International Dairy Journal* 16: 1306–14.

Popineau, Y., Huchet, B., Larrè, C., and Bèrot, S. 2002. Foaming and emulsifying properties of fractions of gluten peptides obtained by limited enzymatic hydrolysis and ultrafiltration. *Journal of Cereal Science* 35: 327–35.

Recio, I. and Visser, S. 1999. Identification of two distinct antibacterial domains within the sequence of bovine alpha(s2)-casein. *Biochimica et Biophysica Acta* 1428: 314–26.

Rival, S. G., Fornaroli, S., Boeriu, C. G., and Wichers, H. J. 2001. Caseins and casein hydrolysates. 1. Lipoxygenase inhibitory properties. *Journal of Agricultural and Food Chemistry* 49: 287–94.

Rutherfurd, K. J. and Gill, H. S. 2000. Peptides affecting coagulation. *British Journal of Nutrition* 84: 99–102.

Sabikhi, L. and Matur, B. N. 2001. Qualitative and quantitative analysis of β-casomorphins in Edam cheese. *Milchwissenschaft* 56: 198–202.

Saiga, A., Tanabe, S., and Nishimura, T. 2003. Antioxidant activity of peptides obtained from porcine myofibrillar proteins by protease treatment. *Journal of Agricultural and Food Chemistry* 51: 3661–67.

Saito, K., Jin, D. H., Ogawa, T., Muramoto, K., Hatakeyama, E., Yasuhara, T., and Nokihara, K. 2003. Antioxidative properties of tripeptide library prepared by the combinatorial chemistry. *Journal of Agricultural and Food Chemistry* 51: 3668–74.

Saito, Y., Okura, S., Kawato, A., and Suginami, K. 1997. Propyl endopeptidase inhibitors in sake and its by-products. *Journal of Agricultural and Food Chemistry* 45: 720–24.

Saito, Y., Wanezaki, K., Kawato, A., and Imayasu, S. 2005. Structure and activity of angiotensin converting enzyme inhibitory peptides from sake and sake lees. *Bioscience, Biotechnology, and Biochemistry* 58: 1761–71.

Schlichtherle-Cerny, H. and Amadó, R. 2002. Analysis of taste-active compounds in an enzymatic hydrolysate of deamidated wheat gluten. *Journal of Agricultural and Food Chemistry* 50: 1515–22.

Schlimme, E. and Meisel, H. 1995. Bioactive peptides derived from milk proteins. Structural, physiological and analytical aspects. *Die Nahrung* 39: 1–20.

Seebah, S., Suresh, A., Zhuo, S., et al. 2007. Defensins knowledgebase: A manually curated database and information source focused on the defensins family of antimicrobial peptides. *Nucleic Acids Research* 35: D265–68.

Settanni, L. and Corsetti, A. 2008. Application of bacteriocins in vegetable food biopreservation. *International Journal of Food Microbiology* 121: 123–38.

Sigh, T. K., Young, N. D., Drake, M., and Cadwallade, K. R. 2005. Production and sensory characterization of a bitter peptide from beta-casein. *Journal of Agricultural and Food Chemistry* 53: 1185–89.

Silva, S. V. and Malcata, F. X. 2005. Caseins as source of bioactive peptides. *International Dairy Journal* 15: 1–15.

Sörensen, N. K. and Jepsen, R. 1998. Assessment of sensory properties of cheese by near-infrared spectroscopy. *International Dairy Journal* 8: 863–71.

Suetsuna, K., Ukeda, H., and Ochi, H. 2000. Isolation and characterization of free radical scavenging activities peptides derived from casein. *Journal of Nutritional Biochemistry* 11: 128–31.

Tanizaki, H., Tanaka, H., Iwata, H., and Kato, A. 1997. Activation of macrophages by sulfated glycopeptides in ovomucin, yolk membrane and chalazae in chicken eggs. *Bioscience, Biotechnology, and Biochemistry* 61: 1883–89.

Turgeon, S. L., Gauthier, S. F., Mollé, D., and Léonil, J. 1992. Interfacial properties of tryptic peptides of β-lactoglobulin. *Journal of Agricultural and Food Chemistry* 40: 669–75.

Vecchi, B. and Añón, M. C. 2009. ACE inhibitory tetrapeptides from *Amaranthus hypochondriacus* 11S globulin. *Phytochemistry* 70: 864–70.

Walstra, P. and De Roos, A. L. 1993. Proteins at air-water and oil-water interfaces: Static and dynamic aspects. *Food Reviews International* 9: 503–25.

Xu, R. J. 1998. Bioactive peptides in milk and their biological and health implications. *Food Reviews International* 14: 1–17.

Yamamoto, N. 1997. Antihypertensive peptides derived from food proteins. *Biopolymers* 43: 129–34.

Yamasaki, Y. and Maekawa, K. 1978. A peptide with delicious taste. *Agricultural and Biological Chemistry* 42: 1761–65.

Yang, J. H., Mau, J. L., Ko, P. T., and Huang, L. C. 2000. Antioxidant properties of fermented soybean broth. *Food Chemistry* 71: 249–54.

Yokomizo, A., Takenaka, Y., and Takenaka, T. 2002. Antioxidative activity of peptides prepared from okara protein. *Food Science and Technology Research* 8: 357–59.

Zheng, X., Li, L., Liu, X., and Wang, X. 2006. Production of hydrolysate with antioxidative activity by enzymatic hydrolysis of extruded corn gluten. *Applied Microbiology and Biotechnology* 73: 763–70.

Zioudrou, C., Streaty, R. A., and Klee, W. A. 1979. Opioid peptides derived from food proteins: the exorphins. *Journal of Biological Chemistry* 254: 2446–49.

Chapter 14

Food-Derived Bioactive Peptides in the Market

Toshio Shimizu

Contents

14.1 Introduction

Peptides and protein hydrolysates are known for their diverse functional properties. In the early 1980s, the Japanese Scientific Academy started a national project on functional foods. The Japanese national project proved that peptides and protein hydrolysates have useful functionalities that are associated with the following functions in the body: (1) control of blood pressure, (2) promotion of the absorption of minerals, (3) maintenance of bone health, (4) improvement of fat metabolism, (5) antioxidation of cells and tissues, and (6) modulation of the immune system.

The scientific support for the benefits of food-derived peptides on human health should be based on human studies as well as animal studies. When new products with bioactive components enter the market, their scientifically substantiated health claims are important for informing the consumers of their health benefits. In 1991, the results of the research and development of the physiological function of foods prompted Japan's Ministry of Health, Labor and Welfare (MHLW) to establish, for the first time in the world, a regulatory system to evaluate the functionalities of food and to approve the health claims as "Foods for Specified Health Use (FOSHU)." Products applying for FOSHU approval are scientifically evaluated for their effectiveness and safety by the MHLW and the Food Safety Commission. The total number of FOSHU-approved products increased to about 840 in March 2009. Peptides and protein hydrolysates are important components for FOSHU products, whose health claims are concerning the abovementioned functions (1), (2), and (4).

In the United States, the health claims for peptides and protein hydrolysates were labeled by the Dietary Supplement Health and Education Act (DSHEA), which was established in 1994, as well as the EU Regulation on Nutrition in 2007. The products from peptides and protein hydrolysates have similar health claims as those approved by the Japanese claims, such as blood pressure reduction and mineral absorption.

This chapter outlines the sources of bioactive peptides, the functionality, the support for the health claims, and the products in the market.

14.2 Sources of Bioactive Peptides

Bioactive peptides and protein hydrolysates are produced after the degradation of the large proteins. These peptides usually consist of 3–20 amino acids, while the protein hydrolysates refer to more than 20 amino acids. These peptides and protein hydrolysates are produced from the proteins of animal, plant, or microorganism origin. The origins of the animal proteins are milk, beef, pork, fish, or egg. Those of the plant and microorganism proteins are derived from soybean, wheat, rice, sesame, seaweed, and mushroom.

Bioactive peptides and protein hydrolysates are produced from these protein sources in the following processes: (a) enzymatic hydrolysis by digestive enzymes, (b) microbial fermentation of proteins with bacteria, and (c) proteolysis by enzymes derived from microorganisms. The most common process used to produce bioactive peptides and protein hydrolysates is through the hydrolysis of raw natural proteins using enzymes such as pepsin, trypsin, chymotrypsin, pancreatin, and thermolysin. Bioactive peptides are also produced industrially by fermentative production with microorganisms, for example, *Lactobacillus helveticus*. After hydrolysis, enriching the peptides from the hydrolysates is necessary for commercial production. Membrane separation systems, including microfiltration, ultrafiltration, and reverse osmosis filtration, are employed to enrich the bioactive peptides with a specific molecular weight range.

14.3 Functionality of Bioactive Peptides

The initial scientific approach to peptides has focused on their nutritional aspects because ingested protein is thought to be absorbed from the intestinal tract as amino acids after degradation by the digestive enzymes. Researchers have revealed that the gastrointestinal tract has transport systems not only for amino acids but also for peptides; therefore, those absorbed peptides have many physiological actions. These have changed the interest in peptides from their nutritional aspects to their role as functional ingredients in foods. Recently, the research in the field of bioactive peptides has been reviewed (Korhonen and Pihlanto 2006; Hartman and Meisel 2007; Korhonen 2009).

The functionalities of peptides are based on their amino acid composition and sequence. In general, the sizes of the active peptides vary from 2 to 20 amino acid residues. The various functionalities for bioactive peptides and protein hydrolysates have been described, including antihypertensive outcome, mineral absorption–enhancing activity, blood lipid–lowering ability, immunomodulating activity, opioid activity, and fatigue-reducing activity.

14.3.1 Antihypertensive Peptides

Hypertension is a major risk factor in cardiovascular disease, which has a multifactorial etiology. Blood pressure is one of the most important risk factors for cardiovascular disease and is influenced by dietary factors such as sodium, calcium, dietary fat, coffee, some proteins, and alcohol. The milk-derived peptides with antihypertensive activity were first found in the early 1990s (Nakamura et al. 1995).

These peptides inhibit the angiotensin 1-converting enzyme (ACE), which plays an important role in regulating the blood pressure. A large number of peptides with the ACE inhibitory activity have been isolated from the hydrolyzed food protein, such as fish, meat, plant seed, and seaweed.

14.3.2 Mineral Absorption–Enhancing Peptides

Minerals are responsible for bone formation. Calcium is one of the major bone-forming minerals, representing about 40% by weight of the mineral in the skeleton. The calcium balance depends on the amount of calcium in the diet and the efficiency of calcium absorption in the gut intestine. The casein-derived phosphorylated peptides, casein phosphopeptides (CPP), can be produced by tryptic hydrolysis of the casein, which contains clusters of phosphorylated seryl residues. Since CPP can bind to minerals and solubilize them, especially calcium, it promotes the bioavailability of calcium.

14.3.3 Blood Lipid–Lowering Peptides

Dietary saturated fatty acids elevate the serum lipid levels, including the triglyceride levels and the LDL cholesterol level. These levels are positively correlated with coronary heart disease, which has become a major cause of death and a public health issue in developed countries. The soybean protein hydrolysates were reported to have a hypocholesterolemic action in the rat (Sugano et al. 1988). The hydrophobic peptides in the soybean protein bind with the bile acids, which helps lower the cholesterol level.

14.3.4 Opioid Peptides

An opioid is a substance that works by interacting with the opioid receptors, which are located principally in the gastrointestinal tract and in the central nervous system. The physiological actions of the opioids modulate the absorption process in the intestine and decrease the perception of pain. The food-derived peptides with opioid activities were identified in various proteins, especially casein. The first opioid peptides were found in 1979 (Zioudrou et al. 1979).

14.3.5 Immunomodulating Peptides

The immune system protects the body against toxins and pathogens by modulating the release of certain cytokines, by proliferating the lymphocyte stimulation of phagocytic activities in the macrophages, and by producing antibodies. Immunomodulating peptides derived from casein and whey proteins were found to enhance the immune cell function (Gil et al. 2000).

14.4 Health Claims on Products with Bioactive Peptides

Health claims on products is one of the most important ways for manufacturers to enter their developed foods with bioactive peptides into the market. The consumers should have enough information about the health claim of the products and then choose the proper foods by using the health claim information given on the label. The consumers' interest in health issues has become a leading factor in their purchasing decision. The manufacturers can emphasize the characteristics of their products and promote the sales by the label or health claim. Therefore, the labeling should be clear and correct and avoid any misunderstanding by the consumers. Also, the labeling of health claims on foods should always be based on the scientific evidences.

A white paper from the European Union (EU) declared in 2000 that, "Consumers have the right to expect information on food quality and constituents that is helpful and clearly presented, so that informed choices can be made (European Union 2000). Proposals on the labeling of foods, building on existing rules, will be brought forward." Confirmation of the scientific support for the efficacy and adverse effects of health claims as well as the total evidence can help the consumers make an "informed Choice."

14.4.1 Regulatory System of Health Claims

In 2008, the draft guidelines (Codex 2008) for scientific substantiation of the Nutrition and Health Claims in Codex (Report of the 30th Session of Codex Committee on Nutrition and Foods for Special Dietary Uses) proposed the following criteria:

1. Health claims should primarily be based on the evidence provided by well-designed human intervention studies. Human observational studies are not generally sufficient per se to substantiate a health claim but where relevant, they may contribute to the total evidence. Animal model studies, ex vivo or in vitro data may be provided in support of the relationship between a food or a food constituent and its health claim, but they cannot be considered alone as sufficient per se to substantiate any type of health claim.
2. The total evidence, including unpublished data where appropriate, should be identified and reviewed, including the evidence to support the claimed effect, evidence that contradicts the claimed effect, and evidence that is ambiguous or unclear.
3. Evidence based on human studies should demonstrate a consistent association between the food or the food constituent and the health claim, with little or no evidence to the contrary.

These statements seem to be in general agreement with the global concepts concerning the approval of health claims.

In the FOSHU approval system, an applicant is required to substantiate the health claim not only by human intervention studies, but also by in vitro metabolism, biochemical studies, and animal studies. These data should demonstrate statistically significant differences. The study should also be well designed, for example, by using an appropriate biomarker, an appropriate sample size, and a sufficient number of subjects to prove statistically significant differences. All the available literature regarding the related functional components, the related foods, and the related function should be reviewed (Shimizu 2003). These requirements are generally similar to the above guidelines in (Codex 2008).

The United States set up the Nutrition Labeling and Education Act (NLEA) in 1990, which allows a claim describing a relationship between a food substance and a disease to be authorized by the Food and Drug Administration (FDA) after reviewing the scientific literature. The Structure/Function Claim, which was defined by the DSHEA in the United States, is a statement that describes the role of a nutrient or a dietary ingredient that is intended to affect the structure or function in humans or that characterizes the mechanism by which a nutrient or a dietary ingredient acts to maintain such structure/function. Under DSHEA, a food with a structure/function claim could be sold by the manufacturers with only a notification and not having been evaluated by the U.S. FDA.

The EU enforced the regulation of health claims on food in 2007, which consists of well-established functional claims, new functional claims, claims regarding disease risk reduction and child development or health. This regulation aims to ensure that any claims made about foods are clear, accurate, and substantiated by scientific evidence. Only products offering genuine health or nutritional benefits are to be allowed to have the claims labeled or marketed. It also aims to enable the consumers to make informed and meaningful choices about the foods they buy and to ensure fair competition and promote innovation in the food industry (EFSA 2008).

14.4.2 Scientific Substantiation of Health Claims

The scientific support for products proved by in vitro and animal studies is insufficient to substantiate the benefit of human health, because the digestion, absorption, and metabolism of the active components in the food are different between animals and humans. There are three essential requirements for health beneficial foods. The first is the functionality based on scientific evidence including human intervention studies. The second is the safety of the product with safety studies in humans, as well as information on the eating history. The third is the analytical determination of the functional component.

The functionality should be proved by evidence from human intervention studies, as well as by in vitro metabolism and biochemical studies, and animal studies. These data should demonstrate statistically significant differences. Basically, a human study should be conducted using the food in question over a reasonably long-term period. The study should also be well designed, for example, using an appropriate biomarker, an appropriate sample size, and a sufficient number of subjects to prove statistically significant differences. The study groups' background diet and other relevant aspects of their lifestyle should be characterized. All the available literature regarding the related functional components, the related foods, and the related claim should be reviewed. Any new scientific evidence used to support the health-related claims must be published in suitably qualified journals with expert referees who can review the evidence and support the credibility of the study.

The safety should be confirmed by human studies at the minimum effective dosage during a proper feeding period, after providing preliminary data from in vivo and in vitro studies confirming safe intake by humans. The literature regarding the related functional components should be reviewed. If the related literature suggests an especially undesirable or adverse health effect, this report should be included as a reference with the scientific explanation or the human study that confirms its safety for humans.

The methods for the analysis of the functional components should be suitable and reliable methods of quantitative/qualitative analytical determinations. These analytical determinations must be provided prior to characterizing the product, carrying out human studies, animal studies, in vitro studies, and stability tests. As additional documentation, the stability of the related functional components should be confirmed. The functional components and other components should be confirmed by suitable methods of analysis. These scientific substantiations of the health benefits to humans are the requirements of the Japanese FOSHU and are similar to the EU Regulation of Nutrition and health claims made on food (EFSA 2008) and the draft guideline for Codex Alimentarius (Codex 2008). The manufacturers should validate the quality, effectiveness, and safety of their products, taking the new scientific studies and the postmarketing research into consideration.

14.5 Products with Bioactive Peptides

The functionalities of bioactive peptides and protein hydrolysates, which were substantiated by human intervention studies for a proper number of subjects, are antihypertensive, enhancing the mineral absorption and reducing the blood lipid. The functional bioactive peptides or protein hydrolysates proved by human intervention studies and approved by the Japanese MHLW were categorized into three groups. The FOSHU-approved products with the clinical studies and the approved health claims are listed in Table 14.1. The products with bioactive peptides or protein hydrolysates, which were entered into the worldwide market, reached about 20 different components. Those products with clinical studies and functions are listed in Table 14.2. The typical peptides and hydrolysates of FOSHU-approved products are as follows:

Blood pressure: Lacto tripeptides, produced from milk fermented with *Lb. helveticus*, have an antihypertensive effect in animal model studies, and fermented milk containing these peptides reduced the blood pressure in subjects with elevated blood pressure. Val-Pro-Pro and Ile-Pro-Pro were isolated as bioactive compounds. It was reported that these peptides were produced from casein by digestion using *Aspergillus oryzae* protease. It was shown that food containing the hydrolysate reduced the blood pressure in normal-high and mildly hypertensive subjects (Mizuno et al. 2005). The fermented milk containing the lacto tripeptides were approved as FOSHU in 1997, and the products including other types of beverages

Table 14.1 FOSHU Products with Bioactive Peptides and Hydrolysates Concerning Blood Pressure

General Name (Peptide Sequence)	Function and Clinical Studies	Manufacturers and Product Name (Form)	Health Claims (Country)	Source (Product)
Lacto tripeptide (VPP and IPP)	Blood pressure *Clinical studies* • 30 persons with borderline hypertension (14 weeks) (Hata et al. 1996) • 30 mild and moderate hypertensive subjects (12 weeks) (Kajimoto et al. 2001a, 2001b) • 18 hypertensive subjects and 26 persons with normal blood pressure (12 weeks) (Itakura, Nakamura, and Takano 2001) • 42 persons with mild and moderate hypertension (14 weeks) (Kajimoto et al. 2001a, 2001b) *Meta-analysis* • 9 studies including 12 trials (total n = 623, 1996–2005) (Xu et al. 2008)	Calpis Co., Japan • Ameal S (beverage) • Ameal S Handy-tab (tablet) • Ameal S Vegetable (juice)	This is suitable for persons with slightly elevated blood pressure. (FS)	Milk, *Lb. helveticus* (fermentation)

Sardine peptide (VY)	Blood pressure *Clinical studies* • 29 mild hypertensive subjects (4 weeks) (Kawasaki et al. 2000) • 88 persons with normal-high blood pressure, mild and moderate hypertension (18 weeks) (Kajimoto et al. 2003) • 24 persons with mild hypertension (14 weeks) (Seki et al. 2000)	Sato Pharmaceutical Co. Ltd., Japan • Sato Marine Super P (tablet) • SenmiEkisu Co. Ltd., Japan • EP Marine Super P (soft drink)	This is suitable for persons with slightly elevated blood pressure. (FS)	Sardine muscle (enzyme degradation)
Casein dodecapeptide (FFVAPFPEVFGK)	Blood pressure *Clinical studies* • 18 mild hypertensive subjects (4 weeks) (Sekiya et al. 1992)	Kracie Pharmaceutical, Japan • Casein DP Peptio Drink (soft drink)	This is suitable for persons with slightly elevated blood pressure. (FS)	Milk casein
Sesame peptide (LVY)	Blood pressure *Clinical studies* • 72 persons with borderline and mild hypertension (16 weeks) (Moriguchi et al. 2004) • 89 persons with borderline and mild hypertension (6 weeks) (Moriguchi et al. 2006)	Suntory Beverage & Food Ltd., Japan • Goma Pepucha (soft drink)	This is suitable for persons with slightly elevated blood pressure. (FS)	Sesame protein (enzyme degradation)

(continued)

Table 14.1 (Continued)   FOSHU Products with Bioactive Peptides and Hydrolysates Concerning Blood Pressure

General Name (Peptide Sequence)	Function and Clinical Studies	Manufacturers and Product Name (Form)	Health Claims (Country)	Source (Product)
Bonito-derived peptides (LKPNM)	Blood pressure *Clinical studies* • 61 persons with borderline and mild hypertension (5 weeks) (Fujita et al. 2001)	Nippon Supplement, Japan • Peptide soup EX (freeze-dried soup) • Peptide Ace tab-type (tablet) • Peptide Tea (drink powder)	This is suitable for persons with slightly elevated blood pressure. (FS)	Dried bonito (enzyme degradation)
Isoleucyl-tyrosine (IY)	Blood pressure *Clinical studies* • 60 persons with normal-high blood pressure and mild hypertension (12 weeks) (Sato et al. 2004) • 60 persons with normal-high blood pressure, mild and moderate hypertension (14 weeks) (Mashiko et al. 2003) • 60 persons with mild and moderate hypertension (18 weeks) (Tsuchida et al. 2001) • 15 persons with mild and moderate hypertension (18 weeks) (Tsuchida et al. 2002)	Yakult Health Foods Co. Ltd., Japan • Bunaharitake (drink powder)	This is suitable for persons with slightly elevated blood pressure. (FS)	Bunaharitake, kind of mushroom (*Mycoleptodonoides aitchisonii*) (water extraction)

Nori peptide (AKYSY) Kind of seaweed (*Porphyra yezoensis*)	Blood pressure *Clinical studies* • 91 persons with normal to high blood pressure (12 weeks) (Kajimoto et al. 2004a, 2004b)	Shirako Co. Ltd., Japan • Mainichi Kaisai Nori Peptide (powder)	This is suitable for persons with slightly elevated blood pressure. (FS)	Nori, kind of seaweed (*P. yezoensis*)
Wakame peptide (FY, VY, IY) Kind of seaweed (*Undaria pinnatifida*)	Blood pressure *Clinical studies* • 54 mild hypertensive subjects (8 weeks) (Kajimoto et al. 2002)	Riken Vitamin Co. Ltd., Japan • Wakame Peptide Jelly (jelly)	This is suitable for persons with slightly elevated blood pressure. (FS)	Wakame, kind of seaweed (*U. pinnatifida*) (enzyme degradation)
Royal jelly peptide (VY, IY, IVY)	Blood pressure *Clinical studies* • 85 persons with normal-high blood pressure and mild hypertension (12 weeks) (Kajimoto et al. 2005) • 87 persons with normal-high blood pressure and mild hypertension (8 weeks) (Kajimoto et al. 2004a, 2004b)	Api Co. Ltd., Japan • StayBalance RJ (soft drink)	This is suitable for persons with slightly elevated blood pressure. (FS)	Royal jelly (enzyme degradation)

(continued)

Table 14.1 (Continued) FOSHU Products with Bioactive Peptides and Hydrolysates Concerning Blood Pressure

General Name (Peptide Sequence)	Function and Clinical Studies	Manufacturers and Product Name (Form)	Health Claims (Country)	Source (Product)
Casein phosphopeptide	Calcium absorption *Clinical studies* • 35 healthy postmenopausal women (single) (Hearney et al. 1994)	Suntory Beverage & Food Ltd., Japan • Tekkotsu Inryou (soft drink) Asahi Soft Drinks Co. Ltd., Japan • Kotsu Kotsu calcium (soft drink)	This product, formulated with CPP and designed to improve absorption of calcium, is suitable for supplementing calcium intake, which tends to be insufficient in normal diets. (FS)	Milk casein (enzyme degradation)
Globin protein hydrolysate (VVYP)	Body fat *Clinical studies* • 12 healthy volunteers (single) (Kamei 2001) • 7 healthy subjects and 6 subjects with normal-high blood triglyceride level (single) (Inagaki et al. 2002)	Morinaga & Co. Ltd., Japan • Seishou-sabou (soft drink)	Product support to improve the dietary life of those whose diets are rich in fat. (FS)	Globin protein derived from red blood cell of bovine or swine (enzyme method)
Soy peptides bound to phospholipids (CSPHP)	Cholesterol *Clinical studies* • 33 male subjects with hypercholesterolemia (8 weeks) (Yamamoto 1998)	Kyowa Hakko, Japan • Remake CholesteBlock (drink powder)	This is suitable for persons with a slightly elevated serum cholesterol level. (FS)	Soy protein, lecithin (enzyme method)

Table 14.2   Products with Bioactive Peptides and Hydrolysates from Japan

General Name (Peptide Sequence)	Function and Clinical Studies	Manufacturers and Product Name (Form)
Lacto tripeptide (VPP, IPP)	Blood pressure *Clinical studies* • 94 hypertensive volunteers (10 weeks) (Jauhiainen et al. 2005) • 89 hypertensive volunteers (24 weeks) (unpublished)	Valio Oy, Finland, Evolus (fermented milk)
Casein dodecapeptide (FFVAPFPEVFGK)	Blood pressure *Clinical studies* • 18 mild hypertensive subjects (4 weeks) (Sekiya et al. 1992) • 10 hypertensive subjects (6 days) (Townsend et al. 2004)	DMV International, the Netherlands, C12 Peption (ingredient)
β-Lactoglobulin fragments	Blood pressure *Clinical studies* • 30 subjects in prehypertensive stage (6 weeks) (Pins 2006)	Davisco, United States, BioZate (hydrolyzed whey protein isolate)

(continued)

Table 14.2 (Continued) Products with Bioactive Peptides and Hydrolysates from Japan

General Name (Peptide Sequence)	Function and Clinical Studies	Manufacturers and Product Name (Form)
κ-Casein f (106–169) (glycomacropeptide)	Anticarcinogenic, antimicrobial, and antithrombotic Blood pressure and blood lipid *Clinical studies* • 30 prehypertensive or stage-1 hypertensive subjects (6 weeks) (Pins et al. 2006)	Davisco, United States, BioPURE-GMP (whey protein isolate)
αs1-Casein f (1–9) αs1-Casein f (1–6) αs1-Casein f (1–7)	Blood pressure (in vitro)	MTT Agrifood Research, Finland, Festivo (fermented low-fat hard cheese)
Casein phosphopeptide	Calcium absorption *Clinical studies* • 35 healthy postmenopausal women (single) (Hearney et al. 1994)	DMV International, the Netherlands, CE90CPP (ingredient)
Casein phosphopeptide	Calcium absorption *Clinical studies* • 28 8-year-old boys (7 days) (Budek et al. 2007)	Arla Foods Ingredients, Sweden, Capolac (ingredient)

Casein-derived dipeptides and tripeptides	Improves athletic performance and muscle recovery after exercise *Clinical studies* • 14 male athletes (single) (Kaastra et al. 2006)	DSM Food Specialties, the Netherlands, PeptoPro (flavored drink)
Milk protein–derived peptide	Aids sleep	DMV International, the Netherlands, Cysteine peptide (ingredient)
Glutamin-rich peptides	Immunomodulatory	DMV International, the Netherlands, Glutamin peptide (dry milk protein hydrolysate)
α-Lactalbumin-rich whey protein hydrolysate	Aids relaxation and sleep *Clinical studies* • 34 moderate-to-heavy social drinkers (single) (Nesic and Duka 2007)	Borculo Domo Ingredients (BDI), the Netherlands, Vivinal Alpha (ingredient)
Lactoferrin-enriched whey protein hydrolysate	Reduce acne *Ex vivo* • Healthy volunteers (topical, single) (Cumberbatch et al. 2003)	DMV International, the Netherlands, Praventin (capsule)
αs1-Casein f (91–100) (YLGYLEQLLR)	Relief of stress symptoms	Ingredia, France, ProDiet F200/ Lactium (flavored milk drink, confectionery, and capsules)

and tablet-type foods were also approved as FOSHU and are in the Japanese markets.

The sardine peptides were derived from the alkaline muscle protein hydrolysate. It was shown in vitro that more than 10 bioactive peptides were isolated from the sardine hydrolysate. Val-Tyr was the most effective compound among them, which showed antihypertensive action. It was demonstrated that food containing this peptide reduced the blood pressure in normal-high and mildly hypertensive subjects (Kajimoto et al. 2003). The tablet-type food containing the sardine peptide was approved as FOSHU in 1999 and approved beverages are also on the market.

The sesame peptides from the thermolysin hydrolysate of the sesame protein were shown to contain Met-Leu-Pro-Ala-Tyr, Val-Leu-Tyr-Arg-Asp-Gly, and Ile-Val-Tyr as antihypertensive compounds. The beverage containing the sesame peptides exhibited antihypertensive action in subjects with high-normal and mild hypertension (Nakano et al. 2006). The beverage containing the sesame peptides was approved as FOSHU in 2005 and is on the market.

It was reported that peptides, other than those previously mentioned, showing antihypertensive results are the dodeca peptides from casein, seaweed peptides such as nori or wakame, mushroom peptides derived from *Mycoleptodonoides*, dried bonito peptides, and soy protein hydrolysates. The products containing these components can be claimed as suitable for people with slightly elevated blood pressure. The Japanese market sales of these FOSHU products containing bioactive peptides that reduce blood pressure have gradually increased since 1997 and are expected to amount to about 15 trillion Yen in 2009 (Figure 14.1).

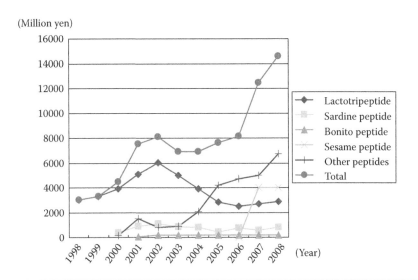

Figure 14.1   Sales amount of FOSHU peptides.

Absorption of mineral: The CPP derived from casein was shown to improve the absorption of calcium by preventing the precipitation of Ca cations and significantly increasing the luminal soluble calcium. The beverage containing CPP was approved as FOSHU in 1997 and is now in the market. The approved products containing CPP can claim that this product helps calcium absorption and supplements calcium to make the bones strong.

Heme-iron derived from bovine blood hemoglobin, which was hydrolyzed with endoproteinase, was shown to have a high bioavailability of iron. The approved product claims containing heme-iron from hemoglobin can state that this product is suitable for people with mild anemia and those who need iron supplementation.

In the Japanese market, the sales of these FOSHU products containing the bioactive peptides related to mineral absorption were estimated at about 1 trillion Yen in 2007.

Blood lipid: Globin hydrolysate could decrease the blood lipid levels after meals. The approved products containing this component can claim that this product helps reduce postprandial serum triglyceride levels and these products are suitable for persons with slightly elevated serum cholesterol level. This product also supports improving the dietary lifestyle of those eating a fat-rich diet. In the Japanese market, the sales of these FOSHU products containing the bioactive peptide for blood lipid control were estimated at 1 trillion Yen in 2007.

14.6 Conclusion

The food-derived peptides have a variety of bioactive functionalities and exist as various kinds of products in the market. The scientific support for most products using in vitro and animal studies is not sufficient in promoting the health claims that benefit human health. The health claim on the product is one of the most important ways for the manufacturers to promote their developed foods with bioactive peptides in the market. The consumers' interest in health issues has become a leading factor in their purchasing decisions. The manufacturers can emphasize the benefit of their products and promote their sales by using health claims. Therefore, the health claims on foods should always be based on scientific evidence. In 1991, Japan's MHLW set up FOSHU as a regulatory system to approve the statements of health claims and foods.

It is desirable for the national governments and international organizations representing both the consumers and the industries to cooperate with Codex in establishing international standards and a regulatory system similar to FOSHU for approving health-related claims on foods as well as promoting the research and development of health foods for world markets and improving the quality of life.

References

Budek, A., Hoppe, C., Michaelsen, K. F., and Mølgaard, C. 2007. High intake of milk, but not meat, decreases bone turnover in prepubertal boys. *European Journal of Clinical Nutrition* 61: 957–62.

Codex. 2008. http://www.codexalimentarius.net/web/archives.jsp?lang=en.

Cumberbatch, M., Bhushan, M., Dearman, R. J., Kimber, I., and Griffiths, C. E. 2003. IL-1 beta-induced Langerhans' cell migration and TNF-alpha production in human skin. *Clinical and Experimental Immunology* 132: 352–59.

European Food Safety Authority. 2008. http://www.efsa.europa.eu/EFSA/efsa_loc ale-1178620753812_1178621456747.htm.

European Union. 2000. http://ec.europa.eu/dgs/health_consumer/library/pub/pub06_en.pdf.

Fujita, H. et al. 2001. Effects of an ace-inhibitiory agent, katsuobushi oligopeptide, in the spontaneously hypertensive rat and in borderline and mildly hypertensive subjects. *Nutrition Research* 21: 1149–58.

Gil, H. S., Doull, F., Rutherfurd, K., and Cross, M. 2000. Immunoregulatory peptides in bovine milk. *British Journal of Nutrition* 84 (Suppl. 1): S111–17.

Hartman, R. and Meisel, H. 2007. Food-derived peptides with biological activity: From research to food applications. *Current Opinion in Biotechnology* 18: 163–69.

Hata, Y., Yamamoto, M., Ohni, M., Nkajima, K., Nakamura, Y., and Takano, T. 1996. A placebo-controlled study of the effect of sour milk on blood pressure in hypertensive subjects. *American Journal of Clinical Nutrition* 64: 767–71.

Heaney, R. et al. 1994. The effect of calcium absorption enhancement of Caseinophospho-peptide. *J Bone Miner Met* 12: 77–81.

Inagaki, H., Fujita, H., Arakawa, K., Sugiyama, K., Hashizume, S., and Okuda, H. 2002. Suppressive effect of tea beverage including globin digest on postprandial serum tri-glyceride elevation. *Journal of Nutritional Food* 5: 131–44.

Itakura, H., Nakamura, Y., and Takano, T. 2001. The effect of sour milk on blood pres-sure in untreated hypertensive and normotensive subjects. *Journal of Japanese Society of Clinical Nutrition* 23: 26–31.

Jauhiainen, T., Vapaatalo, H., Poussa, T., Kyrönpalo, S., Rasmussen, M., and Korpela, R. 2005. Lactobacillus helveticus fermented milk lowers blood pressure in hyperten-sive subjects in 24-h ambulatory blood pressure measurement. *American Journal of Hypertension* 18: 1600–5.

Kaastra, B., Manders, J., Van Breda, E., Kies, A., Jeukendrup, E., Keizer, A., et al. 2006. Effects of increasing insulin secretion on acute postexercise blood glucose disposal. *Medicine & Science in Sports & Exercise* 38: 268–75.

Kajimoto, O., Aihara, K., Hirata, H., and Nakamura, Y. 2001a. Hypotensive effects of the tablets containing "lactotripeptides". *Journal of Nutritional Food* 4: 51–61.

Kajimoto, O., Maruyama, H., Tokunaga, K., and Hirata, H. 2004a. Antihypertensive effect of protease-treated royal jelly in subjects with high-normal or mild hypertension. *Journal of Nutritional Food* 7: 53–71.

Kajimoto, O., Maruyama, H., Yoshida, T., Araki, Y., Mishima, T., Sakamoto, A., et al. 2005. Antihypertensive effects and safety of drink containing protease-treated royal jelly in subjects with high-normal or mild hypertension. *Health Science* 21: 229–45.

Kajimoto, O., Nakamura, Y., Yada, H., Moriguchi, S., Hirata, H., and Takahashi, T. 2001b. Hypotensive effects of sour milk in subjects with mild or moderate hypertension. *Journal of Japanese Society of Nutrition and Food Science* 54: 347–54.

Kajimoto, O., Nakano, T., Kato, T., and Takahashi, T. 2002. Hypotensive effects of jelly containing Wakame peptides on mild hypertensive subjects. *Journal of Nutritional Food* 5: 67–81.

Kajimoto, O., Seki, E., Osajima, K., and Kawasaki, K. 2003. Hypotensive effects of food containing peptide from sardine muscle hydrolyzate, in subjects with high-normal, or hypertension. *Journal of Nutritional Food* 6: 65–82.

Kajimoto, O., Sito, M., and Ogino, K. 2004b. The effect of oligo-peptides derived from Nori (Porphyra yezoensis) on high-normal or mild hypertensive subjects. *Journal of Nutritional Food* 7: 43–58.

Kamei, M. 2001. The effect of globin digest on serum triglyceride of human subjects. *Journal of Nutritional Food* 4: 1–11.

Kawasaki, T., Seki, E., Osajima, K., Yoshida, M., Asada, K., Matsui, T., et al. 2000. Antihypertensive effect of Valyl-Tyrosine, a short chain peptide derived from sardine muscle hydrolyzate, on mild hypertensive subjects. *Journal of Human Hypertension* 14: 519–23.

Korhonen, H. 2009. Bioactive milk protein and peptides: From science to functional application. *Australian Journal of Dairy Technology* 64(1): 16–25.

Korhonen, H. and Pihlanto, A. 2006. Bioactive peptides: Production and functionality. *International Dairy Journal* 16: 945–60.

Markus, C. R., Jonkman, L. M., Lammers, J. H. C., Deutz, N. E. P., Messer, M. H., and Rigtering, N. 2005. Evening intake of alpha-lactalbumin increases plasma tryptophan availability and improves morning alertness and brain measures of attention. *American Journal of Clinical Nutrition* 81: 1026–33.

Mashiko, K., Hiratsuka, H., Itagaki, Y., Samejima, K., and Sato, T. 2003. Anti hypertensive effect of an aqueous extract from *Mycoleptodonoides aitchisonii* in high-normal and mild hypertensive group. *Jpn Pharmacol Ther* 31: 239–46.

Mizuno, S., Matsuura, K., Gotou, T., Nishimura, S., Kajimoto, O., Yabune, M., et al. 2005. Antihypertensive effect of casein hydrolysate in a placebo-controlled study in subjects with high-normal blood pressure and mild hypertension. *British Journal of Nutrition* 94: 84–91.

Moriguchi, S., Iino, T., Kusumoto, A., Kusumoto, A., Shibata, H., Ohta, H., et al. 2006. Effect of the tea beverage added with peptides derived from sesame protein hydrolysates on blood pressure. *Journal of Nutritional Food* 9: 1–14.

Moriguchi, S., Iino, T., Kusumoto, A., Shibata, H., Ohta, H., Kiso, Y., et al. 2004. Effect and safety of the tea beverage added with peptides derived from sesame protein hydrolysates on blood pressure in high normal blood pressure and mild hypertensive subjects. *Journal of Nutritional Food* 7: 49–64.

Nakamura, Y., Yamamoto, N., Sakai, K., Okubo, A., Yamazaki, S., and Takano, T. 1995. Purification and characterization of angiotensin I: Converting enzyme inhibitors from a sour milk. *Journal of Nutritional Food* 7(1): 123–27.

Nakano, D. et al. 2006. Antihypertensive effect of angiotensin I-converting enzyme inhibitory peptides from a sesame protein hydrolysate. *Biosci. Biotechnol. Biochem.* 70(5): 1118–26.

Nesic, J. and Duka, T. 2007. Effects of stress on emotional reactivity in hostile heavy social drinkers following dietary tryptophan enhancement. http://alcalc.oxfordjournals.org/cgi/content/full/agm179v1.

Pins, J. 2006. Clinical study of lactoglobulin fragment on prehypertensive stage. *Journal of Clinical Hypertension* (Greenwich) 10: 775–82.

Sato, T., Shimada, T., Shimowada, K., Kanematsu, T., and Mashiko, K. 2004. Antihypertensive effect of an aqueous extract powder from Mycoleptodonoides *aitchisonii* in high-normal and mild hypertensive group. *Japanese Journal of Pharmacology and Therapeutics* 32: 761–71.

Seki, E., Asada, K., Osajima, K., Matsui, T., Osajima, Y., and Kawasaki, T. 2000. Antihypertensive effect of peptide derived from sardine muscle hydrolyzate on mild hypertensive and high normal blood pressure subjects. *Journal of Nutritional Food* 3: 1–13.

Sekiya, S., Kobayashi, Y., Kita, E., Imamura, Y., and Toyama, E. 1992. Antihypertensive effects of tryptic hydrolysate of casein on normotensive and hypertensive volunteers. *Journal of Japanese Society of Nutrition and Food Science* 45: 513–17.

Shimizu, T. 2003. Health claims on functional foods: The Japanese regulations and an international comparison. *Nutrition Research Reviews* 16: 241–52.

Sugano, M., Yamada, Y., Yoshida, K., Hashimoto, Y., Matsuo, T., and Kimoto, M. 1988. The hypocholesterolemic action of the undigested fraction of soybean protein in rats. *Atherosclerosis* 72: 115–22.

Takahashi, Y., Ide, T., and Fujita, H. 2001. Dietary gamma-linolenic acid in the form of borage oil causes less body fat accumulation accompanying an increase in uncoupling protein 1 mRNA level in brown adipose tissue. *Nutrition Research* 21: 1149–58.

Townsend, R. R., McFadden, C. B., Ford, V., and Cadée, J. A. 2004. A randomized, double-blind, placebo-controlled trial of casein protein hydrolysate (C12 peptide) in human essential hypertension. *American Journal of Hypertension* 17: 1056–58.

Tsuchida, T., Mashiko, K., Nagata, H., and Sato, T. 2001. Effect of an aqueous extract from Mycoleptodonoides *aitchisonii* on human blood pressure. *Japanese Journal of Pharmacology and Therapeutics* 29: 899–906.

Tsuchida, T., Mashiko, K., Nagata, H., and Sato, T. 2002. Effect and safety of an aqueous extract from Mycoleptodonoides *aitchisonii* in the long term administration. *Japanese Journal of Pharmacology and Therapeutics* 30: 31–36.

Xu, J.-Y., Qin, L.-Q., Wand, P.-Y., Li, W., and Hang, C. 2008. Effect of milk tripeptides on blood pressure: A meta-analysis of randomized controlled trials. *Nutrition* 24: 933–40.

Yamamoto, S. 1998. The hypocholesterolemic effect of soybean peptides on human hypertention. *Journal of Nutritional Food* 3/4: 51–58.

Zioudrou, C., Streaty, A., and Klee, A. 1979. Opioid peptides derived from food proteins. *Journal of Biological Chemistry* 254: 2446–49.

Chapter 15

Large-Scale Fractionation of Biopeptides

Kenji Sato and Kaori Hashimoto

Contents

15.1 Introduction

Over the last few decades, it has been demonstrated that oral ingestion of peptide mixtures, which can be produced from food proteins by enzymatic and chemical hydrolysis or fermentation, can improve the human health beyond their inherent nutritional values (Arai 1996; Korhonen and Pihanto 2003; Meisei 2004; Möller et al. 2008), as also reviewed in Book II. Based on these facts, some peptide preparations with beneficial activities have been prepared on an industrial scale and are commercially available worldwide. For elucidation of their mode of action, the isolation of the active peptides from such peptide mixtures is a crucial step. In addition, the identification of peptides is necessary for the quality control of peptide-based products.

Without doubt, high-performance liquid chromatography (HPLC) has been used most frequently for the laboratory-scale fractionation and isolation of peptides. In particular, reversed-phase (RP)-HPLC is involved in most peptide isolation procedures. Ion-exchange chromatography and size-exclusion (also called gel filtration or gel permeation) chromatography are also preferably used. By optimizing the elution conditions and using a combination of these chromatographic modes, the peptides of interest can be isolated from food protein hydrolysates with high resolution at microgram and milligram levels (see Chapter 2). Coupled with in vitro assay systems, the "active" peptides have been isolated and identified from food protein hydrolysates. However, unlike other functional food ingredients, food-derived peptides may be rapidly and extensively degraded by peptidases during digestion, absorption, and circulation in animal and human bodies. Therefore, the peptide with in vitro activity may be degraded and may lose its apparent biological activity during the digestion and absorption processes. On the other hand, some peptides with no in vitro activity might be converted to their active forms by partial digestion in the body. Hence, the activity of a peptide for use as a food ingredient should be evaluated by feeding experiments. For the evaluation of peptide fractions by feeding experiments, relatively high amounts (gram or kilogram orders) of peptides are necessary for even small-scale animal and/or human trials. However, the high initial and production costs of large-scale preparative liquid chromatography systems have hampered the preparation of large amounts of peptide fractions for feeding experiments to identify the real active peptides by ingestion. In addition, some solvents and chemicals frequently used for RP liquid chromatography, such as acetonitorile, methanol, and trifluoroacetic acid (TFA), are harmful to animals and humans. It is necessary to completely remove these toxic substances from the final peptide preparations before the trial. Even nontoxic substances, such as ethanol and salts, should be removed before the feeding experiments, which increases the cost of these experiments. Alternatively, filtration and selective precipitation techniques have been used for both small- and large-scale peptide fractionations. These techniques are generally used to remove the undigested proteins and proteases from the oligopeptide fractions (see Chapter 2). Thus, crude protein hydrolysates or peptide fractions prepared by these low selective methods have been used for feeding trials. These crude preparations are still a mixture of numerous peptides. Thus, these trials can only confirm the efficacy of the peptide preparations by ingestion, but cannot identify the real active peptides. In some cases, peptides with in vitro activity have been chemically synthesized and used in feeding experiments, demonstrating that some single peptides show significant activity by ingestion (Kagawa et al. 1996; Miguel and Aleixandre 2006). However, more potent peptides may be overlooked by this approach, as the candidate peptide for the feeding trial has been selected by in vitro activity-guided fractionation without considering the bioavailability. To overcome this situation, a large-scale, biocompatible, and low-cost fractionation method for the peptides from the proteolytic digest of a food protein is required, which facilitates the identification of the active peptide by ingestion.

Even after identifying the active peptide by in vivo activity-guided fractionation, a crude peptide mixture has been used for nutraceutical and functional food ingredients due to the lack of suitable industrial fractionation procedures for peptides (Korhonen and Pihlanto 2007). Relatively large doses of the peptide mixture are necessary to exert the beneficial activity. These crude peptide preparations frequently show bitter and odd tastes. The lack of a suitable industrial peptide fractionation has hindered the commercialization of peptide-based nutraceuticals and functional foods.

In this chapter, a new approach for the large-scale fractionation of peptides based on their amphoteric nature is introduced. Some application of this approach for the identification of the active peptides by ingestion is also described. In addition, the potential of this approach for industrial application is discussed.

15.2 Development of Large-Scale Peptide Fractionation Method on the Basis of Diversity of Isoelectric Points of Peptides

As described previously, liquid chromatography has been used for separating peptides and proteins with high resolution. In some cases, liquid chromatography has been used for the industrial fractionation of peptides to prepare functional food ingredients (Osajima et al. 1993; Tanabe, Arai, and Watanabe 1999; Kitts and Weiler 2003). To process particle-containing feedstock, the expanded-bed chromatography has been developed. This technique enables proteins and peptides to be recovered directly from the crude extract without the removal of the suspended solids (Anspach et al. 1999). However, liquid chromatography systems have an inherent disadvantage in their scale-up cost. Membrane technology has been successfully used for the industrial fractionation of peptides. However, this technique cannot fractionate the peptides with similar molecular mass. Alternatively, electrophoresis has been used to fractionate proteins and peptides on the basis of their differences in the molecular masses, charges, and isoelectric points. It is difficult to scale up gel electrophoresis based on its sieving effect. On the other hand, isoelectric focusing does not require a gel matrix with sieving effects and can be used for the resolution of small peptides. For conventional isoelectric focusing, soluble and hydrophilic amphoteric polymers with different isoelectric points have been developed (Rilbe 1976), which are referred to as ampholines and can be obtained commercially as Ampholyte and BioLyte. By applying a direct electric current on the ampholine solution through an anode and a cathode filled with acid and base solutions, the ampholine molecules start to migrate to their isoelectric points and stabilize the pH. Thus, a pH gradient can be formed. By using the pH gradient, the peptides and proteins migrate to their own isoelectric points. The isoelectric focusing is frequently used for the resolution of small amounts of proteins and peptides in a gel matrix without the sieving effect, such as agarose gel, for the first resolution in a two-dimensional electrophoresis system. However, this technique can be used for preparative purposes. A preparative isoelectric-focusing apparatus, which can process a 50–100 mL sample, is commercially available from Bio-Rad Laboratories (Rotofor cell) and Nihon Eido (NA-1720). When the peptide is small and hydrophilic, it is difficult to separate it from the chemically synthesized ampholine molecules. Therefore, preparative isoelectric focusing has been generally limited to applications for fractionating large molecular peptides and proteins.

In the early studies on the development of isoelectric focusing, protein hydrolysates, namely peptide mixtures, were used as ampholines (Haglund 1967). However, it was not acceptable to add peptides to purify proteins. Subsequently, synthetic ampholines were developed. Sato and coworkers have noticed the amphoteric nature of the peptides and have assumed that the fractionation of the peptides might occur by just applying a direct electric current on the aqueous solution of a peptide mixture (Yata et al. 1997; Akahoshi et al. 2000; hashimoto et al. 2005). On the basis of this assumption, aqueous solutions of some enzymatic hydrolysates from food proteins were applied to a laboratory-scale (50 mL sample cell volume) preparative isoelectric-focusing apparatus (Rotofor, Bio-Rad Laboratories). As shown in Figure 15.1, after 2 h of focusing, a pH gradient was formed and each fraction consisted of different

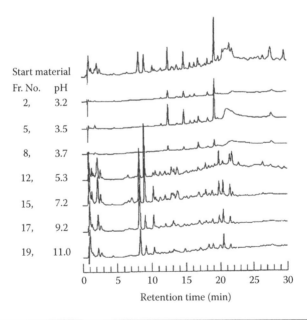

Figure 15.1 **The formation of a pH gradient and the resolution of peptides in a tryptic digest of hen-egg lysozyme by autofocusing using a laboratory-scale preparative isoelectric-focusing apparatus (Rotofor, Bio-Rad Laboratories). See the paper by Yata, M., et al.,** *Journal of Agricultural and Food Chemistry,* **44, 76–79, 1996 for experimental details.**

peptides, which clearly indicated that peptide fractionation occurs by applying a direct electric current to the peptide solutions (Yata et al. 1996). The pH gradient profile developed by autofocusing differs not only by the origin of the sample but also by the sample concentration. However, if fractions with a similar pH are collected, reproducible fractionation can be obtained (Yata et al. 1996).

This technique requires only water as solvent and requires a base (0.1 N NaOH) and an acid (0.1 N phosphate), for the electrode solutions. All the chemicals involved in this technique can be used for food processing, and high pressure is not generated during fractionation. This approach, therefore, has the potential for further scale-up and has the advantages of cost and biocompatibility over the conventional liquid chromatographic technique. We have proposed referring to this technique as autofocusing and believe that it is useful as a first step in in vivo activity-guided fractionation.

15.3 Design and Testing of Large-Scale Fractionator

For fractionating a peptide in sufficient amounts for animal and human trials, a large-scale peptide fractionator based on autofocusing has been developed (Akahoshi et al. 2000; Hashimoto et al. 2005). The most critical point for the scale-up is suppression of the diffusion of the sample in the large focusing chamber (Haglund 1967). In the established laboratory-scale preparative fractionators, the diffusion of the sample is suppressed by a nylon mesh screen (Egen et al. 1988) and/or a glycerin or sucrose gravity gradient (Haglund 1967; Rilbe 1976; Acevedo 1993). In the large-scale chamber, it is difficult to form a stable gravity gradient. In addition, the additives necessary to form a gravity gradient should be removed before an in vivo evaluation. The nylon screen (100 mesh) was first used to suppress the diffusion in our prototype apparatus consisting of 500 mL sample cells. The nylon screen was fixed in the center of the joint tubes connecting the sample cells, as shown in Figure 15.2. However, this type of apparatus achieved no significant resolution of the peptide. Extensive improvement of the resolution was achieved by forming a thin agarose gel layer on the nylon screen. The thin agarose gel layer (less than 1 mm) can be prepared by wetting the nylon screen with 1% hot agarose solution and standing it in air for a few minutes (Figure 15.2c). The cell assembly is illustrated in Figure 15.3. An RP-HPLC analysis revealed that the fractionation of peptides in a casein hydrolysate occurred by introducing the agarose gel layer between the sample chambers, as shown in Figure 15.3 (Akahoshi et al. 2000). This apparatus, which can process up to a 5 L sample, gives a resolution of the peptides corresponding to that of the small-scale apparatus (50 mL). The time necessary for the fractionation is proportional to the sample volume and inversely proportional to the electric power, which suggests that large-scale fractionation can be achieved in a shorter time by just applying high electric power. On the basis of this finding, a larger focusing apparatus, which can process a sample up to 50 L, has been developed (Hashimoto et al. 2005). The schematic drawing of the 50 L apparatus

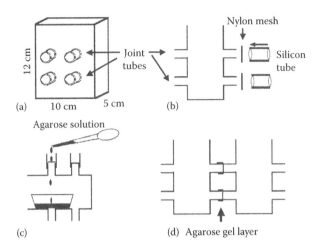

Figure 15.2   A schematic drawing of a sample cell for large-scale autofocusing. (a) Aerial view of the sample cell. The sample cell has four joint tubes on two sides. (b) Fixing the nylon mesh on the end of the joint tubes by inserting a silicon tube. (c) Wetting the nylon mesh with a hot agarose solution. After a few minutes, the agarose gel is formed, supported on the mesh. (d) Connecting the two sample cells by silicon tubes. The agarose gel layers are formed between the two sample cells. (From Akahoshi, A., et al., *Journal of Agricultural and Food Chemistry*, 48, 1955–59, 2000.)

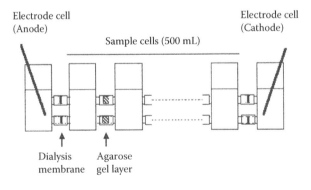

Figure 15.3   A schematic drawing of the cell assembly for large-scale autofocusing. Agarose gel layers are formed between the sample cells. The electrode cell and the sample cell are partitioned by a dialysis membrane. (From Akahoshi, A., et al., *Journal of Agricultural and Food Chemistry*, 48, 1955–59, 2000.)

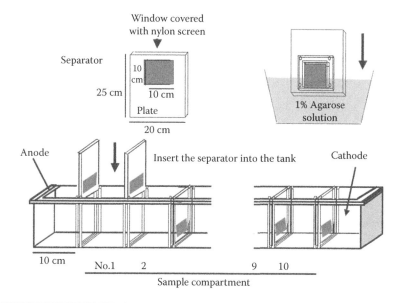

Figure 15.4 A schematic drawing of a 50 L scale autofocusing apparatus. An acryl plate with a window is prepared. The window is covered with a nylon screen. The nylon screen is wetted with a hot agarose solution and is left to stand for few minutes in air to allow gel formation. The plate with the agarose gel layer supported on the screen is referred to as a separator. The separators are inserted into each slit in the tank. Then the tank is partitioned into 12 compartments. Both compartment ends are used as electrode compartments while all others are sample compartments. (From Hashimoto, K., et al., *Journal of Agricultural and Food Chemistry*, 53, 3801–6, 2005.)

is shown in Figure 15.4. A tank is partitioned into 10 sample compartments and 2 electrode compartments by separators with a thin agarose gel layer supported by a nylon screen. As shown in Figure 15.5, the peptides in a casein hydrolysate (500 g) can be fractionated on the basis of their isoelectric points by using this apparatus (Hashimoto et al. 2005).

Two types of sample application methods can be selected for the batch-type apparatus. In method A, a diluted (approximately 1%) sample is applied to all the sample compartments. In method B, a concentrated (approximate 5%) sample is applied to some compartments in the center. The other sample compartments are filled with deionized water. In general, method B gives a better resolution of the peptides than method A. The substance with no electric charge remains in its original position. Thus, the noncharged sample cannot be resolved by method A, but it can be separated from the acidic and basic substances by method B. On the other hand, in method A, a significant concentration of peptides having an electric charge occurs after focusing (Hashimoto et al. 2005).

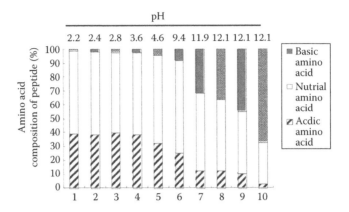

Figure 15.5 The fractionation of peptides in a commercial casein hydrolysate by the 50 L batch-type autofocusing apparatus, as illustrated in Figure 15.4. Hydrolysate (500 g) was applied into sample compartments Nos. 5 and 6 (method B). After 24 h, aliquots from the sample compartments were hydrolyzed and subjected to amino acid analysis. The peptides in the acidic and basic fractions are rich in acidic and basic amino acids, respectively. (From Hashimoto, K., et al., *Journal of Agricultural and Food Chemistry,* **53, 3801–6, 2005.)**

As demonstrated in the following sections, this apparatus enables in vivo activity-guided fractionation for identifying the active peptides using animal models. In addition, it is worth mentioning that this technique requires no harmful chemicals for fractionation. The autofocusing fraction can be evaluated by human trials not only for its potential biological activity, but also for sensory attributes such as taste, flavor, aroma, and other physicochemical functions, such as the emulsifying and foaming activities in the food systems.

If the sample contains a relatively high concentration of salts (approximately more than 1%), a low electric voltage is generated by applying a relatively high electric power. In such a case, the resolution of peptides is generally not as good. To solve this problem, the sample can be desalted by electrodialysis using small pore ion exchange membranes. Alternatively, an ion exchange membrane can be placed between the electrode and the sample compartments, which can improve the resolution by suppressing the migration of the inorganic ions into the electrode chambers. A further problem with this apparatus is caused by the electro-osmosis current, which is probably caused by the negatively charged agarose. During focusing, the sample volume in each sample compartment may change due to the electro-osmosis current. The extent of the electro-osmosis current depends on the sample peptide. To solve this problem, further studies on optimizing the membrane material and improving the apparatus are necessary.

15.4 Identification of Biopeptides After Large-Scale Fractionation

The autofocusing apparatus has been successively used to identify the active peptides in animal experiments. It has been demonstrated that oral ingestion of a protease digest of shark cartilage at 1 g/kg body weight can decrease the serum uric acid level in an oxonate-induced rat model (Murota et al. 2010). To detect the antihyperuricemic activity with an in vitro experiment, the inhibitory activity against xanthine oxidase, a key enzyme for uric acid synthesis, was checked. However, the digest has no inhibitory activity against xanthine oxidase. Subsequently, an in vivo activity-guided fractionation based on autofocusing and animal experiments was carried out. A water extract of shark cartilage was subjected to autofocusing using an apparatus consisting of 10 sample cells (500 mL). Acidic, weak acidic, and basic fractions were obtained. The protease digest of a basic fraction (pI more than 7) showed a significant uric acid–lowering effect in smaller dose (100 mg/kg body weight) than the crude digest (1 g/kg body weight). The basic fraction was further fractionated by preparative RP chromatography using a preparative grade resin with a diameter of 50 μm. The protease digest of the hydrophobic fraction showed significant activity. The peptides in the active fraction were resolved by a series of size exclusion and RP-HPLCs and were identified by a peptide sequencer based on the Edman degradation. Based on the sequence information, some peptides in the active fraction were synthesized and evaluated for their uric acid–lowering effect in the animal model. Consequently, Tyr-Leu-Asp-Asn-Tyr was identified as the uric acid–lowering peptide by ingestion. This is one of the first cases of identifying an active peptide by in vivo activity-guided fractionation.

It is difficult to isolate a single peptide from a highly complex mixture of peptides after the enzymatic digestion of a food protein for the feeding experiments by autofocusing. However, one can prepare the active fraction, which may consist of fewer numbers of peptides than the original digest. When additional fractionation using preparative liquid chromatography is necessary to obtain the active fraction, autofocusing would be useful in concentrating the target peptide for subsequent liquid chromatographic separation, especially in the RP mode, due to the differences in the modes of action. Modern HPLC systems and sequence tools allow the isolation and identification of the peptides in the active fraction. Based on the sequence data, the candidate peptides in the active fraction can be chemically synthesized. By using the synthetic peptide, the active peptide could be identified by feeding experiments, as demonstrated here. This approach enables the identification of the active peptide by ingestion from a highly complex mixture. This approach has great advantages in identifying the active component, which is susceptible to digestion, over the conventional in vitro activity-guided fractionation.

15.5 Industrial Standpoint

To demonstrate the potential of autofocusing for the application in food processing, a continuous-type fractionator has been developed (Hashimoto et al. 2006). As shown in Figure 15.6, the drain tube is fixed through the bottom of each compartment. The tank, consisting of 12 compartments with drain tubes, is referred to as an autofocusing unit. Three sets of autofocusing units are prepared and arranged in tandem, as shown in Figure 15.6. The sample is delivered to the center of the two sample compartments. Deionized water is delivered to the other sample compartments of the first unit. The acid (0.1 N phosphate) and base (0.1 N NaOH) solutions are delivered, respectively, to the electrode compartments at both ends. The solutions drained from the first unit are successively delivered into the corresponding compartments of the second and third units. A direct electric current is applied to the electrode compartments of all the autofocusing units. In the first unit, no significant resolution of the peptides is observed. A better resolution can be obtained in the second and third units (Figure 15.7). By collecting the effluents from the third unit, the fractionation of the peptides can be obtained. Hashimoto et al. (2006) demonstrated that the peptide fractionation of a casein hydrolysate continues for at least 1–6 h with a resolution corresponding to that obtained by using the batch system apparatus after 24 h. After 8 h, the resolution of the peptides became poorer in comparison to that before 6 h. Part of the sample peptides

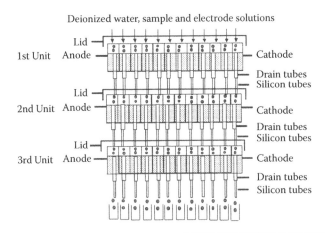

Figure 15.6 Schematic drawing of the assembly of a continuous-type autofocusing apparatus. Three sets of autofocusing units are arranged in tandem. Each compartment has a drain tube to transfer the sample from one unit to the corresponding compartment of the next unit. The sample, the deionized water, and the electrode solutions are delivered to the compartments of the first unit. Effluents from the third unit are collected as a sample. (From Hashimoto, K., et al., *Journal of Agricultural and Food Chemistry*, 54, 650–55, 2006.)

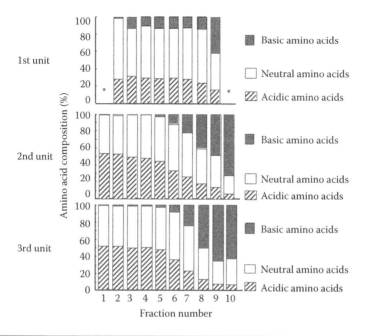

Figure 15.7 **The fractionation of the peptides in a commercial casein hydrolysate by the continuous-type autofocusing apparatus, as illustrated in Figure 15.6. Five percent of the casein hydrolysate was delivered to sample compartments Nos. 5 and 6. The aliquots of the effluents from the third unit were hydrolyzed and subjected to amino acid analysis. The peptides in the acidic and basic fractions are rich in acidic and basic amino acids, respectively. (From Hashimoto, K., et al., *Journal of Agricultural and Food Chemistry,* 54, 650–55, 2006.)**

migrated into the anode compartment after prolonged focusing. These facts suggest that deterioration of the agarose gel layer might occur after prolonged focusing. The effect of prolonged focusing on a resolution also might depend on the sample peptide and the sample loading. The mechanism of this phenomenon remains to be solved.

As summarized in Table 15.1, the currently used continuous-type apparatus shows higher performance by lower electric power in comparison with the 50 L batch type, thereby demonstrating the advantage of the continuous system over the batch system in large-scale fractionation. If the large focusing units (e.g., 50 L) could be used in the continuous system, the peptides might be processed at the kilogram per hour level. Then, the present approach might have the potential for industrial application. Crude enzymatic hydrolysis of food proteins frequently shows a bitter or odd flavor, and most of these peptides are inactive. Thus, separating the active peptide by autofocusing would improve the taste of the peptide-based product by decreasing the peptide dose. This technique coupled with the

Table 15.1 Summary of Performance of Batch- (50 L) and Continuous-Type Autofocusing Apparatuses for Fractionation of the Peptides in a Commercially Available Casein Hydrolysate

	Batch-type[a]	*Continuous-type*[b]
Time (h)	24	6
Total processed sample (g)	400	288
Performance (g/h)	16.6	57.6
Consumed electric power (W · h) to fractionate 1 g sample	20.4	1.3

[a] Hashimoto, K., et al., *Journal of Agricultural and Food Chemistry*, 53, 3801–6, 2005.
[b] Hashimoto, K., et al., *Journal of Agricultural and Food Chemistry*, 54, 650–55, 2006.

conventional fractionation procedures, for example, membrane technology and preparative liquid chromatography, would be useful in preparing food ingredients with enhanced bioactivity and lower adverse effects.

15.5 Conclusion

A peptide mixture dissolved in water can be fractionated in a preparative isoelectric-focusing apparatus based on the amphoteric nature of the sample peptides. This approach is referred to as autofocusing and has an advantage in the processing cost and biocompatibility over liquid chromatography; additionally, it can be scaled up. By using a thin agarose gel layer to prevent the diffusion of the sample, batch and continuous types of autofocusing apparatus have been developed. The batch-type autofocusing apparatus enables in vivo activity-guided fractionation for identifying digestible peptides. The continuous-type autofocusing has the potential for preparing peptide fractions with enhanced activity as functional food ingredients.

References

Acevedo, F. 1993. Isotachophoresis of proteins in sucrose density gradients. *Electrophoresis* 14: 1019–22.

Akahoshi, A., Sato, K., Nawa, N., Nakamura, Y., and Ohtsuki, K. 2000. Novel approach for large-scale, biocompatible, and low-cost fractionation of peptides in proteolytic digest of food protein based on the amphoteric nature of peptides. *Journal of Agricultural and Food Chemistry* 48: 1955–59.

Anspach, F. B., Curbelo, D., Hartmann, R., Garke, G., and Deckwer, W. D. 1999. Expanded-bed chromatography in primary protein purification. *Journal of Chromatography* 865: 129–44.

Arai, S. 1996. Studies in functional foods in Japan – State of art. *Bioscience Biotechnology & Biochemistry* 60: 9–15.

Egen, N. B., Bliss, M., Mayersohn, M., Owens, S. M., Arnold, L., and Bier, M. 1988. Isolation of monoclonal antibodies to phencyclidine from ascites fluid by preparative isoelectric focusing in the Rotofor. *Analytical Biochemistry* 172: 488–94.

Haglund, H. 1967. Isoelectric focusing in natural pH gradients – A technique of growing importance for fractionation and characterization of proteins. *Science Tools* 14: 17–23.

Hashimoto, K., Sato, K., Nakamura, Y., and Ohtsuki, K. 2005. Development of a large-scale (50 L) apparatus for ampholyte-free isoelectric focusing (Autofocusing) of peptides in enzymatic hydrolysates of food proteins. *Journal of Agricultural and Food Chemistry* 53: 3801–6.

Hashimoto, K., Sato, K., Nakamura, Y., and Ohtsuki, K. 2006. Development of continuous type apparatus for ampholyte-free isoelectric focusing (autofocusing) of peptides in protein hydrolysates. *Journal of Agricultural and Food Chemistry* 54: 650–55.

Kagawa, K., Matsutaka, H., Fukuhama, C., Watanabe, Y., and Fujino, H. 1996. Globin digest, acidic protease hydrolysate, inhibits dietary hypertriglyceridemia and Val-Val-Tyr-Pro, one of its constituents, possesses most superior effect. *Life Sciences* 58: 1745–55.

Kitts, D. D. and Weiler, K. 2003. Bioactive proteins and peptides from food sources. Applications of bioprocesses used in isolation and recovery. *Current Pharmaceutical Design* 9: 1309–23.

Korhonen, H. and Pihlanto, A. 2003. Food-derived bioactive peptides – opportunities for designing future foods. *Current Pharmaceutical Design* 9: 1297–1308.

Korhonen, H. and Pihlanto, A. 2007. Technological options for the production of health-promoting proteins and peptides derived from milk and colostrums. *Current Pharmaceutical Design* 13: 829–43.

Meisei, H. 2004. Multifunctional peptides encrypted in milk proteins. *Biofactors* 21: 55–61.

Miguel, M. and Aleixandre, A. 2006. Antihypertensive peptides derived from egg proteins. *Journal of Nutrition* 136: 1457–60.

Möller, N. P., Scholz-Ahrens, K. E., Ross, N., and Schrezenmeir, J. 2008. Bioactive peptides and proteins from foods: Indication for health effects. *European Journal of Nutrition* 47: 171–82.

Murota, I., Tamai, T., Baba, T., Sato, R., Hashimoto, K., Park, E.-Y., et al. 2010. Uric acid lowering effect by ingestion of proteolytic digest of shark cartilage and its basic fraction. *Journal of Food Biochemistry* 34: 182–94.

Osajima, K., Sakakibara, S., Sawabe, T., Kitamura, K., Matsuda, H., and Osajima, Y. 1993. Study on separation system of angiotensin I converting enzyme inhibitory peptides originating from sardine meat. *Nippon Shokuhin Kogyo Gakkaishi* 40: 568–76.

Rilbe, H. 1976. Isoelectric focusing – Development from notion to practically working tool. *Science Tools* 23: 18–22.

Tanabe, S., Arai, S., and Watanabe, M. 1999. Large-scale production of a high-glutamine peptide. *Journal of Japanese Society of Nutrition and Food Science* 52: 103–6.

Yata, M., Sato, K., Ohtsuki, K., and Kawabata, M. 1996. Fractionation of peptides in protease digests of proteins by preparative isoelectric focusing in the absence of added ampholyte: A biocompatible and low-cost approach referred to as autofocusing. *Journal of Agricultural and Food Chemistry* 44: 76–79.

Chapter 16

Industry Perspectives and Commercial Trends for Food Proteins and Biopeptides

Se-Kwon Kim and Isuru Wijesekara

Contents

16.1 Introduction

Food proteins and biopeptides play a vital role in the growth and development of the body's structural integrity and regulation, as well as other functional properties. Additionally, food proteins are used as essential raw materials in most industries. Currently, however, many biopeptides are underutilized. Both food proteins and biopeptides have the potential in novel commercial trends, as they are widely commercialized in the food, beverage, pharmaceutical, and cosmetic industries, in addition to other fields, such as photography, textiles, leather, electronic, medicine, and biotechnology. The industrialists are eager to embrace a novel product if it can deliver what the consumers want; however, the industry needs to balance its involvement against the perceived market potential for a new trend.

Recently, the consumers' demand for products with functional ingredients has increased, underlining the need to guarantee the safety, traceability, authenticity, and health benefits of such products. Therefore, commercially available food protein products have also prompted newer challenges. The bioactive food proteins and biopeptides have been isolated from various sources, including terrestrial food crops, animals, microorganisms, and marine-derived foods including fish and algae by-products (Kim and Mendis 2006). They are potential candidates as functional ingredients for new commercial trends by the industries. In this sense, this chapter presents an overview of the current status, future industrial perspectives, and commercial trends of bioactive food proteins and biopeptides.

16.2 Industry Perspectives

16.2.1 Food Industry

Proteins, from sausage making to bread baking, contribute to the physical/chemical characteristics, such as texture, color, and flavor, as well as the nutritional values of the food products. Proteins can be isolated from a range of biological materials, including legumes, cereals, milk, meat, and marine organisms. While proteins and peptides have excellent antioxidant activity and great potential as food additives, they may not be suitable in all food applications due to their less solubility and bitterness. Therefore, improving the solubility and foaming properties and removing the bitterness of the food proteins are needful for novel commercial trends with consumer appeal. To improve a protein's solubility and foaming properties, a mild hydrolysis can be first used to remove the amide nitrogen group, which increases the number of negatively charged groups. Further hydrolysis then reduces the protein's average molecular weight. As a treatment for the bitterness, Shahidi, Han, and Synowiecki (1995) treated a fish protein hydrolysate with activated carbon, which removed the bitterness. The challenge for food technologists will be to develop functional foods and nutraceuticals without the undesired side effects of the added peptides.

Chemical modification techniques have been the subject of much research. The gelation of the milk protein casein may be enhanced by introducing the functional groups that are hydrophilic and can form disulfide bonds. The solubility of casein in the presence of calcium has been altered by the addition of chemical groups through succinylation and thiolation. The succinylation of whey protein for use in coffee whiteners has been patented. The protein is less likely to coagulate at high temperatures. In addition to chemical and enzymatic modifications, genetically modified proteins are available, but global acceptance of these has limited their use.

The perspective of improving the functional characteristics of the food proteins could be the way to novel food products. The antifreeze proteins of polar fishes and the bacterial proteins that act as nucleating agents could have useful applications in frozen food systems. The former could be used to lower the temperature and retard crystallization during frozen storage, and the latter could be used to give the desirable crystal size and texture when added to ice-cream formulations.

Furthermore, over 100 million tons of fishes are harvested annually and approximately 30% of the total catch is used for fish meal and animal feed because of its poor functional properties. In addition, more than 25% of the total catch is discarded as processing waste or as by-products (Kim and Mendis 2006). Among the by-products from fish processing plants are fish frames, including the head, bone, and tail, which are very important protein sources. Therefore, researchers have shown interest in developing techniques to convert or upgrade fish processing waste into value-added by-products. One of the approaches for effective protein recovery from these by-products is enzymatic hydrolysis, which is widely applied to improve and upgrade the functional and nutritional properties of proteins.

In the food industry, lipid oxidation and the rancidity of fats and oils influence the consumer appeal for these high-fat products, and novel bioactive peptides can be incorporated as antioxidants to prevent these undesirable issues. Currently, the food and pharmaceutical industries use many synthetic antioxidants, such as butylated hydroxytoluene (BHT), butylated hydroxyanisole (BHA), tert-butylhydroquinone (TBHQ), and propyl gallate (PG) to retard the oxidation and peroxidation processes. However, the use of these synthetic antioxidants is under strict regulation due to their potential health risk (Hettiarachchy et al. 1996; Park et al. 2001). Hence, the search for natural antioxidants as safe alternatives is important to the food industry. Furthermore, natural bioactive food proteins, such as phycobiliproteins, can be derived from blue-green algae and marine red algae, which also have the potential as natural food colorants. Therefore, bioactive food proteins and biopeptides have important functional properties that could be scaled up and economically favorable as nutraceutical ingredients for the food industry.

16.2.2 Pharmaceutical Industry

Food-derived proteins and biopeptides are important ingredients in pharmaceutical products. Some are currently used as drugs, for example, insulin, while others are used to provide filler for other drugs and therapeutic agents in the pharmaceutical industry. Thus, novel food proteins and biopeptides have the potential to be used as novel drugs. In the last two decades, several studies have been performed on caseino-phosphopeptides and marine-derived bioactive peptides to act as carriers for different minerals, especially calcium (Jung et al. 2006; Jung and Kim 2007).

In addition, food- and marine-derived bioactive peptides have biological activities that benefit health, such as antioxidant, antihypertensive, anticoagulant, antimicrobial, antihypercholesterol, and anticancer activities (Clare and Swaisgood 2000; Elias, Kellerby, and Decker 2008; Kim and Wijesekara 2010). Therefore, new applications are appearing as novel natural pharmaceuticals. Recently, it has been observed that most synthetic drugs have several side effects. For example, several side effects of heparin have been identified, such as development of thrombocytopenia, hemorrhagic effect, ineffectiveness in congenital or acquired antithrombin deficiencies, and incapacity to inhibit thrombin bound to fibrin (Pereira, Melo, and Mourao 2002). In addition, synthetic antihypertensive drugs have also shown several side effects (Atkinson and Robertson 1979). Therefore, bioactive peptides have the potential as novel drugs in the pharmaceutical industry, without undesirable side effects.

Moreover, the need to discover new antimicrobial substances is important due to the progressive development of resistance by pathogenic microorganisms against conventional antibiotics. Antibacterial peptides purified from food proteins, such as marine organisms, play a significant role as antimicrobial agents and have shown the potencies that make them useful in the pharmaceutical industry.

16.2.3 Cosmetic Industry

Bioactive food proteins and biopeptides are biologically active substances with the potential to act as cosmetics (Kim et al. 2008). Recently, the young generation has paid much attention to beauty as well as maintaining a young appearance with novel cosmetics containing natural bioactive ingredients. Therefore, the search for safe and cheap natural sources of bioactive food proteins and biopeptides as cosmetic ingredients is promising.

The phycoerythrin pigment protein in red algae can be used as a pigment in cosmetics. In addition, food proteins have the potential as functional ingredients in cosmetics, such as sunscreen lotions, shampoos, conditioners, hair gels, nail polishes, and lipsticks.

Recently, cosmetic application of collagen and gelatin, derived from terrestrial animals such as cows and pigs, is declining because of animal diseases and some ethnic or religious barriers. New trends are pushing to use other food sources, such as marine fish-derived collagen and gelatin. They are excellent functional ingredients for the cosmetic industry. Collagen and gelatin have a high moisturizing property and can be produced as novel cosmetic creams and gels.

With the invention of UV radiation protection compounds, antiwrinkling agents, and antiaging compounds in the cosmetic industry, new trends have focused on the manufacture of sunscreen lotions, creams, and other cosmetics. In this case, bioactive food proteins and biopeptides are promising functional ingredients for these new cosmeceuticals with pharmaceutical benefits.

16.2.4 Other Industries

The perspective for the use of bioactive food proteins and biopeptides is promising in other industries, such as printing and photography, textiles, electronic, and leather.

16.3 Food Proteins and Biopeptides with Commercialization Trends

16.3.1 Collagen

Collagen is the most abundant animal protein polymer, representing nearly 30% of the total protein in the animal body. It is the main structural element of the bones, cartilages, skin, tendons, ligaments, blood vessels, teeth, cornea, and all other organs of the vertebrates. The molecular structure of collagen contains three polypeptide α-chains wound together in a tight triple helix. Each polypeptide, called an α-chain, consists of repeated sequence of triplet $(Gly-X–Y)_n$, where X and Y are often proline (Pro) and hydroxyproline (Hyp). For industrial purposes, collagen is mainly extracted from the skin and bones of cattle and pigs. However, due to highly infectious and contagious animal diseases and some religious barriers, the industrial use of collagen from these sources is becoming limited. Therefore, marine-derived collagen is an alternative with safe and economic advantages (Senaratne, Park, and Kim 2006).

On the industrial front, most of the applications of collagen are a direct result that collagen can be synthesized into gelatin, which turns out to be a highly useful raw material in the food industry. It is now well-known and scientifically proved that hydrolyzed collagen significantly contributes to maintaining the bone and joint health and prevents osteoporosis and osteoarthritis. Additionally, collagen improves the elasticity of the skin by stimulating the production of collagen by the skin cells themselves, thereby maintaining youthful and vibrant skin and neutralizing the skin's continuous aging and thereby deterioration by redensifying the dermis. To take advantage of all these healthy claims, a number of functional foods or nutraceuticals have been introduced in the market (Mendis et al. 2005).

Collagen has also found application in the manufacture of photography aids, as well as in the production of cosmetic and pharmaceutical products. In addition, polythene–collagen hydrolysate blends have been shown to be a successful alternative in the thermoplastic industry as environmentally friendly biodegradable plastics. Blends of collagen hydrolysate and low-density polyethylene with a content of collagen hydrolysate up to 20%–30% produce transparent, cohesive, and flexible films that are characterized by satisfactory thermal and mechanical responses and can be applied in the packaging and agriculture fields.

In the medical field, collagen can be used to promote healing of the skin that has been damaged or removed as a result of skin grafting, ulceration, burns, cancer, excision, or mechanical trauma. This type of artificial skin prevents moisture and heat loss from the wounded skin and also prevents microbial infiltration to the body. Artificial skin is used not only to limit the entry of foreign matter into the body and prevent mass and heat transfer out of the body, but also to provide a continuous cellular layer over the skin. Moreover, collagen can be applied to restructure dental damage and for the manufacture of artificial bones in medicine (Karim and Bhat 2009).

16.3.2 Gelatin

Gelatin is derived from the fibrous protein collagen and is one of the most popular biopolymers, widely used in the food, pharmaceutical, cosmetic, and photographic industries, because of its unique functional properties, such as foam stabilization, water binding, creaminess, and capability to act as a matrix (Karim and Bhat 2009). Nowadays, consumer demand has arisen for marine-derived gelatins due to animal diseases and religious issues, as discussed for collagen. In the food industry, gelatin is utilized in confections (mainly to provide chewiness, texture, and foam stabilization), low-fat spreads (to provide creaminess, fat reduction, and mouth feel), dairy products (to provide stabilization and texturization), baked goods (to provide emulsification, gelling, and stabilization), and meat products (to provide water-holding capacity). Gelatin has a tendency to form complexes with other proteins and hydrocolloids. This property makes gelatin useful for precipitating materials that cause haze or cloudiness in wine, beer, cider juice, and vinegar. Hence, gelatin can be used in the food industry as a clarification agent. In addition, the use of gelatin in the nutraceutical industry is widespread. It not only serves as an excipient, but it is also an excellent and economical source of multiple amino acids. Therefore, gelatin can serve as a functional ingredient in medicinal food formulas. It is also widely utilized in nutritional bars, sports drinks, and energy drinks, and it reduces the carbohydrate content in foods formulated for diabetic patients. Gelatin, being low in calories, is normally recommended for use in foodstuffs to enhance the protein levels and is especially useful in body-building foods.

In the pharmaceutical and medical fields, gelatin is used as a matrix for implants, in injectable drug delivery microspheres, and in intravenous infusions. There are also reports in which live attenuated viral vaccines, used for immunization against measles, mumps, rubella, rabies, and tetanus toxin, contain gelatin as the stabilizer. Moreover, gelatin is widely used in the pharmaceutical industry for the manufacture of hard and soft capsules, plasma expanders, and in wound care. As a protein, gelatin is low in calories, and it melts in the mouth to give excellent sensory properties resembling fat, making it ideal for use in low-fat products. The low-gelling temperature of gelatin offers new potential applications, such as use in dry products (for microencapsulation), and in fact, one of its major applications is in the encapsulation of vitamins and other pharmaceutical additives such as azoxanthine. Moreover, gelatin can be used to microencapsulate food flavors.

Gelatin could be utilized as the base for light-sensitive coatings, which are important to the electronics trade, because of its low-temperature gelling property. In addition, it is a good medium for precipitating silver halide emulsions, since this process can be carried out at a lower temperature. The photography industry uses large quantities of gelatin in several applications due to this phenomenon. In the cosmetic industry, gelatin has been used for many years as hydrolyzed animal protein in shampoos, conditioners, lipsticks, and nail formulas. Recently, additional uses of gelatin have been found as a collagen source in topical creams and other value-added cosmetic products (Kim et al. 2008).

16.3.3 Seaweed Proteins

The protein content of seaweeds differs according to the species. Generally, the protein fraction of brown seaweeds is low (3%–15% of the dry weight) compared with that of green or red seaweeds (10%–47% of the dry weight). A few reports are available on the nutritional value of algal proteins, and some perspectives on the potential applications and commercialization trends of algal proteins for the development of new foods or additives for human consumption are promising (Fleurence 1999). With respect to their high protein content and amino acid composition, the red seaweeds appear to be an interesting potential source of food proteins and in the development of novel functional foods. In addition, red seaweeds contain a particular protein called phycoerythrin, which is already used as a dye in the immunofluorescence reactions in biotechnology applications.

16.3.4 Lectins

Lectins are carbohydrate-binding proteins or glycoproteins that are finding valuable commercial trends in the biomedical industry. Thus, lectins derivatized with fluorescent dyes, gold particles, or enzymes, are employed as histochemical and cytochemical reagents for the detection of glycoconjugates in the tissue sections, on the cells and subcellular organelles, and in the investigations of the intracellular pathways of protein glycosylation. Lectin binding has been used to demonstrate that the membrane receptors for hormones, growth factors, neurotransmitters, and toxins are glycoconjugates. Another clinical application of lectins is in blood typing. The lectins from *Lotus tetragonalobus* and *Ulex europaeus*, both specific for fucose, are employed to identify type O blood cells and to identify the secretors of blood group substances. It has been reported that a 30 kDa β-galactose–specific lectin derived from a marine worm, *Chaetopterus variopedatus*, has shown anti–HIV-1 activity in vitro (Wang et al. 2006). Moreover, some marine algae lectins can be developed as antibiotics against marine vibrios. In the food industry, calcium-binding lectins can be incorporated into nutraceutical or functional foods to prevent calcium deficiency (Jung, Park, and Kim 2003).

16.3.5 Enzymes

Enzymes are proteins, and food grade enzymes are widely used in the food industry. New enzyme trends have arisen in the fat and oil industries for the processing of margarine and the removal of phospholipids in vegetable oils (degumming), using a highly selective microbial phospholipase. In addition, enzymes are potential candidates in the textile industry, beverage industry, animal feed industry, detergent industry, and organic chemical synthesis industry. Due to advances in modern biotechnology, enzymes can be developed today for processes where no one would have expected an enzyme to be applicable just a decade ago. Common to most applications, the trends for enzymes due to cost-effective catalysts working under mild conditions result in significant savings in resources, such as energy and water, for the benefit of both the industry in question and the environment (Kirk, Borchert, and Fuglsang 2002). New commercial trends show promise in using underutilized food or fish waste extract enzymes, which can be applicable for industrial purposes (Kim et al. 2003).

16.3.6 Soy Proteins

Sales of soy protein products have increased markedly over the past decade. The most dramatic rise has been in the many recently created convenience products, such as energy bars, flavored soy nuts, chips, meat substitutes, and beverages. Soy protein–based coatings and films are being researched as coatings for meat products. In the meat industry, some meat products such as sausages can be coated by spraying or dipping in soy protein mixtures and this retards moisture loss during short-term storage. In addition, soy protein coatings could help maintain the original flavor of fat-containing foods. The coating with soy proteins will boost the nutritional value of the products by adding extra protein to the foods, than the current cellulose coatings (Emmert and Baker 1995).

16.3.7 Whey Proteins

Whey protein is a pure, natural, and high-quality protein from cow's milk. It is a rich source of essential amino acids that the body requires on a daily basis. High nutritional quality, potent biological activity, and unique functional properties are the foremost attributes of whey proteins that help sustain interest in their utilization, not only in the food industry but also in the pharmaceutical and biomedical fields (Bhattacharjee, Bhattacharjee, and Dutta 2006; Marcelo and Rizvi 2008). Whey protein is an excellent protein choice for individuals of all ages, and it provides a number of benefits in areas including sports nutrition, weight management, immune support, bone health, and general wellness; hence, it is a potential candidate in functional food sources. In addition, it can be used as a filler in lipsticks and as a moisture-retaining agent in cosmetics.

16.3.8 Bioactive Peptides

Recently, a great interest has arisen in the study of the structural, compositional, and sequential properties of bioactive peptides. They can be produced by either one of the three methods: solvent extraction, enzymatic hydrolysis, or the microbial fermentation of food proteins. Bioactive peptides are inactive within the sequence of their parent protein and can be released by enzymatic hydrolysis. Moreover, bioactive peptides usually contain 3–20 amino acid residues and their activities are based on their amino acid compositions and sequences. They have been detected in many different food sources. Furthermore, depending on the amino acid sequences, they can be involved in various biological functions, such as antihypertensive, opioid agonists or antagonists, immunomodulatory, antithrombotic, antioxidant, anticancer, and antimicrobial activities, in addition to nutrient utilization (Byun and Kim 2001; Jo, Jung, and Kim 2008; Kim, Je, and Kim 2007; Liu et al. 2008; Rajapakse et al. 2005; Sheih, Fang, and Wu 2009). Some bioactive peptides have shown multifunctional activities based on their structures and other factors, including hydrophobicity, charge, or microelement-binding properties.

Recent research has shown that milk proteins can yield bioactive peptides with opioid, mineral binding, cytomodulatory, antihypertensive, immunostimulating, antimicrobial, and antioxidative activities in the human body. A variety of naturally formed bioactive peptides have been found in fermented dairy products, and some of these peptides have been commercialized in the form of fermented milk. Furthermore, marine bioactive peptides have been isolated widely by enzymatic hydrolysis of marine organisms. However, fermented marine food sauces in which enzymatic hydrolysis has already been done by microorganisms and bioactive peptides can be purified without further hydrolysis. In addition, several bioactive peptides have been isolated from marine processing by-products or wastes. Marine-derived bioactive peptides have been shown to possess many physiological functions, including antihypertensive or angiotensin-I-converting enzyme inhibition, antioxidant, anticoagulant, and antimicrobial activities.

Some of these bioactive peptides have been identified as possessing nutraceutical properties that are beneficial in human health promotion, and recently, it has been shown that food-derived bioactive peptides can reduce the risk of cardiovascular diseases. Furthermore, the increasing consumer knowledge of the link between diet and health has raised the awareness and demand for functional food ingredients and nutraceuticals. This leads to the avoidance of the undesirable side effects associated with organically synthesized chemical drugs and also the avoidance of the high cost of drug therapies. Bioactive peptides derived from marine organisms as well as fish processing by-products have the potential in the development of functional foods, and they can act as potential physiological modulators of metabolism after absorption. Hence, marine-derived bioactive peptides can be used as versatile raw materials to produce nutraceuticals and pharmaceuticals for human beings.

Recent studies have provided evidence that marine-derived bioactive peptides play a vital role in human health and nutrition. The possibilities of designing new functional foods and pharmaceuticals that support reduced diet-related chronic malfunctions are promising. Biologically active marine peptides are food-derived peptides that function beyond their nutritional values, having physiological, hormone-like properties with possible roles in reducing the risk of cardiovascular diseases by lowering the plasma cholesterol levels and/or anti-cancer activity by reducing the cell proliferation of human breast cancer cell lines, as shown previously.

Moreover, calcium-binding bioactive peptides derived from the pepsin hydrolysates of marine fish species, such as the Alaska pollack (*Theragra chalcogramma*) and hoki (*Johnius belengerii*) frame, can be introduced to oriental people with a lactose intolerance as calcium-fortified supplements such as fruit juices or calcium-fortified foods as alternatives to dairy products (Jung et al. 2006; Jung and Kim 2007). These calcium-binding peptides have shown a high affinity to calcium and are also suitable for reducing the risk of osteoporosis. It has been proven that small peptides (dipeptides and tripeptides) generated in the diet can be absorbed across the brush-border membrane by a specific peptide transport system, thereby producing diverse biological effects. Hence, novel pharmaceuticals developed from marine bioactive peptides can be given as an oral administration instead of an intravenous administration. For that reason, bioactive peptides have the potential to be used in the formulation of health-enhancing nutraceuticals and cosmetics and as potent drugs with well-defined pharmacological effects.

16.4 Conclusions

Recently, much attention has been focused by consumers toward natural bioactive compounds as functional ingredients, and hence, it can be suggested that food proteins and biopeptides are excellent alternative sources of novel functional foods and pharmaceuticals that can contribute to the consumers' well-being by replacing synthetic ingredients and cosmetics. The possibilities of designing new functional foods and pharmaceuticals using food proteins and biopeptides to reduce diet-related chronic malfunctions are promising. Furthermore, fish processing by-products such as food proteins can be easily utilized for producing bioactive peptides. These evidences suggest that due to their valuable biological functions with health beneficial effects, bioactive proteins and peptides have a great potential as active ingredients for the preparation of various functional foods and nutraceutical and pharmaceutical products.

References

Atkinson, A. B. and Robertson, J. I. S. 1979. Captopril in the treatment of clinical hypertension and cardiac failure. *Lancet* 2: 836–39.

Bhattacharjee, S., Bhattacharjee, C., and Dutta, S. 2006. Studies of the fractionation of β-lactoglobulin from casein whey using ultrafiltration and ion-exchange membrane chromatography. *Journal of Membrane Science* 275: 141–50.

Byun, H. G. and Kim, S. K. 2001. Purification and characterization of angiotensin I converting enzyme (ACE) inhibitory peptides from Alaska Pollack (*Theragra chalcogramma*) skin. *Process Biochemistry* 36: 1155–62.

Clare, D. A. and Swaisgood, H. E. 2000. Bioactive milk peptides: A prospectus. *Journal of Dairy Science* 83: 1187–95.

Elias, R. J., Kellerby, S. S., and Decker, E. A. 2008. Antioxidant activity of proteins and peptides. *Critical Reviews in Food Science and Nutrition* 48: 430–41.

Emmert, J. L. and Baker, D. H. 1995. Protein quality assessment of soy products. *Nutrition Research* 15(11): 1647–56.

Fleurence, J. 1999. Seaweed proteins: Biochemical, nutritional aspects and potential uses. *Trends in Food Science and Technology* 10: 25–28.

Hettiarachchy, N. S., Glenn, K. C., Gnanasambandan, R., and Johnson, M. G. 1996. Natural antioxidant extract from fenugreek (*Trigonella foenumgraecum*) for ground beef patties. *Journal of Food Science* 61: 516–19.

Jo, H. Y., Jung, W. K., and Kim, S. K. 2008. Purification and characterization of a novel anticoagulant peptide from marine echiuroid worm, *Urechis unicinctus*. *Process Biochemistry* 43: 179–84.

Jung, W. K., Karawita, R., Heo, S. J., Lee, B. J., Kim, S. K., and Jeon, Y. J. 2006. Recovery of a novel Ca-binding peptide from Alaska pollack (*Theragra chalcogramma*) backbone by pepsinolytic hydrolysis. *Process Biochemistry* 41: 2097–2100.

Jung, W. K. and Kim, S. K. 2007. Calcium-binding peptide derived from pepsinolytic hydrolysates of hoki (*Johnius belengerii*) frame. *European Food Research and Technology* 224: 763–67.

Jung, W. K., Park, P. J., and Kim, S. K. 2003. Purification and characterization of a new lectin from the hard roe of skipjack tuna, *Katsuwonus pelamis*. *International Journal of Biochemistry and Cell Biology* 35: 255–65.

Karim, A. A. and Bhat, R. 2009. Fish gelatin: Properties, challenges, and prospects as an alternative to mammalian gelatins. *Food Hydrocolloids* 23: 563–76.

Kim, S. Y., Je, J. Y., and Kim, S. K. 2007. Purification and characterization of antioxidant peptide from hoki (*Johnius balengerii*) frame protein by gastrointestinal digestion. *Journal of Nutritional Biochemistry* 18: 31–38.

Kim, S. K. and Mendis, E. 2006. Bioactive compounds from marine processing byproducts – A review. *Food Research International* 39: 383–93.

Kim, S. K., Park, P. J., Byun, H. G., Je, J. Y., Moon, S. H., and Kim, S. H. 2003. Recovery of fish bone from hoki (*Johnius belengeri*) frame using a proteolytic enzyme isolated from mackerel intestine. *Journal of Food Biochemistry* 27: 255–66.

Kim, S. K., Ravichandran, Y. D., Khan, S. B., and Kim, Y. T. 2008. Prospective of the cosmeceuticals derived from marine organisms. *Biotechnology and Bioprocess Engineering* 13: 511–23.

Kim, S. K. and Wijesekara, I. 2010. Development and biological activities of marine-derived bioactive peptides: A review. *Journal of Functional Foods* 2: 1–9.

Kirk, O., Borchert, T. V., and Fuglsang, C. C. 2002. Industrial enzyme applications. *Current Opinion in Biotechnology* 13: 345–51.

Liu, Z., Dong, S., Xu, J., Zeng, M., Song, H., and Zhao, Y. 2008. Production of cysteine-rich antimicrobial peptide by digestion of oyster (*Crassostrea gigas*) with alcalase and bromelin. *Food Control* 19: 231–35.

Marcelo, P. A. and Rizvi, S. S. H. 2008. Physicochemical properties of liquid virgin whey protein isolate. *International Dairy Journal* 18: 236–46.

Mendis, E., Rajapakse, N., Byun, H. G., and Kim, S. K. 2005. Investigation of jumbo squid (*Dosidicus gigas*) skin gelatin peptides for their in vitro antioxidant effects. *Life Sciences* 77: 2166–78.

Park, P. J., Jung, W. K., Nam, K. D., Shahidi, F., and Kim, S. K. 2001. Purification and characterization of antioxidative peptides from protein hydrolysate of lecithin-free egg yolk. *Journal of American Oil Chemists Society* 78: 651–56.

Pereira, M. S., Melo, F. R., and Mourao, P. A. S. 2002. Is there a correlation between structure and anticoagualant action of sulfated galactans and sulfated fucans? *Glycobiology* 12(10), 573–80.

Rajapakse, N., Jung, W. K., Mendis, E., Moon, S. H., and Kim, S. K. 2005. A novel anticoagulant purified from fish protein hydrolysate inhibits factor XIIa and platelet aggregation. *Life Sciences* 76: 2607–19.

Senaratne, L. S., Park, P. J., and Kim, S. K. 2006. Isolation and characterization of collagen from brown backed toadfish (*Lagocephalus gloveri*) skin. *Bioresource Technology* 97: 191–97.

Shahidi, F., Han, X.-Q., and Synowiecki, J. 1995. Production and characteristics of protein hydrolysates from capelin (*Mallotus villosus*). *Food Chemistry* 53: 285–93.

Sheih, I. C., Fang, T. J., and Wu, T. K. 2009. Isolation and characterization of a novel angiotensin I-converting enzyme (ACE) inhibitory peptide from the algae protein waste. *Food Chemistry* 115: 279–84.

Wang, J. H., Kong, J., Li, W., Molchanova, V., Chikalovets, I., Belogortseva, N., et al. 2006. A β-galactose-specific lectin isolated from the marine worm *Chaetopterus variopedatus* possesses anti-HIV-1 activity. *Comparative Biochemistry and Physiology C* 142: 111–17.

Chapter 17

Protein Nanotechnology: Research, Development, and Precaution in the Food Industry

Masami Matsuda, Geoffrey Hunt, Yoshinori Kuboki, Toshio Ogino, Ryuichi Fujisawa, Fumio Watari, and Rachel L. Sammons

Contents

17.1 Introduction

Geoffrey Hunt and Masami Matsuda

Nanotechnology is gradually revolutionizing the food industry, in processing additives, functionalized or bioactive ingredients, packaging, storage, and traceability. The use of nanotechnology in animal feed, fertilizers, and pesticides is also relevant, from the point of view of both the benefits and the risks in the food sector. Nanotechnology will also play a role in environmental sustainability by reducing packaging, extending the shelf-life of food product, and reducing spoilage and waste. A study in 2009 from the U.S. Institute for Agriculture and Trade Policy, appropriately named "Identifying our Climate Foodprint [*sic*]," points out that food processing and packaging is currently both energy-intensive and waste-intensive, even if not as severely so as modern agricultural methods generally. Nanotechnology could make a large contribution to addressing this issue (U.S. Institute for Agriculture and Trade Policy 2009).

Nanoscale materials and techniques are already being used in the food-related areas of nanosensors, nanoparticles in polymers and barriers, nanostructured films, nanofiltration, nanodispersions, and nanoemulsions, and research in these areas is intensifying each year. To give two examples, we are seeing nanocoatings of aluminium and nanoceramics in polyethylene terephthalate (PET) bottles to prevent oxygen entry. However, as we shall see in the following sections, concerns are being increasingly raised about the possible health, safety, and environmental hazards; the risks of this new technology; and the adequacy of current global and local regulations. Few studies have been dedicated to this matter (Pusztai and Bardocz 2006).

One specific area of research that will have a high relevance for the food and agriculture industries is the interface between biomolecules, such as proteins, and solid surfaces, which presents challenging and promising avenues of investigation for applications, as well as new questions of health and safety. In this chapter, three nanoscience researchers explain some of the progress and questions in this particular field, and the general developments are briefly examined from the point of view of risk, ethics, and regulation.

In Section 17.2, Kuboki et al. explain that the discovery of carbon nanotubes (CNTs) has opened a new avenue of questions, including: How do proteins interact with solid clusters with a certain geometry at the same nanometer scale? Protein molecules have their own definite geometry of conformation. The same is true with clusters of CNTs. Despite the fact that CNTs are devoid of specific charged groups, they interact well with various proteins, as exemplified by the strong sustainability of various insoluble collagens with CNTs.

Kuboki et al. assume that the solid geometry of CNTs may function as receptors for protein molecules. Thus, an analysis of the mode of interaction between the CNT and various proteins will give us fruitful insights concerning the biological applications of CNTs. They consider the example of the interaction between the collagen triple helix and the CNTs.

In Section 17.3, Ogino explains that interfaces between solid surfaces and biomolecules play an important role in the development of high performance and highly reliable biosensors and biosolid hybrid devices. In this section, the adsorption and immobilization processes of protein molecules on solid surfaces are described. Protein adsorption sites on a sapphire surface can be controlled by an atomic-step arrangement. The selective adsorption of protein molecules is realized by a domain structure formed during the step rearrangement process. Specific pattern formations of lipid bilayers also take place on such an atomically controlled sapphire surface. Both of these theoretical contributions will ultimately raise questions at the level of applications. Other concerns may be raised about the nanotechnology of proteins in food, food packaging, storage, and distribution that are not manifested in these two sections, and some are outlined next.

17.2 Impact of Carbon Nanotubes on Protein Chemistry

Yoshinori Kuboki, Ryuichi Fujisawa, Fumio Watari,
Rachel L. Sammons

17.2.1 Interacting Forces of Proteins

The kinds of forces interacting between proteins themselves and other molecules are traditionally categorized into four groups: (1) van der Waals, (2) electrostatic,

(3) hydrophobic, and (4) hydrogen bonds. These bonds are important for creating the fundamental binding forces responsible for the three-dimensional geometric structures of proteins. However, if we look at a higher level of protein interactions, we will find the complex forms of the above-mentioned bonds, in which each amino acid in a certain sequence may interact with other amino acids or sequences through different forces, and these are far more important in the protein interactions, such as in enzymes and antibodies.

Using the four fundamental forces, the side chains and the backbone of the polypeptide chains interact to construct the secondary, tertiary, and quaternary structures of the protein molecules themselves. At the same time, these fundamental forces mainly play a role in the interactions with other proteins and biomolecules.

On the other hand, the CNTs are not equipped with any functional group, but with a rigid rod-like structure, which may interact with other molecules with a bulk geometrical structure, which is equivalent to the quaternary structure of proteins.

17.2.2 Hierarchy of Binding between Proteins and Nanomaterials

When we consider the interactions of proteins with CNTs, the tertiary and quaternary structures, or furthermore, the bulk shape of the protein, such as the globular, rod-like, or fibrillar shape of the whole molecule, must be taken into consideration. In addition, both the molecules of the protein and the clusters of CNTs are generally liable to assemble and form self-aggregation products. In the case of certain proteins, these aggregation products are formed in a strictly regular manner, typically exemplified by collagen and fibrinogen molecules. The CNTs can also easily form large longitudinal or random aggregates ranging from a fine suspension state to insoluble clots.

From the four fundamental binding forces to the aggregating forces of the molecular clusters, we see a hierarchy of binding, when we investigate the interaction between the protein and the CNT. One example of the interaction between collagen and CNTs is discussed in detail next.

17.2.3 Impact of Carbon Nanotubes Discovery on Protein Chemistry

Ever since Iijima and Ichihashi (Iijima, Ajayan, and Ichihashi 1993) discovered single-walled carbon nanotubes (SWNT) in 1993, the exceptionally high mechanical strength, the electrical conductivity, and the unique geometrical structure of CNTs have attracted attention from the scientists in various fields, including those interested in the biological and biochemical applications of this new material. In the field of protein chemistry, the discovery of CNTs is considered to have a strong potential due to the following reasons.

Table 17.1   Comparison of Geometrical Scales of Proteins and Carbon Nanotubes

	Diameter/ Amplitude (nm)	*Length (nm)*
SWNT	1–3	Up to several micrometers
MWNT	20–30	Several nanometers to several hundred micrometers
α-Helix	0.5	Up to 3–4
β-Sheet	0.2	Up to 3–4
Collagen helix	0.3	Up to 300
Collagen molecule	1.4	300
Albumin (BSA)	4.0	14
Myoglobin	4.4	2.5
Lysozyme	3.0	4.5
Fibrinogen	2–9	45

1. CNTs have a definite chemical structure based on graphite, which can be chemically functionalized with various side chains.
2. The definite rod-like, three-dimensional, geometric structure of CNTs, whose diameters are typically 1–3 nm, and a minimum of 0.7 nm in an SWNT, is almost the same as those of the fundamental building stones of proteins. The diameters of the α-helix, the collagen helix, and the width (amplitude) of the β-sheet are 0.5, 0.3, and 0.2 nm between the α-carbons, respectively (Table 17.1).
3. By attaching the side chains with the functional groups to the carbon atoms of the CNT, we may create new protein-mimicking substances, since the molecular sizes of the CNT and the proteins are almost similar, as shown in Table 17.1.
4. By hybridizing a specific protein with a CNT, we can increase or modify the function of the proteins.

17.2.4 Interactions between Proteins and Carbon Nanotubes: Previous Studies

Although there have been numerous attempts to combine CNTs with various proteins to fabricate new hybrid materials, the detailed mechanism of the interactions between CNTs and important biomolecules, in particular, protein molecules, has not been studied extensively. Various hybrid materials are composed of CNTs and proteins, including albumin (Shen et al. 2008), hemoglobin (Wu et al. 2008), and immunoglobulin G (Cid et al. 2008).

Many proteins have been shown to adhere to or bind with nonmodified CNTs (SWNT or MWNT) via noncovalent bonds. Furthermore, when the CNTs are modified with the functional groups, such as the carboxyl or amino groups, they are reasonably expected to bind with almost all proteins. This situation will naturally lead us to several new aspects of protein chemistry.

17.2.4.1 Conformational Studies

Shen et al. (2008) showed that human serum albumin underwent stepwise conformational changes on the CNT. These kinds of conformational changes are likely to occur in many other proteins on binding with the CNT and are worth further studies in detail.

17.2.4.2 Sensors

Another important application of a protein–CNT hybrid is the result of its electroconductivity. It was proven that electron transfer from the hybridized redox enzyme glucose oxidase to the CNT supporter can be detected by amperometry (Cid et al. 2008). The device was shown to be useful in detecting the glucose concentration in a biological fluid. Also, specific IgG–CNT hybrids change their electroconductivity as a result of binding with a specific antigen, and this can also be detected amperometrically (Lyons and Keeley 2008).

17.2.4.3 Artificial ECM for Regenerative Medicine

Another aspect is the creation of new biomedical materials. In the field of regenerative medicine (tissue engineering), the reconstruction of specific tissues often requires scaffolds to support the stem cells or the developing cells. These scaffolds are called artificial extracellular matrices (ECM) and a conclusively powerful one has not yet been developed. In searching for the most suitable artificial ECM, various forms of CNT and CNT-containing hybrids have been tested for their potential to permit cell attachment, growth, and differentiation (Cao et al. 2007; MacDonald et al. 2005; Harrison and Atala 2007; Firkowska et al. 2006).

17.2.4.4 Cell Attachment to CNT

Perhaps one of the promising indications of biocompatibility is the fact that many types of cells in culture attached firmly (Aoki et al. 2007a; Zanello et al. 2006) and spread (Aoki et al. 2006) on the CNT-coated surfaces of the substrate. This means that the CNT interacts strongly with the cell surface proteins, probably with integrins, cadherins, and other adhesion proteins and that the CNT-coated cell culture devices will provide useful tools for effective cell culture systems in tissue engineering (Li et al. 2008, 2009). However, the details of the interaction mechanism between the CNT and the adhesion proteins are still to be further studied.

The combination of the CNT with known biomaterials, that is, collagen-related products, may improve their physicochemical and biological properties. MacDonald et al. (2005) reported a hybrid biomaterial that was made by mixing type I collagen with a solution of carboxylated SWCT at various concentrations. They found that a 2% wt CNT-containing collagen gel showed a lower shrinkage than the pure collagen control. They stated that the interaction of both substances is a physical one, as observed by electron microscopy, and they did not detect a strong molecular interaction, as observed by Raman spectroscopy.

17.2.5 Aggregation Caused by the Interaction between CNTs and Proteins

One of the difficulties in studying the interaction between CNTs and biomolecules is partly due to the fact that CNTs are solid entities, while most biomolecules can be prepared in solution. Accordingly, the conventional methods of analysis for protein conformation in solutions, such as optical rotatory dispersion, circular dichroism, and viscometry, are difficult to apply. On the other hand, analytical technologies for suspensions of solids, such as turbidity measurement, morphological observations, x-ray diffraction analysis, and Raman spectroscopy, are useful. In this section, we will introduce the studies performed by microscopical methods, precipitation analyses, and turbidity measurements (Kuboki et al. 1979).

17.2.5.1 Effects of CNTs on Protein Fibrillation

Linse et al. (2007) reported an interesting example of protein fibrillation caused by CNTs. They showed that nanoparticles, including multiwalled CNTs, of 6 nm diameter, increased the rate of fibrillation of β-microglobins into amyloid-like fibrils by shortening the lag phase. They proposed an interesting mechanism that there is an exchange of proteins between the solution and the particle surface (CNT) and the β-microglobins formed multiple layers on the particle surface, providing a locally increased protein concentration promoting oligomer formation. This process was assumed to form a critical nucleus for the nucleation of the amyloid-like fibrils of the protein. The nature of the binding forces between the CNTs and β-microglobins was not fully described, but they gave an important model of the mechanism by which the CNT promotes protein fibrillation. One possibility is that the rod-like geometry of the CNT may be related to the fibrillation of this protein. This possibility is further discussed in the next section.

17.2.6 Collagen: A Rod-Like Molecule Induces Aggregation of Carbon Nanotubes

Another important example of aggregation caused by the protein–CNT interaction was found in collagen, which is the most abundant protein in the human body, distributed among all of the tissues and organs. If we are to use the CNT as a biomaterial, which is implanted into the animal body in any form, the interaction between the collagen molecules and the CNT is one of the most essential problems to be clarified. In order to challenge this problem, we used turbidity as a means of evaluating the interaction between the CNT and the collagen molecules.

Recently, it was found that the addition of only a slight amount of collagen to a stable suspension of CNT caused a dramatic aggregation by hybridization between both components (Figure 17.1). By contrast, with heat-denatured collagen (gelatin), which has lost the natural triple-helical structure characteristic of the rod-like geometry of this protein, no aggregation occurred (Kuboki et al. 2009).

These results indicate that the rigid rod-like structure of the native collagen triple helix is essential for the interaction with the CNT, in order to cause aggregation. The mechanism is considered to be dependent on the geometric properties of the rod-like collagen molecules, where the like-to-like interaction at nanoscale level is proposed (Figure 17.2). These findings will open a new avenue to clarify the details of the mechanism of interaction between the collagen molecules and the CNT.

By discovering CNTs, protein chemistry has entered into a new era, where proteins of all kinds are able to reside in a rigid "house" and possibly express new functions. This has potential applications in the field of protein engineering for medical

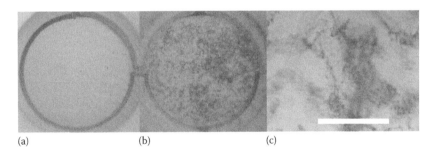

(a)   (b)   (c)

Figure 17.1 **Aggregation of a CNT with collagen observed in culture wells. To a clear suspension of 10 ppm CNT in 0.1% Triton, a collagen solution (0.1%) was added to obtain a final concentration of 25 ppm. Before the addition of collagen (a), the CNT suspension was clear. After the addition of collagen, the aggregation of the CNT occurred within 10 min (b). The diameter of the wells is 10 mm. An enlarged view of the aggregation after the addition of collagen shows that the aggregation seems to have a somewhat fibrous appearance (c). The bar in (c) indicates 0.1 mm.**

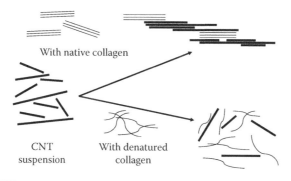

Why native collagen induces aggregation with CNT, but denatured collagen does not: a hypothesis

With native collagen

CNT suspension

With denatured collagen

Figure 17.2 **A schematic explanation of the mechanism of aggregation of CNT with collagen. One set of thin triple lines indicates the intact collagen molecules, and the thin-curved lines indicate the heat-denatured collagen chains (gelatin). The rigid triple helical structure of the rod-like collagen molecules can induce the aggregation of another rod-like clustered CNT, while the denatured random structure of gelatin can neither induce nor support the aggregation of the CNT.**

and biological applications. Further studies of the interaction between the CNT and proteins will not only be of practical value, but will also give new opportunities to clarify the fundamental mechanisms of the structure–function relationships of various proteins.

17.3 Interfaces between Solid Surfaces and Biomolecules

Toshio Ogino

The interdisciplinary research field between nanotechnology and biotechnology attracts much interest, partly because the development of nanotechnology will be able to promote biotechnology on a molecular level due to their similar dimensions. This research field may be classified into two groups. One is the application of biomolecules to electronics and mechanical engineering, and the other is the utilization of nanotechnology in biodevices, such as biosensors. Some of these will have relevance to food storage, distribution, and packaging. In any case, the interface control between the biomolecules and the solid materials used in nanotechnology plays an important role (Tiefenauer and Ros 2003; Kasemo 2002). Recently, research groups (Whaley et al. 2000; Sano and Shiba 2003) reported that peptides with specific amino acid sequences may be selectively bound to a range of semiconductor or metal surfaces. These reports suggest the possibility of molecular recognition by a solid surface.

In this section, we focus on selective adsorption of the protein molecules on sapphire surfaces. Sapphire is a very hard crystal variety of aluminium oxide (Al_2O_3) and can be found naturally or manufactured. It is used, among many applications, in wafers for the deposition of semiconductors, such as gallium nitride. Sapphire is chemically stable and its atomic structures on the surface, which consist of flat terraces and abrupt steps with one atomic layer height—atomic steps—can be preserved in a solution because it is an oxide material. Therefore, sapphire can be applied to biosensor substrates utilizing the surface nanostructures.

The protein adsorption sites on a sapphire surface can be controlled by an atomic-step arrangement. An overview of protein adsorption is described here, and then the atomic structure control of sapphire surfaces is shown. Subsequently, we show that selective adsorption of the protein molecules is realized by a domain structure formed on well-defined surfaces.

17.3.1 Interfaces between Biomolecules and Solid Surfaces

Figure 17.3 shows the typical interfaces between the solid surfaces and the biomolecules. A technique to immobilize the protein molecules on the inorganic surfaces through chemical bonds has been established. The primary requirements are an efficient immobilization of the molecules without denaturing of the protein and blocking of nonspecific adsorption. Figure 17.3a shows a representative process of protein immobilization using a silane-coupling agent. In this process, the protein molecules are immobilized on amino- or carboxyl-terminated alkane molecules fixed on the substrate through amide bonds. Figure 17.3b shows the physical adsorption of the proteins on a solid surface. In this type of adsorption, the van

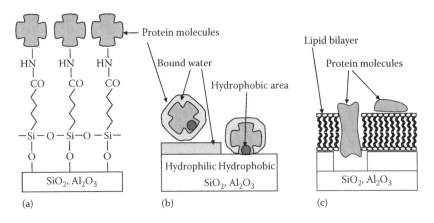

Figure 17.3 Typical interfaces between solid surfaces and biomolecules: (a) chemical bond, (b) physical adsorption, and (c) incorporation into a membrane.

der Waals force (attractive), hydrophobic interaction (attractive), hydration force (repulsive), and electrostatic force (attractive or repulsive) are involved. It should be emphasized that water molecules between the solid surfaces and the biomolecules often play a crucial role in protein adsorption.

Nonspecific adsorption is one of the serious problems in biosensor development because such adsorption may produce an error signal or degrade the sensor surfaces. The bioactivity of the membrane proteins is retained only when they are buried in a biomembrane. To immobilize a membrane protein on a solid surface, an artificial lipid bilayer is used (Castellana and Cremer 2006), as shown in Figure 17.3c.

17.3.2 Atomic-Structure Control on Sapphire Surfaces

When biomolecules are immobilized on a solid surface, the substrates should be immersed in a buffer solution to retain their bioactivity. Silicon substrates are one of the most perfect single crystals and are widely used in many solid-state devices. The silicon surface, however, is quickly oxidized in air or in a buffer solution and its features as a single crystal are lost. Sapphire, on the other hand, is an oxide single crystal and its atomic structures can be utilized in air or in a buffer solution. There have been many reports (Yoshimoto et al. 1995; Shiratsuchi et al. 2002) on the arrangement control of the atomic steps on the sapphire surfaces. The step arrangement depends on the annealing temperature after wafer cutting. Below about 800°C, the atomic steps are randomly distributed, as can be observed in the cross-sectional view of an atomic force microscopy (AFM) image, shown in Figure 17.4a.

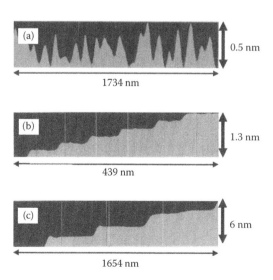

Figure 17.4   Step arrangement on sapphire surfaces: (a) randomly stepped surface, (b) single-stepped surface, and (c) multistepped surface.

After annealing at 1000°C–1200°C, the steps are separated into single atomic steps, as shown in Figure 17.4b. After annealing above 1300°C, several steps are gathered to form a step bunch, as shown in Figure 17.4c. Here, we call the surface shown in Figure 17.4c, a "multistepped surface." Well-ordered atomic steps can be used as templates for protein immobilization, as shown in Figure 17.3a (Aoki et al. 2007b).

On a multistepped surface, there are two domains that exhibit different hydrophilicities (Binggeli and Mate 1994). Figure 17.5 shows (a) the topographical and (b) the frictional images on a sapphire (0001) surface miscut by about 0.15°. The wafers were annealed at 1400°C and then chemically treated with an H_2SO_4–H_2O_2 solution. The misorientated direction of this surface was tilted slightly from the direction of a stable atomic step. We refer to this angle as the "tilting angle." On the surface shown in Figure 17.5, the tilting angle is small and the density of the crossing steps is also small. The topographical image after the thermal treatment at 700°C was the same as that after the acid treatment. As shown in Figure 17.5b and c, however, the frictional images are different between the treatment methods. In Figure 17.5b, the elliptic domains with less frictional force are clearly observed, whereas there was no specific pattern after the thermal treatment, as shown in Figure 17.5c. Generally, the difference in the frictional force is generated by the amount of adsorbed water (Binggeli and Mate 1994). The sapphire

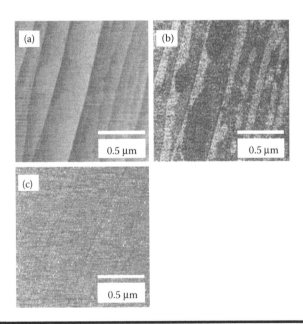

Figure 17.5 A domain structure on a sapphire surface: (a) topography after acid treatment, (b) frictional force image after acid treatment, and (c) frictional force image after annealing at 700°C.

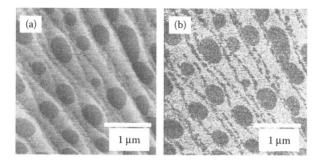

Figure 17.6 **A domain structure on a sapphire surface with a high crossing step density: (a) topography after acid treatment and (b) frictional force image after acid treatment.**

surface is hydrophilic after the acid treatment because the surface is terminated with OH groups.

To investigate the spatial distribution of the OH group, an OTS (octadecyltrichlorosilane, $C_{18}H_{37}SiCl_3$) self-assembled monolayer (SAM) was deposited on the surface. Since the OTS molecules are chemically fixed only on the OH-terminated areas, we can examine whether the surface is OH-terminated or not. The experimental results clearly demonstrated that the OH group density on the elliptic areas is much smaller than that on the outside areas. In other words, a phase separation into the hydrophilic and hydrophobic domains takes place on the acid-treated surface. Figure 17.6 shows (a) the topographical and (b) the frictional images on another sapphire (0001) surface with a larger tilting angle after a chemical treatment using an H_2SO_4–H_2O_2 solution. On this surface, the crossing step density is high, and 100 nm scaled terraces are grown on the areas surrounded by the bunched steps and the crossing steps. The formation process of this step arrangement is similar to that observed on an Si surface (Ogino, Hibino, and Homma 1996).

17.3.3 Selective Adsorption of Protein Molecules on Phase-Separated Surfaces

Figure 17.7 shows AFM images of the sapphire surfaces after ferritin molecule adsorption (a) on the acid-treated surface and (b) on the thermally treated surface (Ikeda et al. 2008). These images were taken in a buffer solution to retain the bioactivity, at least before the adsorption. On the acid-treated surface, the ferritin molecules are preferentially adsorbed on the hydrophobic domains, whereas no specific pattern is observed on the thermally treated surface. This pattern formation shows that the acid-treated surface consists of two domains and that the ferritin molecules are preferentially adsorbed on the hydrophobic domains. Generally, the protein molecules are more often adsorbed on a hydrophobic surface rather than on a

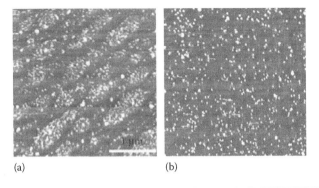

(a) (b)

Figure 17.7 The adsorption patterns of ferritin molecules on (a) the acid-treated [3 × 3 (μm²)] and (b) the thermally treated sapphire surfaces [5 × 5 (μm²)].

hydrophilic surface. The result shown in Figure 17.7a coincides with such a general phenomenon. However, on the thermally treated surface shown in Figure 17.7b, the hydrophilicity is uniform over the surface. Therefore, a uniform adsorption of the ferritin molecules occurs.

Figure 17.8 shows the adsorption patterns for various proteins (Ikeda et al. 2008). In these images, the hydrophobic (smaller density of the OH group) domains are almost circular and are isolated from each other. Therefore, the adsorption patterns can be more clearly demonstrated. In Figure 17.8a and b, the albumin and ferritin molecules are adsorbed preferentially on the hydrophobic domains. The avidin molecules, however, are preferentially adsorbed on the hydrophilic domains, as shown in Figure 17.8d, which is the opposite of the albumin and ferritin cases. In the case of fibrinogen, no clear adsorption pattern is observed. Although the adsorption patterns should be explained by a difference in the electrostatic interactions between the substrate surface and the protein, they

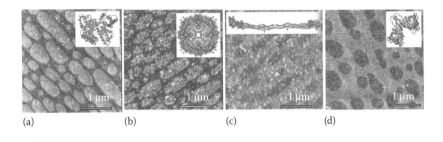

(a) (b) (c) (d)

Figure 17.8 The adsorption patterns of various protein molecules on the acid-treated sapphire surfaces. (a) Albumin, (b) ferritin, (c) fibrinogen, and (d) avidin. The protein molecules are preferentially adsorbed on the bright domains. Insets are the molecule structures from the Protein Data Bank of Japan (PDBj).

can be roughly classified by the molecule structures (Protein Data Bank Japan) that are shown in the insets in Figure 17.8. The albumin and ferritin molecules are mainly composed of α-helixes, whereas avidin is composed of β-sheets. The fibrinogen molecules have both parts. Generally, an α-helix prefers hydrophobicity to hydrophilicity if compared with β-sheets, as demonstrated in the configuration of the membrane proteins incorporated into the cell membrane. The adsorption patterns shown in Figure 17.8, therefore, can be roughly explained by the molecule structure.

Although the origin of the selectivity should be quantitatively described by considering the hydrophobic attractive force, the hydrophilic repulsive force, the van der Waals force, and the electrostatic force generated by the charges on the substrate surface and the protein molecules, the observed selectivity suggests a possibility of protein selection by a simple method.

In conclusion, a domain structure is self-organized on step-controlled sapphire surfaces where hydrophilic and hydrophobic areas appear after acid treatment. On this surface, protein molecules exhibit specific adsorption patterns. The present surface can be obtained simply by an acid treatment if the step arrangement is carefully designed. Therefore, this surface is promising for protein selection chips. The present finding also suggests the possibility of molecular recognition by a solid surface.

17.4 Policy and Regulatory Aspects of Protein Nanotechnology

Geoffrey Hunt and Masami Matsuda

The findings of Kuboki et al. and Ogino, and the similar bionanoscience work now progressing apace, holds out many opportunities in the fields of nutrition, food processing, storage, packaging, distribution, and traceability, as well as in agriculture. However, the same kinds of discoveries, when applied in the public and environmental domain, may have poorly understood the hazards and risks, which must be understood and controlled at an early stage.

17.4.1 Biointeractivity and Safety

The relatively biochemical inertness of the CNTs as "bulk geometries" may not always rule out the issues of concern about unintended or undesirable physical interaction, fibrillation, and bioaccumulation at some level in the organism. Functionalized CNTs, in which various molecular groups are bonded as side chains, may raise many more issues about adverse bioreactivity. In particular, CNT–protein hybrids that mimic or partially mimic other proteins and molecules, being of a similar size, will always raise the possibility of unintended consequences for networked biological processes, such a protein folding and hormonal reactions, elsewhere in the organism's physiology and development.

Ogino's experimental contribution in this chapter does not in itself suggest any particular concerns, but it does show the surprising things that *can* now be done, and we may envisage many applications in the future in which proteins and other biomolecules are adsorbed onto solid surfaces for applications in the food industry, especially sensors for the purposes of traceability and detection of impurities and pathogens. However, mainstream food nanotechnology research is largely focused elsewhere at the time of writing.

For example, extensive research is continuing on reducing the weight and gas absorption of drinks containers and improving their strength and thermal properties. This accords with sustainability principles, by extending the shelf-life and lowering the food transport energy use. For example, one research group found that nanolayered polyethylene oxide (PEO) crystallized into a single layer, thereby reducing the gas permeability by about 100 times in polymer packaging (Wang et al. 2009). Another intense area of research is the antipathogenic behavior of certain manufactured nanoparticles in food and packaging. One study looked at zinc oxide as a powder and as a suspension in a polyvinylprolidone gel (ZnO-PVP). It found significant antimicrobial activity against *Escherichia coli* O157:H7, *Salmonella* spp., and *Listeria monocytogenes* (Jin et al. 2009). Hazard and risk research will be needed in such areas too.

17.4.2 Protein Folding and Misfolding

It is worthwhile emphasizing that proteins are ubiquitous and fundamental in biological systems and systems of systems. There are tens of thousands of different proteins at work in the human physiology. Their interlocked functions within these systems are highly complex and far from fully understood and predictable in their interactions, despite the great advances in the last two decades. They are instrumental in almost every cell process and function as enzymes, structural components, cell signaling agents, and immunological agents.

The subtlety and complexity of the role of proteins have recently come to the fore in the matter of protein folding and the serious health impacts of some forms of misfolding, such as Alzheimer's, Parkinson's, and CJD (Selkoe 2004). As Dobson (2003) has written (see also Radford and Dobson 1999):

> The manner in which a newly synthesized chain of amino acids transforms itself into a perfectly folded protein depends both on the intrinsic properties of the amino acid sequence and on multiple contributing influences from the crowded cellular milieu. Folding and unfolding are crucial ways of regulating biological activity and targeting proteins to different cellular locations. Aggregation of misfolded proteins that escape the cellular quality-control mechanisms is a common feature

of a wide range of highly debilitating and increasingly prevalent diseases…

Based on the complexity principles (Hunt 2006), this implies the reasonable hypothesis that some manufactured proteins, protein fragments, biotechnology protein products, nanoparticle–protein hybrids, and protein-scale particles (possibly including CNTs) may, in some contexts, have an adverse effect, short or long term, on the normal physiological and metabolic systems and processes. They might interfere with the normal protein interactions of the cellular and subcellular environment. It is, for example, possible that since the endoplasmic reticulum plays a critical role in normal protein folding, manufactured nanoparticles may disrupt these processes if they enter the cell, damaging the reticulum and interacting in undesired or unexpected ways (Sitia and Braakman 2003). Studies are beginning to indicate the relevance of nanoparticle bioreactivity to the issue of protein folding and the chaperone molecules involved (Akiyoshi et al. 1999).

These possibilities are a matter for a balanced benefits–risks investigation by means of convergent research programs bringing together medical research into specific disease conditions and toxicology with research in protein chemistry, nutrition, nanoscience, and other disciplines.

17.4.3 Regulation and Precaution

The discussion of a new technology assessment will generally involve the Precautionary Principle, especially in Europe. This principle was first formulated as a "nonpreclusion" principle, for instance,

> In order to protect the environment, the precautionary principle shall be widely applied by the States according to their capabilities. Where there are threats of serious or irreversible damage, lack of full scientific certainty shall not be used as a reason for postponing cost-effective measures to prevent environmental degradation. (United Nations 1992)

This was aimed at defensive statements from the authorities of the form "There is no evidence that, e.g. Bovine spongiform encephalopathy (BSE) is connected with variant Creutzfeltd Jakob Disease (vCJD), therefore we shall continue with the *status quo*." The principle has also been interpreted as the proper observance of a margin of error or as requiring that potentially harmful activities should be subject to the "best available technology" tests. In the politics of technology, it has been seen as shifting the onus (burden of proof) from the consumers showing potential harms, to the producers (and/or regulators) showing safety, as far as reasonably possible.

In the European formulation of the principle, precautionary actions should be proportionate, nondiscriminatory, and consistent with the previous action, considering both the costs and benefits, and be subject to review (European Commission 2009). A possible criticism is that the Precautionary Principle does not properly take into account the potential benefits of a technology or an intervention. It assumes zero "opportunity cost," that is, that there is no cost associated with not acting or not developing the technology. This point is often made by the adherents of the "Proactionary Principle," who emphasize that opportunity costs should be accounted for in evaluating a technology. In the case of protein nanotechnology, the technology is still at a theoretical and experimental stage, but it is clearly wise to balance, as far as possible, the potential benefits and harms from the beginning, putting aside special interests.

17.4.4 Some Concerns

From 2009, concerns about the potential harms of nanomaterials were raised at a significant level in official governmental bodies, especially in Europe. Although the European Commission has not yet argued for an amendment of the chemicals regulation—Registration, Evaluation, Authorisation (and restriction) of Chemicals (REACH) (European Commission 2002)—to take account of the *novelty* of nanomaterials, the European Parliament appears to be moving in another direction. In April 2009, the parliament's environment committee even recommended that consumer products, including food-related items, be withdrawn until more adequate safety tests are performed. The parliament adopted a report calling for tighter controls and adherence to the REACH principle of "no data, no market" (http://www.euractiv.com/en/science/data-market-nanotechnologies-meps/article-180893). This followed on a parliamentary vote on novel food regulations, calling for an adequate definition, labeling, and targeted risks assessments for foods containing manufactured nanoscale entities (http://www.europarl.europa.eu/news/expert/briefing_page/51585-082-03-13-20090311BRI51584-23-03-2009-2009/default_p001c010_en.htm/).

In Switzerland, the Technology Assessment Centre (TA-SWISS) issued a report in 2009—"Nanotechnology in the Food Sector"—which recommends that regulations be revised to take account of the novel properties of nanomaterials and that the Precautionary Principle and a life cycle analysis serve as a basis for such revisions (http://www.ta-swiss.ch/e/them_nano_nafo.html). In the UK, a House of Lords select committee consulted widely on the use of nanotechnology in the food industry in 2009 (House of Lords: http://www.foodproductiondaily.com/Quality-Safety/Nanotechnology-in-UK-food-sector-under-review).

There is no suggestion that all food additive nanoparticles are harmful, of course; for example, nanoscale carotenoids and various micelles that have been in use for years are well tested. However, some products containing nanoparticles of greater uncertainty, especially metallic, may appear on the market, with some produced outside Europe. In Japan, a yogurt containing platinum "microparticulated by means of nanotechnology" is on the market (www.nipponham.co.jp/en/).

17.5 Conclusions

Geoffrey Hunt and Masami Matsuda

The discovery of CNTs and other nanomaterials and their functionalization have meant that protein chemistry has also entered into a new era. Proteins of all kinds may be housed within or attached to a variety of rigid structures to express their new functions of utility in the food and other industries. The impacts will affect the field of protein engineering, as well as the medical and biotechnology applications of proteins. Further studies of the interaction between nanomaterials and proteins will not only demonstrate the practical value of proteins but also give new opportunities to clarify the fundamental mechanisms of their structure–function relationships.

However, given the high level of complexity and interactivity on the nanoscale, such research must go hand-in-hand with properly resourced and coordinated health, safety, and environmental research and regulatory development on the basis of the principles of complexity, precaution, and a life cycle analysis.

References

Akiyoshi, K., Sasaki Y., et al. 1999. Chaperone-like activity of hydrogel nanoparticles of hydrophobized pullulan: Thermal stabilization with refolding of carbonic anhydrase B. *Bioconjugate Chemistry* 10: 321–24; Billsten, P., Freskgard, P. O., et al. 1997. Adsorption to silica nanoparticles of human carbonic anhydrase II and truncated forms induce a molten-globule-like structure. *FEBS Letters* 402: 67–72; Ishii, D., Kinbara, K., Ishida, Y., Ishii, N., Okochi, M., Yohda, M., and Aida, T. 2003. Chaperonin-mediated stabilization and ATP-triggered release of semiconductor nanoparticles. *Nature* 423: 628–32.

Aoki, N., Yokoyama, A., Nodasaka, Y., Akasaka, T., Uo, M., Sato, Y., Tohji, K., and Watari, F. 2006. Strikingly extended morphology of cells grown on carbon nanotubes. *Chemistry Letters* 35: 508–509.

Aoki, N., Akasaka, T., Watari, F., and Yokoyama, A. 2007a. Carbon nanotubes as scaffolds for cell culture and effect on cellular functions. *Dental Materials Journal* 26(2): 178–85.

Aoki, R., Arakawa, T., Misawa, N., Tero, R., Urisu, T., Takeuchi, A., and Ogino, T. 2007b. Immobilization of protein molecules on step-controlled sapphire surfaces. *Surface Science* 601: 4915–21.

Binggeli, M. and Mate, C. M. 1994. Influence of capillary condensation of water on nanotribology studied by force microscopy. *Applied Physics Letters* 65: 415–17.

Cao, Y., Zhou, Y. M., Shan, Y., Ju, H. X., and Xue, X. J. 2007. Preparation and characterization of grafted collagen-multiwalled carbon nanotubes composites. *Journal of Nanoscience and Nanotechnology* 7(2): 447–51.

Castellana, E. T. and Cremer, P. S. 2006. Solid supported lipid bilayers: From biophysical studies to sensor design. *Surface Science Reports* 61: 429–44.

Cid, C. C., Riu, J., Maroto, A., and Rius, F. X. 2008. Carbon nanotube field effect transistors for the fast and selective detection of human immunoglobulin G. *Analyst* 133(8): 1005–1008.

Dobson, C. M. 2003. Protein folding and misfolding. *Nature* 426: 884–90. See also: Radford, S. E. and Dobson, C. M. 1999. From computer simulations to human disease: Emerging themes in protein folding. *Cell* 97: 291–98.

European Commission. 2000. *Communication on the Precautionary Principle.* COM(2000) 1 Final, Brussels.

European Commission. 2002. *Registration, Evaluation and Authorisation of Chemicals.* Brussels.

Firkowska, I., Olek, M., Pazos-Perez, N., Rojas-Chapana, J., and Giersig, M. 2006. Highly ordered MWNT-based matrixes: Topography at the nanoscale conceived for tissue engineering. *Langmuir* 22(12): 5427–34.

Harrison, B. S. and Atala, A. 2007. Carbon nanotube applications for tissue engineering. *Biomaterials* 28(2): 344–53.

House of Lords: http://www.foodproductiondaily.com/Quality-Safety/Nanotechnology-in-UK-food-sector-under-review.

http://www.euractiv.com/en/science/data-market-nanotechnologies-meps/article-180893.

http://www.europarl.europa.eu/news/expert/briefing_page/51585-082-03-13-20090311BRI51584-23-03-2009-2009/default_p001c010_en.htm/.

http://www.nipponham.co.jp/en/.

http://www.ta-swiss.ch/e/them_nano_nafo.html.

Hunt, G. 2006. Nanotechnoscience and complex systems. In Nanotechnology: Risk, Ethics & Law, eds. G. Hunt and M. Mehta, pp. 43–56. London: Earthscan.

Iijima, S., Ajayan, P. M., and Ichihashi, T. 1993. Growth model for carbon nanotubes. *Physical Review Letters*, 69(21): 3100–103.

Ikeda, T., Isono, T., Aoki, R., and Ogino, T., 2008. Selective adsorption of protein molecules on atomic-structure-controlled sapphire. Extended Abstracts of the *International Conference on Solid State Devices and Materials*, pp. 948–949. Tsukuba, Japan.

Jin, T., Sun, D., Su, J. Y., Zhang, H., Sue, H. J. 2009. Antimicrobial efficacy of zinc oxide quantum dots against *Listeria monocytogenes*, *Salmonella Enteritidis*, and *Escherichia coli* O157:H7. *Journal of Food Science* 74(1): M46–52.

Kasemo, B. 2002. Biological surface science. *Surface Science* 500: 656–67.

Kuboki, Y., Fujisawa, R., Aoyama, K., and Sasaki, S. 1979. Calcium-specific precipitation of dentin phosphoprotein: A new method of purification and the significance for the mechanism of calcification. *Journal of Dental Research* 58(9): 1926–32.

Kuboki, Y., Terada, M., Kitagawa, Y., Abe, S., Uo, M., and Watari, F. 2009. Interaction of collagen triple-helix with carbon nanotubes: Geometric property of rod-like molecules. *Biomedical and Biomaterials Engineering* 19(1): 3–9.

Li, X., Gao, H., Uo, M., Sato, Y., Akasaka, T., Abe, S., Feng, Q., Cui, F., and Watari, F. 2009a. Maturation of osteoblast-like SaoS2 induced by carbon nanotubes. *Biomedical Materials* 4: 015005 (8 pp.). doi: 10.1088/1748-6041/4/1/015005.

Li, X., Gao, H., Uo, M., Sato, Y., Akasaka, T., Feng, Q., Cui, F., Liu, X., and Watari, F. 2009b. Effect of carbon nanotubes on cellular functions in vitro. *Journal of Biomedical Materials Research A* 91A: 132–39.

Linse, S., Cabaleiro-Lago, C., Xue, W. F., Lynch, I., Lindman, S., Thulin, E., Radford, S. E., and Dawson, K. A. 2007. Nucleation of protein fibrillation by nanoparticles. *Proceedings of the National Academy of Science of the United States of America* 104(21): 8691–96.

Lyons, M. E. and Keeley, G. P. 2008. Immobilized enzyme-single-wall carbon nanotube composites for amperometric glucose detection at a very low applied potential. *Chemical Communication (Cambridge)* 14(22): 2529–31.

MacDonald, A., Laurenzi, B. F., Viswanathan, G., Ajayan, P. M., and Stegemann, J. P. 2005. Collagen–Carbon nanotube composite materials as scaffolds in tissue engineering. *Journal of Biomedical Materials Research* A 74(3): 489–95.

Ogino, T., Hibino, H., and Homma, Y. 1996. Step arrangement design and nanostructure self-organization on Si(111) surfaces by patterning-assisted control. *Applied Surface Science* 107: 1–5.

Protein Data Bank Japan. http://www.pdbj.org/index.html. Protein Data Bank (in English), http://www.rcsb.org/pdb/home/home.do.

Pusztai, Á. and Bardocz, S. 2006. The Future of Nanotechnology in Food Science and Nutrition: Can Science Predict Its Safety? In *Nanotechnology: Risk, Ethics and Law*, eds. G. Hunt and M. Mehta. London: Earthscan.

Sano, K. and Shiba, K. 2003. A hexapeptide motif that electrostatically binds to the surface of titanium. *Journal of American Chemical Society* 125: 14234–35.

Selkoe, D. J. 2004. Folding proteins in fatal ways. *Nature* 428: 900–904.

Shen, J. W., Wu, T., Wang, Q., and Kang, Y. 2008. Induced stepwise conformational change of human serum albumin on carbon nanotube surfaces. *Biomaterials* 29(28): 3847–55.

Shiratsuchi, Y., Yamamoto, M., and Kamada, Y. 2002. Surface Structure of Self-Organized Sapphire (0001) Substrates with Various Inclined Angles. *Japanese Journal of Applied Physics* 41: 5719–25.

Sitia, R. and Braakman, I. 2003. Quality control in the endoplasmic reticulum protein factory. *Nature* 426: 891–94.

Tiefenauer, L. and Ros, R. 2003. Biointerface analysis on a molecular level, New tools for biosensor research. *Colloids and Surfaces B: Biointerfaces* 23: 95–114.

United Nations. 1992. Rio Declaration, Principle 15.

US Institute for Agriculture and Trade Policy, Identifying our Climate Foodprint: http://www.iatp.org/iatp/publications.cfm?accountID=258&refID=105667.

Wang, H., Keum, J. K., Hiltner, A., Baer, E., Freeman, B., Rozanski, A., and Galeski, A. 2009. Confined crystallization of polyethylene oxide in nanolayer assemblies. *Science* 323(5915): 757–60.

Whaley, S. R., English, D. S., Hu, E. L., Barbara, P. F., and Belche, A. M. 2000. Selection of peptides with semiconductor binding specificity for directed nanocrystal assembly. *Nature* 405: 665–68.

Wu, X. C., Zhang, W. J., Sammynaiken, R., Meng, Q. H., Yang, Q. Q., Zhan, E., Liu, Q., Yang, W., and Wang, R. 2008. Non-functionalized carbon nanotube binding with hemoglobin. *Colloids and Surface B: Biointerfaces* 65(1): 146–49.

Yoshimoto, M., Maeda, T., Ohnishi, T., Koinuma, H., Ishiyama, O., Shinohara, M., Kubo, M., Miura, R., and Miyamoto, A. 1995. Atomic-scale formation of ultrasmooth surfaces on sapphire substrates for high-quality thin-film fabrication. *Applied Physics Letters* 67: 2615–17.

Zanello, L. P., Zhao, B., Hu, H., and Haddon, R. C. 2006. Bone cell proliferation on carbon nanotubes, *Nano Letters* 6(3): 562–67.

Index

443

For Product Safety Concerns and Information please contact our EU representative GPSR@taylorandfrancis.com Taylor & Francis Verlag GmbH, Kaufingerstraße 24, 80331 München, Germany

Printed and bound by CPI Group (UK) Ltd, Croydon, CR0 4YY

01/05/2025

01858478-0002